T0179707

Drought and Aquatic Ecosystems:
Effects and Responses

Drought and Aquatic Ecosystems:
Effects and Responses

P. Sam Lake

WILEY-BLACKWELL

A John Wiley & Sons, Ltd., Publication

Library of Congress Cataloguing-in-Publication Data

Lake, P. Sam,
  Drought and Aquatic Ecosystems: Effects and Responses / P. Sam Lake.
     p. cm.
  Includes bibliographical references and index.
  ISBN 978-1-4051-8560-8 (cloth)
  1. Biotic communities. I. Title.
  QH541.M574 2011
  577.8′2–dc22
                                        2011000108

A catalogue record for this book is available from the British Library.

This book is published in the following electronic formats: ePDF 9781444341782; Wiley Online Library 9781444341812; ePub 9781444341799; MobiPocket 9781444341805.

Set in 10.5/12.5pt Photina by Thomson Digital, Noida, India
Printed and bound in Malaysia by Vivar Printing Sdn Bhd

1   2011

For Marilyn, Katherine and Jessica

Contents

Colour plate section appears between pages 210 and 211

Acknowledgements

I wish to thank Land and Water Australia (now abolished) for a Senior Fellowship and the Commonwealth Environment Research Facilities (CERF) of the Australian Federal Government for a Senior Research Fellowship, both of which substantially supported me in the tasks of reading the literature and writing this book. Further invaluable support came from the School of Biological Sciences, Monash University.

I am greatly indebted to Professor Andrew Boulton of East Fremantle for reading all of the draft chapters and offering wise, helpful, critical and sometimes trenchant comments, as well as helping to shape and improve some of the figures. Patrick Baker of Monash University provided much-needed and helpful criticism of the first four chapters. Both provided new leads and alerted me to new material, as well as having many discussions with me on drought, its ecological impacts and its poor management.

For encouraging me to pursue research on drought in aquatic systems, I wish to thank the late Professor Peter Cullen when he was Director of the Cooperative Research Centre for Freshwater Ecology. I am also very grateful for comments and discussions to Nick Bond, Rob Hale, Danny Spring, Gillis Horner, Ralph Mac Nally, Ross Thompson, Hania Lada, Greg Horrocks and Shaun Cunningham of Monash University, Paul Reich and Dave Crook of the Victorian Department of Sustainability and Environment, Paul Humphries of Charles Sturt University, Michael Douglas of Charles Darwin University, Barbara Downes of the University of Melbourne, Darren Baldwin of the Murray-Darling Research Centre, Fran Sheldon of Griffith University, Dr. Dale McNeil of SARDI in Adelaide, Athol McLachlan of the Isle of Mull, Scotland, Mary Power of the University of California, Berkeley, and Margaret Palmer of the University of Maryland.

For great help in dealing with computer problems and in critically evaluating and producing the figures, I wish to thank Tom Daniel and Matthew Johnson of Monash University.

I am very grateful for the help that I received from the Docdel unit of the Main Library, Monash University, and from the staff of the Hargrave Library. With arranging the various permutations of semi-retirement and providing office space for the past four years, I wish to thank Anne Fletcher and Jodie Weller of the School of Biological Sciences, Monash University.

For initially encouraging me to embark on the task of writing this book and for continued and cheerful support, I wish to thank Alan Crowden, and for continued support and assistance, I am very grateful to Ward Cooper of Wiley-Blackwell.

1

Introduction: the nature of droughts

Living in Australia, a land of 'droughts and flooding rains' and the most drought-prone continent in the world, it is not surprising that, like many Australians, I have an acute awareness of the perils of drought. Drought comes with images of crops wilting, livestock being destroyed, dust storms, bushfires, dry farm dams, empty reservoirs, dying trees and drastic restrictions on water use. As a freshwater ecologist, I have become only too aware of the damaging and lasting effects of drought on freshwater systems. Projects planned and premised on the availability of sufficient water have been compromised, if not halted. Thus, drought has moved from being a matter of concern for me to becoming a hazard to research, and it has now grown to become a major research interest of mine. This interest has been heightened by the realization that of the two major flow-generated hazards to freshwater ecosystems – floods and droughts – our ecological understanding of floods is much more comprehensive and deeper than our understanding of droughts (Giller, 1996; Lake, 2000, 2003, 2007).

The literature on the ecology of drought and freshwater systems is limited in quantity in comparison with that on floods and other disturbances (e.g. pollution). It is also very scattered across different types of publications, is uneven in quality, and some of it is quite difficult to access (Lake *et al.*, 2008). Following the international conference on the 'Role of Drought in the Ecology of Aquatic Systems' in Albury, Australia in 2001 (Humphries & Baldwin, 2003), I read much of the literature on drought and freshwater ecosystems and produced an interim report (Lake, 2008). This present book is the culmination of this extended research effort.

Drought is a ubiquitous climatic hazard. It is a recurring climatic phenomenon and its frequency, duration, intensity, severity and spatial extent all vary with locality and with time at any one location. As a hazard, it is determined relative to the prevailing normal conditions of a locality. Thus, partly because of this variation, it has been difficult to find a universal

---

Drought and Aquatic Ecosystems: Effects and Responses, First Edition. P. Sam Lake.
© 2011 P. Sam Lake. Published 2011 by Blackwell Publishing Ltd.

definition of drought; indeed, 'a universal definition is an unrealistic expectation' (Wilhite, 2000). This lack of generality makes the effects of drought difficult to evaluate and compare among localities and regions across the world.

The numerous definitions of drought can be split into two forms: those that define it as a natural climatic phenomenon and those that define it as a hazard to human activities (especially agriculture). The latter type of definition is understandably much more common. Examples of drought definitions focused on human impacts include:

- 'a deficiency of rainfall from expected or normal that, when extended over a season or longer period of time is insufficient to meet the demands of human activities' (Tannehill, 1947);
- 'drought is a persistent and abnormal moisture deficiency having adverse impacts on vegetation, animals, or people' (National Drought Policy Commission (USA), 2000);
- 'a drought is a prolonged, abnormally dry period when there is not enough water for users' normal needs' (Bureau of Meteorology, Australian Government, 2006).

This type of definition leads to an imprecise determination of drought, as it depends on the nature of human activities that are judged to be impaired by drought. However, it is nevertheless perfectly understandable, as the declaration of drought at a locality can have serious economic and social implications.

In looking at the effects of drought on freshwater ecosystems, it is above all necessary to define drought as a natural phenomenon, whilst recognizing the many interactions between human activities and drought. Following Druyan (1996b), drought can be defined as 'an extended period – a season, a year or several years – of deficient rainfall relative to the statistical multiyear mean for a region'. It should be noted that 'rainfall' is usually the major form of precipitation, but other forms such as snow, and even fog, can be important. This definition relies on the availability of lengthy data sets (25–30 years) to determine the 'multiyear mean'. Furthermore, the determination of the 'multiyear mean' may be incorrect when there is a long-term trend in the climate – a move away from the assumption of no significant change in long-term mean values or stationarity (Milly *et al.*, 2008).

In this work, I will be regarding drought as a phenomenon affecting ecosystems and their constituents rather than one affecting human activities. Defining drought this way must, however, recognize that human activities can either create conditions that increase the likelihood of drought or may exacerbate natural drought. For example, the clearing of vegetation may render land more prone to drought (Glantz, 1994), and extraction of water for human use from waterways can exacerbate the low flow conditions generated by natural drought (Bond *et al.*, 2008). Thus, there will be many instances in which the drought affecting biota and ecosystems will be exacerbated by humanity's use of water and land.

Drought must be distinguished from aridity. Aridity occurs where it is normal for rainfall to be below a low threshold for a long and indeterminate duration, whereas drought occurs when rainfall is below a low threshold for a fixed duration (Coughlan, 1985). In arid areas, provided there is a good long-term rainfall record, it is possible to distinguish drought when it occurs in spite of the prevailing regime of low rainfall. Aridity in a region means that there is an overall negative water balance due to the potential evapotranspiration of water exceeding that supplied by precipitation, with precipitation being low, usually less than 20 cm per year (Druyan, 1996a) and highly variable. At some times in arid regions, precipitation may exceed potential evapotranspiration, but in the long run there is a continual deficit in precipitation. In drought, precipitation is less than potential evapotranspiration for an extended period, but not permanently. Again, the assumption of stationarity is challenged if extended droughts are part of the onset of a drying phase, a climate change or a move toward aridity.

As stressed in Wilhite (2000) and Wilhite *et al.* (2007), drought is a very complex phenomenon and it remains a poorly understood climatic hazard. Bryant (2005) ranked 31 different natural hazards, ranging from drought to rockfalls, in terms of nine hazard characteristics: degree of severity; length of event; area extent; loss of life; economic loss; social effect; long-term impact; suddenness; and occurrence of associated hazards. Drought scored the most severe on all characteristics except for the last two, and it is the most severe natural hazard in terms of duration, spatial extent and impact.

Surprisingly, drought did not score as severe in terms of the occurrence of associated hazards. Droughts in many parts of the world, from North America to Indonesia, can be associated with severe and very extensive bushfires. In drier areas, severe dust storms, such as in the Great Plains of the USA in the 1930s (Worster, 1979) or in eastern Australia, are produced during drought. Most other natural hazards are of short duration, of limited spatial extent, and are due to an excess of forces (e.g. cyclones) or of material (e.g. floods). However, drought is an unusual hazard as it is generated by a deficit; out of 31 different types of natural hazard, it only shares this critical characteristic with subsidence (Bryant, 2005).

1.1 The social and economic damage of drought

The range of impacts of drought on human economic and social activities is immense. This is perfectly understandable, as water is essential for life and for the sustainable operation of natural and human-dominated ecosystems, both aquatic and terrestrial. Drought can reduce agricultural production, with direct losses of both crops and livestock, as well as causing the cessation of both cultivation and livestock population maintenance. Land may be lost to future production by dust storms, loss of vegetation and erosion. Forest production may be damaged both by severe water stress to trees and by severe and extensive bushfires. Water restrictions may reduce energy production (e.g. hydro-electricity), industrial production and the availability of clean water for human consumption. Water loss in rivers may even limit water transport; for example, in the 1987–1988 drought in the USA, barge traffic on the Mississippi river was limited by the low depths of the channel (Riebsame *et al.*, 1991). Economic losses can be incurred across a range of activities from agricultural and industrial to tourism and recreation. In addition, costs during drought may rise sharply, as reflected in food prices, water prices for industry, agriculture and human consumption, and in costs for drought relief to farmers and rural communities.

Drought is a natural hazard that humans cannot modify meteorologically. However, with forethought it may be possible to modify some of its impacts on natural ecosystems and on human society. Drought 'has both a natural and social dimension' (Wilhite & Buchanan-Smith, 2005); the human responses to deal with drought may vary from being hasty and reactive to being well-planned and proactive.

These responses are encompassed in the concept of vulnerability. The four essential components of vulnerability to drought are: capacity to predict drought; effective monitoring of drought with the capacity to provide early warning of drought attributes (e.g. extent, severity); effective mitigation and preparedness; and a readiness in society for the need to have a coordinated strategy to deal with drought. Various societies in different regions have different levels of vulnerability to drought, and thus there are 'drought-vulnerable' and 'drought-resilient' societies (Wilhite & Buchanan-Smith, 2005). While there are many drought-vulnerable societies, there are very few examples of drought-resilient societies, though in some regions, such as the USA and Australia, resilience at the societal level is improving (Wilhite, 2003; Wilhite *et al.*, 2007).

In the south-west of what is now the USA, the Anasazi people in the Four Corners region developed a complex society, starting about 650 AD, based on the cultivation of maize supported by extensive and intricate systems of water harvesting, that lasted until the 13th century (Diamond, 2005; Benson *et al.*, 2007). Two severe and lengthy droughts (megadroughts – droughts lasting longer than 10 years: Woodhouse & Overpeck, 1998) in the middle 12th and late 13th centuries greatly reduced maize yields, causing the abandonment of settlements (Diamond, 2005; Benson *et al.*, 2006, 2007).

To the hazard of extended drought, Anasazi society had a high vulnerability and a very low resilience – little capacity to recover.

In drought-vulnerable societies, drought may be linked with famine, disease and social upheaval – both now, as in the Sahel region of Africa (Dai *et al.*, 2004a), and in the past. In the case of colonial India, the two severe droughts of 1876–1879 and 1896–1902 are estimated to have killed 12.2 to 29.3 million people, and in China the death toll was estimated to be 19.5–30 million people (Davis, 2001). Indeed, the failure of the monsoon in 1876–79 that caused drought over much of Asia caused a famine that 'is the worst ever to afflict the human species. The death toll cannot be ascertained, but certainly it exceeded 20 million' (Hidore, 1996).

The high death toll from the two late Victorian droughts in India was no doubt linked to the great increase in drought vulnerability in rural India due to the commodification of village agriculture by Britain. A switch to growing crops for export swept away traditional and local means of storage and support to contend with drought (Davis, 2001). Indeed, the catastrophic impacts of drought on societies high in drought vulnerability and low in preparedness in India and China at that time (Davis, 2001), and in the 'Dustbowl' in the 1930s in the USA (Worster, 1979) can be seen as significant historical events that had major effects on the futures of the affected societies.

Economic losses, mainly through reduction of agricultural production, can be immense; droughts are costly. For example, the drought years of 1980 and 1988 in the USA are estimated to have cost $48.8 billion and $61.6 billion (2006 dollars) respectively (Riebsame *et al.*, 1991; Cook *et al.*, 2007), while the very severe drought of 2002–03 in Australia (Nicholls, 2004) is estimated to have cost $A7.4 billion in lost agricultural production (Australian Bureau of Statistics, 2004).

As droughts usually cover a large spatial extent and are invariably of considerable duration, they slowly produce ecological, economic and social deficiencies. These deficiencies, such as high mortality of biota (plant and animal, natural and domestic) and the poor condition and health of organisms, including humans, do not allow a rapid recovery once a drought breaks; there may be a long lag in recovery.

In human societies, the damaging social and economic effects can persist for a long time. For example, if drought gives rise to famine, children may become seriously malnourished and the effects of malnutrition on health and mental well-being may be lifelong (Bryant, 2005). The replenishment of seed for crops and of livestock numbers from remnant survivors are also lengthy and costly processes. Moreover fire, dust storms and overgrazing may severely damage pastures and croplands and even prevent full recovery (Bryant, 2005).

1.2   Major characteristics of drought

As suggested by Tannehill (1947), when he labelled droughts 'creeping disasters', it can be difficult to detect the beginning of a drought, as the deficiency of moisture in a region takes time to emerge (e.g. Changnon, 1987). As drought is a form of disturbance that steadily builds in strength, Lake (2000, 2003) suggested that it constitutes a ramp type of disturbance, which steadily builds in severity with time. For the same reason, it can also be difficult to detect the end of a drought as it gradually fades away (inverse ramp). However, if the drought is linked with an El Niño event, it may be broken by severe flooding (Whetton, 1997) – a pulse disturbance.

As a form of disturbance – a hazard – droughts are distinctive in not causing major geomorphological changes or damaging or destroying human structures. However, droughts may cause some smaller geomorphological changes, such as those due to accompanying dust storms with consequent wind erosion and deposition of soil and sand, exemplified by the 'Dustbowl' drought in the USA (Worster, 1979). Droughts are distinctive in occurring over large areas. They differ from floods in usually being drawn-out ramp disturbances rather than rapid pulses, and in being a type of disturbance from which ecological recovery can be a long, drawn-out process.

Most droughts consist of abnormal extended periods of hot and dry weather that inexorably deplete water availability across regions. Such droughts may also have long spells of dry winds, dust storms and wildfires. In some regions, notably those parts of the world that have severe winters, there may be winter droughts (e.g. McGowan *et al.*, 2005; Werner & Rothhaupt, 2008), in which poor seasonal precipitation and freezing conditions greatly reduce runoff, reducing flows in downstream rivers and depleting levels and volumes in lakes. Such winter droughts may then lead on to severe supra-seasonal droughts.

Droughts have four major characteristics (Bonacci, 1993; Wilhite, 2000; see Figure 1.1):

Intensity or magnitude;
Duration;
Severity (water deficiency); and
Spatial extent.

Other important characteristics include probability of recurrence and time of initiation and termination (Yevjevich, 1967).

Intensity refers to the average water deficiency (i.e. severity/duration) and is a measure of the degree of reduction in expected precipitation (or river flow) during the drought.

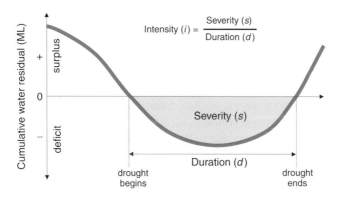

Figure 1.1 Depiction of the characteristics of drought as illustrated by hydrological drought with severity (*s*) (cumulative water deficit), duration (*d*) and intensity (*i*), which is severity divided by duration.

Duration refers to the length of the drought and is entirely dependent for its determination on the thresholds used to define the onset and the end of drought. The duration of a drought is strongly correlated with the severity. Depending on the indices used to detect drought, it usually takes 2–3 months as a minimum for drought to become established.

Severity refers to the cumulative deficiency in precipitation or in water (Bonacci, 1993).

Spatial extent refers to the area covered by and in mapping the areas, such as in the continually updated US Drought Monitor (Svoboda *et al.*, 2002). The areas are delineated in terms of drought intensity, from D_0 (abnormally dry) to D_4 (exceptional) (see Figure 1.2).

Large-scale droughts of long duration can have within them regional droughts of shorter duration (Stahle *et al.*, 2007). Droughts occur in some regions more than others; for example, both severe annual droughts and pluvials (high rainfall events) in the USA 'occur more frequently in the central United States' (Kangas & Brown, 2007).

1.3   The formation of droughts

Droughts develop almost imperceptibly and insidiously and, depending on the drought indicator used, it usually takes at least three months of abnormally low rainfall to detect a drought. The almost imperceptible onset of drought is sensitively recounted by Barry Lopez: 'In the years we have been here I have trained myself to listen to the river, not in the belief that I could understand what it said, but only from one day to the next know

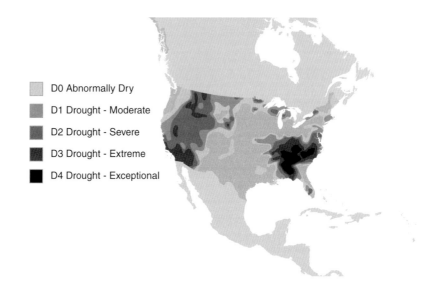

D0 Abnormally Dry

D1 Drought - Moderate

D2 Drought - Severe

D3 Drought - Extreme

D4 Drought - Exceptional

Figure 1.2 An example of the output from the Drought Monitor, showing the extent and severity of drought in central and southern USA on October 9, 2007. (See the colour version of this figure in Plate 1.2.)

its fate . . . It was in this way that I learned before anyone else of the coming drought. Day after day as the river fell by imperceptible increments its song changed.' (Lopez, 1990)

Droughts may gradually finish with the return of normal rainfall. However, they can also end with heavy rains such as those of tropical storms (e.g. Churchell & Batzer, 2006) or with sudden and very damaging floods (Whetton, 1997), such as the floods that ended the 1982–83 drought in south-eastern Australia and the recent floods (2010) that abruptly ended the long drought in southern Australia.

Droughts arise from a lack of precipitation that is due to the development of stationary or slow-moving weather systems – a subsidence of moisture-depleted, high-pressure air over a region. The development of slow-moving high-pressure systems has been proposed to occur due to two different basic causes – changes in solar activity and sea surface temperature fluctuations.

For a considerable time, droughts were thought to arise from sunspot activity (e.g. Tannehill, 1947). Sunspots are due to intense magnetic activity on the sun reducing convectional activity and causing cooling of the area affected. Sunspot activity is correlated with solar activity and, during periods of high activity, the production of cosmogenic isotopes (e.g. ^{14}C and ^{10}Be) decreases. Changes in the concentrations of these isotopes in lake sediments (e.g. Yu & Ito, 1999) and in the polar icecaps (e.g. Ogurtsov, 2007) may be used to detect changes in solar activity. Low levels of these isotopes from high solar activity are held possibly to indicate drought (Hodell *et al.*, 2001).

Mensing *et al.* (2004), in analyzing pollen from cores taken from Pyramid Lake, Nevada, detected prolonged droughts going back 7,600 years. They found that the periods of prolonged droughts coincided with periods of reduced drift ice activity in the north Atlantic Ocean (Bond *et al.*, 2001). These periods of reduced drift ice activity were correlated with periods of increasing solar activity as revealed by reduced levels of cosmogenic isotopes (Bond *et al.*, 2001). In contrast, high levels of the cosmogenic isotope ^{14}C correlated with low solar activity have been linked with drought and dry periods in the northern Great Plains of the USA (Yu & Ito, 1999).

Whether solar forcing is a major force producing drought appears debatable. It is worth noting that the changes in 0.1–0.25% of total radiation in solar forcing (Crowley, 2000) may appear to be slight, but so are the oceanic sea surface temperature changes associated with El Niño/La Niña events (Cook *et al.*, 2007).

Although droughts have been suggested to be caused by various forces, including sunspot activity (Tannehill, 1947) and solar forcing (Hodell *et al.*, 2001), recent studies suggest that the primary cause for severe droughts is small fluctuations in sea surface temperatures over a large area. These fluctuations, linked with changes in air pressure, alter winds carrying moisture onto land. A major driving force for marked fluctuations of long duration in sea surface temperatures, air pressure and onshore moisture-laden winds is the oceanic oscillation of the El Niño/Southern Oscillation (ENSO) system. As this system was the first such oscillatory system to be unravelled, it is worth a brief account of the history of the discovery of the system and its behaviour.

1.4   El Niño Southern Oscillation (ENSO) and drought

A powerful, worldwide and persistent creator of droughts and floods resides in the El Niño-Southern Oscillation (ENSO) phenomenon that produces the El Niño and La Niña events. ENSO is a major climatic event, creating not only year-to-year climate variability (Gergis *et al.*, 2006) but also extreme events or indeed disasters – floods and droughts (e.g. Dilley & Heyman, 1995; Bouma *et al.*, 1997; Davis, 2001). This phenomenon is now relatively well understood (e.g. Allan *et al.*, 1996; Cane, 2005) and is clearly a very powerful force driving the world's climate.

The identification of ENSO, and coming to understand how it operates and the nature and spread of its effects, is a fascinating story involving many investigators in many parts of the world (Allan *et al.*, 1996; Davis, 2001).

As recounted by Davis (2001), in seeking to explain droughts in India and China in the late 19th century, meteorologists initially placed a strong reliance on sunspots, solar activity and air pressure. In 1897, the Swedish meteorologist Hugo Hildebransson described an inverse relationship in mean air pressure between Iceland and the Azores and recognized that this was connected to rainfall. This relationship is now known as the North Atlantic Oscillation (NAO). He also recognized two other oscillations – one between Siberia to India and one across the Pacific from Buenos Aires to Sydney.

Aware of this discovery, Sir Gilbert Walker, director-general of observatories in the India Meteorological Office (1904–1924), embarked on a programme involving many Indian clerks to identify, through a multitude of hand-calculated regressions, patterns of air pressure and rainfall relationships from a mountain of data collected around the world. In 1924, he identified three systems of long-distance atmospheric oscillation – the Southern Oscillation (SO) across the Pacific, the North Atlantic Oscillation (NAO) and the North Pacific Oscillation (now called the PDO). The Southern Oscillation involved an air pressure oscillation linked with rainfall between India, Indonesia and Australia in the west, and the Pacific including Samoa, Hawaii, South America and California in the east. Elaborate equations were used to calculate summer and winter SOI values (Allan *et al.*, 1996). However, no clear mechanism was identified to account for the Southern Oscillation. Progressively, the SOI was refined and simplified, so that now the SOI refers to mean sea level pressure differences between Darwin and Tahiti.

In the late 1950s and 1960s, the Dutch meteorologist Hendrik Berlage linked the SOI with sea surface temperatures (SST) and related an increase in SST in the tropical eastern Pacific to El Niño events, producing drought in Australia and floods in western South America. From this, Jacob Bjerknes, in a key paper in 1969, linked the low pressure and warm pool of the western Pacific (WPWP) with the cold water and high air pressures of the eastern Pacific. Sea level winds (easterly Trades) flow from the high pressure system to the low pressure WPWP. As they flow, these winds are heated and gain moisture so that the moisture-laden air rises, releasing heat and rain. This upper level air then moves eastward across the Pacific to descend in the eastern Pacific. Bjerknes called this circulation the Walker circulation. The winds from the east Pacific cause the WPWP to gain more warm water and to increase in level up to 40–60 cm above the east Pacific (Wyrtki, 1977). This is a positive feedback – the Bjerknes feedback. When the south-east trade winds fail, the warm water of the WPWP expands eastward and the upwelling of cold water off Peru weakens. This is reflected in the SOI as pressures decline in the east Pacific and rise across the west Pacific, centred on Australia and Indonesia.

Thus, what happens in an El Niño event was deduced, but the mechanisms producing the phenomenon remained uncertain.

In the 1970s, Klaus Wyrtki (1976, 1977) examined the oceanography of El Nino events. In the Pacific, as in many large bodies of water, there is stratification, with a warm layer of surface water separated from a cooler much deeper layer by a boundary layer called the thermocline. Wyrtki posited that the easterly trade winds build up the waters of the WPWP, deepening the thermocline. An El Niño event was marked by a relaxation of the trade winds, or even a pulse of westerly winds, and consequently the thermocline would decrease in depth and the accumulated water of the WPWP would move eastward across the Pacific. Near South America, this warm water mass would suppress the Humboldt Current upwelling.

In turn, the warm water off South America serves to further weaken trade winds. The winds and the SSTs are closely linked in phase, but it is the delayed changes in the depth of the thermocline altering the heat content of the WPWP that serves to create the oscillation (Cane, 2005). ENSO was so named by Rasmusson and Carpenter in 1982. Furthermore, Wyrtki (1976) explained why as an El Niño event ceases: there may be an overshoot of conditions to generate a colder WPWP and a return of the Humboldt Current upwelling, producing La Niña events (Philander, 1985).

From the work of Walker and others, it was realized that droughts across the world were linked in time, but the mechanism was unknown. Bjerknes proposed that forces arising in ENSO events in the tropical Pacific were transmitted away to interact with other climate systems. These connections he called teleconnections, a term originally coined by Ångström (1935). Teleconnections, for example, exist between ENSO and Indian droughts (Whetton & Rutherfurd, 1994) due to the failure of the Asian monsoon (Wahl & Morrill, 2010; Cook et al., 2010a, 2010b), and between ENSO and the North Pacific Oscillation, affecting North China rainfall (Whetton & Rutherfurd, 1994), and they serve to create floods and droughts in many parts of the world.

The tropical region of the Pacific Ocean, with its considerable length, the Humboldt Current upwelling in the east and the warm pool of water in the west, appears to be a very suitable area for an oscillator with the great strength of ENSO to be generated and, through teleconnections, to exert extreme events on sub-tropical and temperate regions. In terms of generating severe droughts of long duration in North America, southern Europe and south-west Asia, the Pacific has been described as 'the perfect ocean for drought' (Hoerling & Kumar, 2003).

El Niño and La Niña events are closely linked to the Southern Oscillation. When the Southern Oscillation Index (SOI) is positive, La Niña events occur;

when it is negative, El Nino events occur. Equatorial sea surface temperature (SSTs) in the Pacific are used to indicate ENSO events.

There are three major Niño regions: Niño 1+2 (0–10°S, 80–90°W), Niño 3 (5°N–5°S, 90–150°W), Niño 4 (5°N–5°S, 160–150°W) with a fourth Niño 3+4 (5°N–5°S, 120–170°W) being added in 1997 (Trenberth, 1997). Initially, the onset of El Niño events was detected by rises in SSTs in Niño 1+2, but Hanley *et al.* (2003) found that SSTs in this region were rather unreliable. Both Allan *et al.* (1996) and Hanley *et al.* (2003) suggested SST readings from Niño 3 were more sensitive and reliable, and today this region 'remains the primary area for climate model prediction of ENSO, (Gergis *et al.*, 2006).

When SOI has high positive values (La Niña) and SSTs are lower than normal in Niño 3, major flooding may occur in Australia, Indonesia, India, southern Africa and north-eastern South America, and droughts may occur in east and north-western Africa, Spain, southern North and South America (Ropeleweski & Halpert, 1989; Whetton & Rutherfurd, 1994; Allan *et al.*, 1996; see Figure 1.3a.). When SOI values are strongly negative (El Niño), with high sea surface temperatures in Niño 3 (Figure 1.3b), drought may occur in Australia, Indonesia, Oceania, central China, northern India, northern South America, Central America and southern Africa, while floods may occur in southern North and South America, southern Europe, east Africa, central and southern China (Ropeleweski & Halpert, 1987; Whetton & Rutherfurd, 1994; Allan *et al.*, 1996). El Niño events may be terminated by the rapid onset of La Niña, sometimes with severe flooding (Whetton, 1997). Clearly, not all droughts in the world are primarily caused by ENSO events, but it is also very evident that ENSO events, with the linked teleconnections, are responsible for many of the severe and damaging droughts.

The age of ENSO is uncertain; biological adaptations to high rainfall variability suggest that 'ENSO has been operating and affecting Australia for millennia' (Nicholls, 1989b). Evidence from lake deposits from Ecuador suggest that ENSO is at least 11,000 years old (Moy *et al.*, 2002), and evidence from fossil coral from northern Indonesia (Hughen *et al.*, 1999) and from peat sediments covering 45,000 years from Lynch's Crater in north Queensland (Turney *et al.*, 2004) suggests that ENSO was active in the last glacial-interglacial period. Cane (2005) contends that ENSO 'has been a feature of earth's climate for at least 130,000 years'. Such a time span would presumably be sufficient for biota to develop adaptations to deal with the extremes of ENSO cycle, as suggested by Nicholls (1989b).

The strength of the ENSO cycle has fluctuated in time with data, suggesting that ENSO events were absent or at least very weak in the early Holocene (10,000–7,000 years BP (before present)) (Moy *et al.*, 2002). Donders *et al.* (2007), in analyzing palynological data across many sites in

(A) El Niño

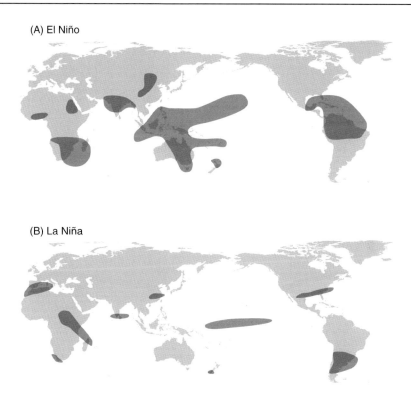

(B) La Niña

Figure 1.3 (a) Map of the world, indicating regions liable to incur drought conditions with an El Niño event. (b) Map of the world, indicating regions liable to incur drought conditions with a La Niña event. (Adapted from Allan *et al.*, 1996.) (See the colour version of this figure in Plate 1.3.)

eastern Australia, have produced strong evidence for an increase from 5,000 to 3,500 years BP in ENSO activity to current levels. In an analysis of ENSO signals from the present back to 1525, Gergis & Fowler (2006) identified 37 major El Niño events, including nine extreme events, four of which occurred in the 20th century, and 46 major La Niña events, including 12 extreme events, five of which occurred in the 16th to mid-17th centuries. As it appears that recent ENSO variability is strong and increasing in the 20th–21st centuries, there is cause for concern. Whether this increase is induced by climate change is quite uncertain (Cane, 2005).

El Niño events cause major changes in rainfall and, consequently, in surface runoff and streamflow – floods or droughts, depending on the region. An El Niño signal causing low streamflow and drought occurs in eastern Australia (e.g. Simpson *et al.*, 1993; Chiew *et al.*, 1998; Chiew & McMahon, 2002). Rainfall and streamflow both have lag correlations with the SOI

(Chiew & McMahon, 2002). Links between low streamflow and ENSO events have been reported for India (Ganges) (Whitaker *et al.*, 2001), New Zealand, Nepal (Shrestha & Kostaschuk, 2005), north-east South America, central America, and to a lesser extent northern (e.g. Nile River) (Eltahir, 1996) and south-eastern Africa (Chiew & McMahon, 2002) (see Figure 1.3A). The occurrence of La Niña events is associated with low flows and droughts in south-western North America (Cayan *et al.*, 1999; Cook *et al.*, 2007), southern South America, Spain and north-east Africa, southern India and central coastal China (Allan *et al.*, 1996) (see Figure 1.3B).

In summary, it is very evident that El Niño and La Niña events exert a powerful influence in generating drought conditions around the world.

Droughts linked to ENSO events can occur in mid- and south-western North America. A closely observed drought was the Dustbowl drought (1931–1939) that devastated the southern parts of the Great Plains in the states of Texas, New Mexico, Colorado, Oklahoma and Kansas (Figure 1.4). This severe drought, which temporally consisted of four droughts (Riebsame *et al.*, 1991), was driven by below-average sea surface temperatures (SSTs) in the tropical eastern Pacific (Schubert *et al.*, 2004; Seager *et al.*, 2005; Cook

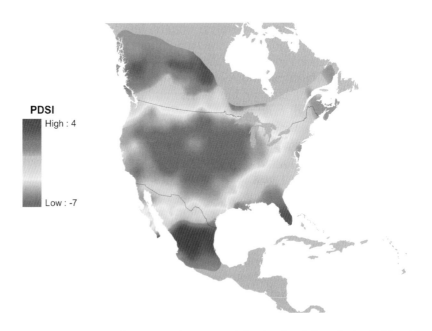

Figure 1.4 The spatial extent and severity of the Dustbowl drought in 1934, with regions in drought depicted by the Palmer Drought Severity Index PDSI (in red with negative values) and wet regions (positive PDSI and in blue). (Drawn using data from Cook, E.R., 2000.) (See the colour version of this figure in Plate 1.4.)

et al., 2007). Such temperature changes are small, being only 0.1–0.4 °C colder than normal (Cook *et al.*, 2007).

The train of events creating and maintaining drought, like the Dustbowl drought, appears to follow a sequence (Seager *et al.*, 2005; Cook *et al.*, 2007). The small sea-surface temperature changes cause the tropical troposphere (lower portion of the atmosphere) to cool, which subsequently causes the subtropical jet streams that flow from west to east to move poleward. This results in causing the weather systems which normally bring rain to the Great Plains to move poleward, which in turn causes moisture-deficient air in the upper troposphere to descend on the Great Plains. As long as this condition persists, there will be reduced rainfall.

Such droughts in North America are linked with La Niña events, in which there is abnormal cooling of the eastern Pacific Ocean. Persistent droughts, such as the extended drought of the 1950s, and the recent 1999–2002 drought, have been regarded as being due to a persistent 'La Niña-like state' (Seager *et al.*, 2005; Seager, 2007; Herweijer *et al.*, 2007; Herweijer & Seager, 2008). Indeed, extending this idea, Herweijer and Seager (2008) have suggested that 'the global pattern of persistent drought appears to be a low-frequency version of interannual ENSO-forced variability'.

Within some drought-affected regions, such as the Great Plains, there is a further phenomenon that may serve to maintain the drought. A coupling between the land and the atmosphere may develop whereby, as precipitation declines, soil moisture and evapotranspiration also decline and thus less moisture goes back into the atmosphere to generate precipitation, which consequently declines even more. Oglesby (1991), Forman *et al.* (2001) and Schubert *et al.* (2004, 2008) have suggested that this phenomenon is important in maintaining the extended droughts of the Great Plains, and Koster *et al.* (2004) have suggested that such land-atmosphere coupling reducing available moisture may occur in 'hot spots' in the Great Plains, central India and the Sahel (Dai *et al.*, 2004a; Foley *et al.*, 2003). Such a factor exacerbating drought is indicated by the research by Cook *et al.* (2008) on the Dustbowl drought. Climate model runs driven by east Pacific sea surface temperature data of the Dustbowl drought resulted in a simulated drought weaker than that observed in reality. However, the addition of data estimating the dust aerosol load increased the intensity and spatial extent of the drought to observed levels (Cook *et al.*, 2008).

1.5   Other important oscillations creating drought

While a major and extended research effort has gone into the discovery and unravelling of the mechanisms of the El Niño/La Niña oscillation, other oscillations have been discovered and have been found to be tied with the creation of drying conditions and drought.

The issue of clearly identifying the climatic factors generating prolonged droughts over North America seems not to be fully resolved. In addition to the concept that 'La Niña-like states' with cool sea surface temperatures in the east Pacific is a major generator of droughts, it is likely that two low-frequency oscillations in sea surface temperatures are also influential in drought generation. These oscillations, linked as teleconnections, are the Atlantic Multidecadal Oscillation (AMO) (Kerr, 2001) and the Pacific Decadal Oscillation (PDO) (Mantua *et al.*, 1997), with the AMO being an oscillation in sea surface temperatures of the north Atlantic Ocean with a recurrence interval of 70–80 years and the PDO being an oscillation in SST in the Pacific Ocean with a recurrence interval of 50–70 years.

It is proposed that both of these oscillations are correlated with hydrologic variability and the occurrence of severe droughts in western USA (McCabe *et al.*, 2004; Hidalgo, 2004). The AMO has been linked with droughts in central and eastern North America (Enfield *et al.*, 2001) and in western Africa (Shanahan *et al.*, 2009). In the latter location, the AMO in its current phase (30 years) has weakened the West African monsoon to possibly produce the severe and continuing 'Sahel Drought' (Foley *et al*, 2003; Dai *et al.*, 2004a; Held *et al.*, 2005).

Linked with the Atlantic Multidecadal Oscillation is the North Atlantic Oscillation (NAO), which operates with a periodicity of 5–10 years (Stenseth *et al.*, 2003). This oscillation is indicated by changes in sea level air pressure between the Azores and Iceland, and it is particularly active in winter. Changes in the oscillation result in major changes in wind speeds, and correspondingly in temperatures and precipitation (Hurrell *et al.*, 2003). Positive NAO index (NAOI) values (high pressure in the Azores) results in wet winters, with strong westerly moisture-laden winds over northern Europe but decreased precipitation in southern Europe. Negative NAOI values lead to weakened westerlies and cold, dry winters over northern Europe and increased precipitation over southern Europe (Hurrell *et al.*, 2003; Yiou & Nogaj, 2004).

Accordingly, extended periods of negative NAOI values are linked with dryness and droughts over northern Europe, and extended positive NAOI values are linked with drought in Mediterranean Europe (Hurrell *et al.*, 2003; Straille *et al.*, 2003). Severe drought and extremely low river flows in northern Europe are linked with negative NAOI values, and in southern Europe hydrological droughts occur when winters are dominated by a positive NAO phase (Shorthouse & Arnell, 1997; Pociask-Karteczka, 2006).

There is an oscillating sea surface temperature gradient between Indonesia and central Indian Ocean called the Indian Ocean Dipole (IOD) (Saji *et al.*, 1999). This oscillation is indicated by changes in sea surface temperatures between the tropical western Indian Ocean (50–70°E, 10°S–10°N) and the tropical south-eastern Indian Ocean (90–110°E, 10°S to equator) (Saji *et al.*, 1999; Saji & Yamagata, 2003). In a 'normal' year, south-east trade winds blow from Indonesia into the oceanic tropical convergence zone, delivering rain to India and Sri Lanka. However, the dipole oscillation is in the positive phase when there is cooling in the tropical eastern Indian Ocean off Indonesia, and a warming of the waters of the north-eastern waters of the Indian Ocean off western India (Saji *et al.*, 1999). Under these conditions, Indonesia and south-western Australia undergo drying and may be in drought. The drying may spread to central and eastern Australia, even exacerbating the effects of an ENSO-created drought (Nicholls, 1989a; Dosdrowsky & Chambers, 2001; England *et al.*, 2006; Barros & Bowden, 2008).

Both around the Arctic and Antarctica, there are annular modes (Thompson & Wallace, 2000), named the Northern Annual Mode (NAM) and the Southern Annual Mode (SAM) respectively. These are large systems that have a strong influence on temperate and subtropical weather systems, as they modulate the circumpolar westerly systems and strongly influence the strength and number of rain-bearing frontal systems moving from sub-Arctic or sub-Antarctic regions into temperate zones. In recent years, the SAM has appeared to become stronger and has been moving polewards. This strengthening may be related to the long-term decline in winter frontal systems and rainfall across southern Australia (Nicholls, 2010). In exerting a strong influence on rainfall in temperate and sub-tropical zones, SAM and NAM may thus interact with phenomena such as the NAO and ENSO to induce drying and droughts.

Droughts being induced by these dynamic oscillations and modes may vary in their severity and in the ways of formation. This is illustrated by research on three strong Australian droughts (Verdon-Kidd & Kiem, 2009). The 'Federation' drought (≈1895–1902) appears to have been primarily caused by ENSO and the PDO (IPO); the 'World War II' drought appears to have been multi-causal, with contributions from the IOD, SAM and PDO (IPO); and the major contribution to the recent 'Big Dry' drought has come from the SAM, along with ENSO (Verdon-Kidd & Kiem, 2009).

Through access to more accurate and comprehensive meteorological data concerning droughts, and through the rapid development of more and more sophisticated modelling, it appears that we are now gaining a more precise understanding of the mechanism(s) that may create, maintain and terminate droughts.

1.6 Drought in Australia

Being in the mid-latitudes, the flattest of the continents and relatively close to the warm pool of the western Pacific (WPWP) – the western dipole of the ENSO phenomenon – it is not surprising that the major part of Australia is arid and that the continent as a whole is drought-prone (Lindesay, 2003). For 82 of the 150 years from 1860, when reliable records began, until 2010, Australia has had severe droughts (McKernan, 2005; Bureau of Meteorology, 2006).

Most droughts, especially those affecting eastern Australia, arise from El Niño events (Allan *et al.*, 1996). In recent years, there has been considerable concern at the decline in rainfall in south-western Western Australia (Allan & Haylock, 1993) and with the possible influence of Indian Ocean conditions in influencing rainfall (and drought) (Nicholls, 1989a; Drosdowsky & Chambers, 2001).

Drought has been a force moulding the patterns of land use and abuse in Australia since European settlement, but this has not been fully recognized by historians (with some exceptions, e.g. Griffiths (2005) and McKernan (2005)). Indeed, shortly after the establishment of the first European settlement in Port Jackson (Sydney) in 1788, there was a severe El Niño drought from 1791 to 1793 (Nicholls, 1988; Stahle *et al.*, 1998b; Gergis *et al.*, 2006) which may account for the penal colony suffering a major setback and severe hardship.

As argued by Heathcote (1969, 1988, 2000), the reality of living in a drought-prone continent has taken a long time to be fully accepted by European settlers. If anything, it appears that Australia as a nation has been locked into the 'hydro-illogical' cycle of drought described by Wilhite (1992); (See Figure 1.5). In this cycle, drought arises and there is alarm, but when the drought breaks, activities return to the pre-drought state without any anticipatory and pro-active measures to contend with the next drought and with a firm belief in the existence of wet and dry cycles.

Both Keating (1992) and McKernan (2005) contend that it was the severe 1982–83 drought that effectively made dealing with drought a central part of Australia's politics and economy. Belatedly, planning for drought and adapting to climate change have become key political and management issues (Connell, 2007). Australia has had a considerable number of major droughts, including the Centennial drought (1888) (Nicholls, 1997), the Federation drought (1895–1903), the droughts of the two World Wars (1911–16, 1939–45) and the drought of 1982–83 (Keating, 1992; McKernan, 2005).

With drought, severe bushfires tend to occur, such as Black Friday of 1939 and Ash Wednesday of 1983 (Keating, 1992). Most, but not all, droughts

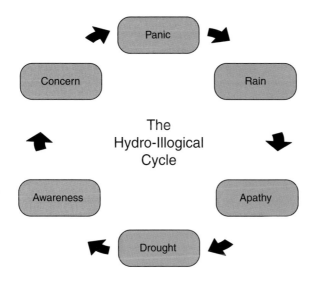

Figure 1.5 The 'hydro-illogical cycle' indicating the social reactions to drought. (Redrawn from Figure 2 in Chapter 1 of Wilhite, 2000.)

are linked with ENSO events (Nicholls, 1985; Whetton, 1997). In drought, river flows have been very low. For example, in the recent drought, the Murray River only received an inflow of 770 GL in 2006–2007, compared with an average annual inflow of 5,400 GL (Cai & Cowan, 2008).

The recent drought (1997–2010) was both severe and long and 'unprecedented in the historical records' (Timbal & Jones, 2008). It was much more severe than the long droughts of 1939–45 and 1946–49 (Watkins, 2005) and the Federation Drought. Nicholls (2004) noted that during this drought, in 2002, temperatures (and evaporation levels) were very high, which suggested that the nature of Australian droughts may be changing, being exacerbated by the enhanced greenhouse effect. This suggestion is supported by Karoly *et al.* (2003), Watkins (2005) and Timbal & Jones (2008).

2

Types of drought and their assessment

There are five recognized forms of drought: meteorological or climatological; hydrological; agricultural; ecological; and socioeconomic drought or operational drought (Tate & Gustard, 2000; Wilhite, 2000; Heim, 2002; NDMC, 2005).

Meteorological drought occurs when there is a deficit between the actual amount of precipitation received and the amount that may normally be expected for an extended duration. It is dependent in its determination on rainfall falling below threshold levels that are determined from long-term rainfall records. This form of drought is regionally specific and, given the need for long-term records to define it, it may be difficult to define in regions with highly variable rainfall (such as arid areas) or in regions with insufficient long-term rainfall data. Meteorological drought is the primary form of drought that leads on to the other four –hydrological, agricultural, ecological and socioeconomic. It is usually perceived as a shortage of rainfall, but it also includes precipitation by snow and even possibly fog, as for example in the case of tropical cloud forests. Snow may accumulate as snowpack and ice, and drought can become evident when the flows from spring thaw are abnormally low due to a poor snowpack.

Hydrological drought occurs when the amount of precipitation in a region is insufficient to maintain surface water:

1 for normally expected flows in streams/rivers (lotic systems); or
2 for normally expected levels or volumes in lakes/wetlands/reservoirs (lentic systems) (Tate & Gustard, 2000; Wilhite & Buchanan-Smith, 2005); and
3 to maintain subsurface water volumes.

It is usually defined at the basin or catchment level.

Drought and Aquatic Ecosystems: Effects and Responses, First Edition. P. Sam Lake.
© 2011 P. Sam Lake. Published 2011 by Blackwell Publishing Ltd.

In the natural state, hydrological drought is induced by shortfalls in precipitation affecting surface runoff and groundwater storage of water. However, it may be induced not only by precipitation deficits, but also by water deficits created by human land use and water storage (NDMC, 2005; Wilhite & Buchanan-Smith, 2005). The state of the catchment in terms of such factors as dryness, plant cover and soil porosity may strongly influence the onset of hydrological drought.

Agricultural drought 'is typically defined as a period when soil moisture is inadequate to meet evapotranspirative demands so as to initiate and sustain crop growth' (Changnon, 1987). As such, it focuses on 'soil moisture deficits and differences between actual and potential evapotranspiration' (Tate & Gustard, 2002) and is primarily centred on the availability of soil moisture in the root zone of crops, though it may also refer to lack of water for plant growth to meet the needs of livestock (Changnon, 1987). Agricultural drought is concerned with soil moisture so, while it dependent on rainfall, it is also strongly influenced by factors that govern water infiltration and soil water-holding capacity (Wilhite & Buchanan-Smith, 2005).

As agricultural drought is primarily concerned with crop growth, drought related to non-agricultural terrestrial biota is not covered by indices for agricultural drought. One could expect the terrestrial biota native to an area to be better able to deal with drought than exotic and domestic plants and animals.

Ecological drought has been recognized only recently. It has been defined as 'a shortage of water causing stress on ecosystems, adversely affecting the life of plants and animals' (Tallaksen & van Lanen, 2004). However, unlike agricultural drought, ecological drought currently lacks specific indices to quantify it.

Socioeconomic or **operational drought** depends basically on the availability of water for human activities, so this form of drought varies greatly with locality, human demand and with the level of infrastructure for water capture, storage and delivery (Mawdsley *et al.*, 1994). For example, socioeconomic drought in Australia is regarded politically as an exceptional circumstance that is defined economically in terms of a 'severe downturn in farm income over a prolonged period' (Botterill, 2003).

In this account, the major concern will be with meteorological and, especially, hydrological drought, though indices developed for agricultural drought may be relevant at times. In many parts of the world, there are seasonal dry periods that are both normally expected and predictable. As they are the outcome of normal climatic variation, these dry periods or 'seasonal droughts' (Lake, 2003), are not unusual in terms of expected patterns of precipitation and hydrology. Such seasonal droughts occur, for example, in those areas with a Mediterranean climate (e.g. Towns, 1985; Resh *et al.*, 1990; Gasith & Resh, 1999; Mesquita *et al.*, 2006) and in those tropical/subtropical areas with distinct wet and dry seasons (Kushlan, 1976a; Rincon & Cressa, 2000; Douglas *et al.*, 2003). While such dry periods do have ecological effects, because they occur predictably they may not be regarded as a damaging disturbance.

The droughts which are the major focus of this account are defined as those events that occur across seasons, and are 'supra-seasonal droughts' (Lake, 2003). Thus, even in areas with normal dry seasons or 'summer droughts', supra-seasonal droughts can occur due to the failure of normally expected wet seasons (e.g. Boulton & Lake, 1992b; Bravo *et al.*, 2001; Power *et al.*, 2008; see Figure 2.1).

In the literature, there is some confusion about the definition of drought that affects hydrological systems. As most studies on drought in freshwater ecosystems do not give comprehensive details of the drought(s) studied, distinguishing dry periods and seasonal droughts from supra-seasonal droughts can be difficult. Even in the case of studies on supra-seasonal droughts, characterizing the nature of the drought(s) may be

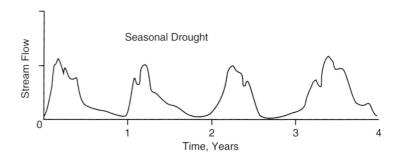

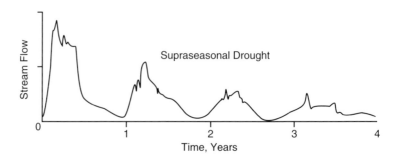

Figure 2.1 Seasonal droughts in comparison with a supra-seasonal drought.

difficult, because drought indices and characteristics of the drought – such as duration, intensity and spatial extent – are rarely given. This makes informed comparisons between different studies difficult.

Hydrological drought has mostly been defined in terms of availability of surface water, especially in terms of streamflow, but it is critical to note that groundwater levels may also undergo drought. Groundwater drought can be defined as occurring when there is a deficit in groundwater storages or heads in relation to normally expected storage levels or heads (van Lanen & Peters, 2000). In the natural state, drought is due to a reduction in groundwater recharge in relation to discharge. In the human-impacted state, groundwater drought can be created by extraction, and with excessive extraction groundwater droughts may be created independent of surface hydrological conditions (Scanlon *et al.*, 2006).

In any one region, as droughts develop, the form may change. Meteorological drought is the initial form of drought. In any one catchment, meteorological drought can be correlated with hydrological drought, which is indicated by cumulative water deficit. However, in some instances, such as in long periods of meteorological droughts, this correlation may weaken (see Figure 2.2). Dependent on soil properties and the type of agriculture, meteorological drought may lead on to agricultural drought.

Hydrological drought, which depends on precipitation, evapotranspiration and human land and water uses, usually takes time to set in. If the drought is relatively short, groundwater drought may not occur. However, when a surface water deficit persists for an extended period, groundwater drought sets in (Peters *et al.*, 2005), and its onset and severity may be accelerated by human groundwater extraction (van Lanen & Peters, 2000; Scanlon *et al.*, 2006; see Figure 2.3).

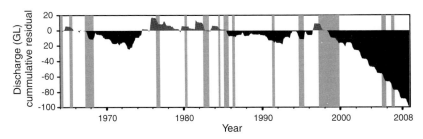

Comparison of discharge cumulative residual deficits at Joyces Creek from 1964 to 2009. Shading indicates periods of meteorlogical drought as defined by BOM indices. Zero is the long-term average.

Figure 2.2 Comparison at a local catchment level between meteorological drought (as determined by the rainfall deciles method) and the cumulative water deficit (severity), which is determined as a deficit from the long-term mean. The locality is Joyce's Creek in central Victoria, Australia. This severe drought was broken by flooding in mid-2010.

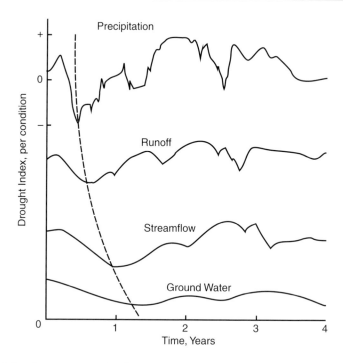

Figure 2.3 The progressive development of a supra-seasonal drought in a region from meteorological drought, to hydrological drought in surface water, and finally in groundwater.

When meteorological drought breaks with increased precipitation, agricultural drought may end shortly afterwards, dependent on the soil moisture deficits and the infiltration rate. The end of a hydrological drought usually takes time, due to considerable lags to recover as the deficits in water volume in catchments may be very substantial. The soil conditions of the catchment may take up the runoff to such extent that little may end up in running or standing waters. If the drought, aided by overgrazing, has created a hard, impermeable soil surface, precipitation inputs may be lost as surface runoff. This also occurs in urbanized areas, with their large areas of impermeable surfaces. Groundwater droughts usually have long lags in recovery.

Socioeconomic (operational) drought is the most difficult to define objectively as it arises from the interaction between meteorological, agricultural and hydrological drought with social and economic drivers of water use in a region. It occurs when water demands for economic goods exceeds the available supply, due to a weather-related shortfall in water supply. The announcement of this type of drought may be strongly influenced by political pressures (Heathcote, 2000).

2.1   Drought monitoring and indices

The World Meteorological Organization (1992) defined a drought index as 'an index which is related to some of the cumulative effects of a prolonged and abnormal moisture deficiency'. There are numerous drought indices, many of which are designed to meet particular needs or purposes. In fact, as with definitions of drought, the number of drought indices is immense. Their common theme is that they all 'originate from a deficiency of precipitation which results in a water shortage for some activity or some group' (Wilhite & Glantz, 1985). If there were standard drought indices, it would make the comparative study of the impacts of drought much more reliable and more comprehensive.

In their review, Tate & Gustard (2000) outline 12 different indices and list 25 different definitions of meteorological drought. Heim (2002) provides a detailed history of the development of numerous drought indices in USA, along with an evaluation of 13 of these indices, from Munger (1916) to the Drought Monitor (Svoboda, 2000). Given the difficulty in producing a universal definition of drought, there has been a proliferation of drought indices.

In an ideal world, drought indices could be used to compare droughts from region to region, to compare current droughts with those of the past, to identify drought-prone areas and to determine whether there are trends in droughts with time. The lack of consistent indices to characterize drought remains a major impediment to such comparative studies of the effects of drought.

In a study of the impacts of drought on aquatic ecosystems, the important indices are those used for detecting and assessing meteorological drought and hydrological drought (surface water and groundwater). Rather than review a wide range of indices, I will deal with several widely used indices.

2.2   Meteorological drought

There are two basic types of meteorological drought indices: those that address rainfall and precipitation deficits, and those that relate to moisture or water availability, with the latter addressing agricultural impacts of drought.

As indicated by Heim (2002) and Tate & Gustard (2000), there are many indices addressing rainfall deficiencies, many of them place-specific. For example, an early definition (1887) of absolute drought in Britain was 'a period of at least fifteen days without a daily total of 0.25 mm or more of rain' (Rodda, 2000). This definition persisted into the 1930s (Tate & Gustard, 2000).

The precipitation-based indices that are used include Rainfall Deciles (Gibbs & Maher, 1967) and the Standardized Precipitation Index (SPI) (McKee *et al.*, 1993). The Rainfall Deciles method overcame a major problem in earlier indices of drought that were based on percentage deviation from normal rainfall, which was usually based on monthly medians or means. As rainfall is not normally distributed, there could be significant differences between the median and the mean for each month.

Gibbs & Maher (1967) overcame this problem by dividing the frequency of rainfall occurrences into tenths or deciles of the overall frequency distribution. Thus, the first decile covers the rainfall total (e.g. for three months) that is not exceeded by the lowest ten per cent of rainfall occurrences. The median value is the 5th decile – that amount of rainfall not exceeded by 50 per cent of rainfall occurrences. The onset of drought is indicated if the total of precipitation over three preceding consecutive months is in the lowest decile (Kininmonth *et al.*, 2000). The end of a drought occurs when either the rainfall of the past month already places the rainfall for the three-month period in or above the fourth decile, or the rainfall for the past three months is above the average for that period (Kininmonth *et al.*, 2000; Keyantash & Dracup, 2002). The Rainfall Deciles method is used in Australia, but it is greatly limited by its dependence on long-term rainfall records for the localities of interest. The index is rated highly by Keyantash & Dracup (2002) in their evaluation of meteorological drought indices, just above the Standardized Precipitation Index (SPI).

The Standardized Precipitation Index (SPI) was devised by McKee *et al.* (1993) to indicate the strength of drought over a range of time periods. Short time periods may indicate soil moisture deficits resulting from low precipitation, while the longer time periods indicate rainfall deficits affecting streamflow, reservoir storages and even groundwater (Hayes, 2006). Like the Rainfall Deciles Index, the SPI requires a lengthy period of rainfall records (> 50 years of data (Guttman, 1999)), so that the long-term record after fitting it to a probability distribution can then be normalized. The mean SPI for a location is zero; if the normalized deficit reaches a level of -1 or less, a drought may have begun. SPI values of -2 or less indicate severe drought for the period under examination. The usual time intervals are 1, 3, 6, 9 and 12 months.

The SPI is valuable as it can clearly indicate the beginning, the end and the severity of a drought (Hayes, 2006; Keyantash & Dracup, 2002). It has proven to be a very useful index and is gaining increased support. For example, to assess the predictability of droughts in the Murray-Darling Basin in Australia, Barros and Bowden (2008) calculated the 12-month SPI for 345 gauge sites for the period from 1973 to 2002.

The Palmer Drought Severity Index (PDSI) was originally developed by Palmer (1965) to monitor meteorological drought as it affects soil moisture levels and hence agricultural production. The soil moisture levels are divided into two layers – the top layer (the 'plough layer') and the bottom layer (the 'root zone'). As it depends on the soil properties of any location, considerable knowledge of soil types is required for the index to be applicable, and this requirement limits its application. At any one location, long-term records (usually monthly averages of precipitation, evapotranspiration, soil moisture loss and recharge and surface runoff) are required to make calculations of coefficients of evapotranspiration, recharge, loss and runoff. These coefficients are standardized to determine the amount of precipitation required to maintain normal soil moisture levels for each month. The values for normal soil moisture are the Climatologically Appropriate for Existing Conditions (CAFEC) quantities and the deviation each month between the recorded precipitation and CAFEC precipitation produce a moisture anomaly index or Palmer Z index (Heim, 2002). The beginning and the end of a drought are then determined, given that the Z index for any one month depends on the moisture condition for that month and that of previous months. The value for a single month is called the Palmer Hydrological Drought Index (PHDI), as it is basically a measure of the likelihood of surface runoff, whereas the value for three months is the Palmer Drought Severity Index (PDSI).

The PDSI is available in the USA for many locations, due to the availability of good records of soil moisture properties across many different soil types. The PDSI for drought ranges from -1 to -2 for a mild drought to < -4 for an extreme drought (Hayes, 2006). In the USA. it has been standardized so that an extreme drought in Michigan may also be equivalent to an extreme drought in Utah.

The Palmer Index is applied widely across North America, but its application is limited in many parts of the world that, for a variety of reasons, do not have the required set of long-term records. As the PDSI is a measure indicating the potential for plant growth, dendrochronological records can be used to calculate past PDSI values, allowing reconstruction of droughts (Cook et al., 1999) and 'megadroughts' (droughts lasting more than ten years) across the continental USA (Stahle et al., 2007), and also of droughts across the entirety of North America (Cook et al., 2007).

However, given the complexity of the calculations and the data requirements, it is not surprising that the PDSI has attracted considerable criticism. Such criticisms include the arbitrary basis of selecting the beginning and the end of a drought, the lack of consideration of forms of precipitation other than rainfall, the assumption that runoff only occurs when the soil layers are full of moisture, and also the fact that the PDSI does not reflect the state of hydrology in long droughts. The PDSI is thus poorly suited to portray droughts in places such as Australia, where there is extreme variability of precipitation and runoff (Hayes, 2006).

The PDSI does not effectively operate in regions that have snowpacks as a significant component of the regional water input. Thus, in Colorado, where much of the water supply is derived from snow in the Rocky Mountains, the Surface Water Supply Index was developed (Shafer & Dezman, 1982) to complement the PDSI. This index requires four inputs: snowpack; stream-flow; precipitation; and reservoir storage, with streamflow replacing snow-pack in summer. Thus, this is actually a hydrological drought index developed for particular winter conditions in which the PDSI does not operate effectively.

In an evaluation of six meteorological drought indices, Keyantash & Dracup (2002) used six different criteria: robustness; tractability; transparency; sophistication; extendability; and dimensionality. They rated the PDSI the lowest, while the Rainfall Deciles index rated highest. However, a shortcoming of the Rainfall Deciles index is that it does not indicate hydrological conditions, whilst the PDSI can provide a soil moisture index as well as provide the PHDI.

2.3   Hydrological drought

Hydrological drought occurs when there are abnormally low discharges in flowing waters, low volumes of water in lakes and reservoirs and little or no water in wetlands and ponds. While surface water or streamflow droughts usually arise and end after meteorological drought, the lag between them means that they are not necessarily well correlated. Hydrological drought also covers groundwater drought, which usually lags well behind surface water drought (e.g. Changnon, 1987). Groundwater drought is indicated by the lowering of water yields from wells or bores, lower flows in springs and streams and 'even the drying-up of wells, brooks and rivers' (van Lanen & Peters, 2000). In the case of short or mild droughts, groundwater drought may not eventuate.

The study of surface water drought has focused almost exclusively on flow in running waters, though the volume of water in storages for human use is a form of surface water drought. The latter is also a form of socioeconomic drought, as it directly affects human economic activities. Hydrological drought is part of a continuum of low flow conditions. Thus, in detecting drought using streamflow records, the detection of the abnormal flow conditions of drought relies upon discriminating between normal low flows (or base flows) and abnormally low flows (Smakhtin, 2001; Hisdal *et al.*, 2004).

Low flows may be defined as those that occur in streams during the dry weather of seasonal low rainfall period (Smakhtin, 2001). The development of low flow conditions in a catchment or region depends on climatic factors (precipitation, temperature, evapotranspiration), catchment morphometry (area, length and slope, drainage density, elevation), catchment morphology (relief, orientation, standing waters), catchment geology and soil types, and patterns and types of human land use (Nathan & Weinmann, 1993; Smakhtin, 2001; van Lanen *et al.*, 2004).

In most situations, the low flows are base flow, which is defined as that 'part of discharge which enters a stream channel from groundwater and other stores, such as large lakes and glaciers' (van Lanen *et al.*, 2004). In regions where, in winter, most of the available water becomes frozen, underneath the ice of frozen waterways there may be low flows, producing a seasonal low flow period. In most cases, base flow consists of water from groundwater storage.

Flows from the catchment into its stream consist of basically three forms: overland flow or surface runoff, throughflow or interflow that moves downslope above the water table, and groundwater discharge (Gordon *et al.*, 2004). In catchments with highly variable rainfall, steep slopes and impermeable surfaces (e.g. rock, lateritic soils, shallow soils), drainage may be rapid; thus, with a rainfall event, there is characteristically a rapid rise and fall in streamflow – the flashy condition (van Lanen. *et al.*, 2004). Such streams due to low catchment water storage may be intermittent.

The flashy condition also occurs in urban areas with greater than 30 per cent of their catchments consisting of impervious surfaces and rapid drainage systems (Walsh *et al.*, 2005). For these system, it may be difficult to define base flow, so droughts may be frequent and relatively short-lived (van Lanen *et al.*, 2004). In the case of streams in arid and semi-arid places, it may be difficult, using streamflow, to discriminate drought from the normal prevailing dry conditions.

Converse to the flashy type of catchment and stream, there are flowing waters that are more predictable and receive a significant component of their flow from groundwater storages. These rivers and streams make up the majority of the watercourses used by humans. Such streams develop in size as catchment area increases, and there is a fusion of the effect of both intermittent and perennial streams.

In stream systems with considerable levels of groundwater storage, even in dry spells of considerable duration, surface water drought may be slow to emerge. However, once a drought emerges in these systems, it may be of considerable duration. Streams may have predictable low flow periods and be buffered from drought if they are draining wetland systems or if they flow out of large permanent lakes or are fed by flows from melting glaciers. In regions with Mediterranean or monsoonal climates, the groundwater recharge in the wet season is critical to the maintenance of the extended base flow of summer/dry season. Failure of winter or wet season rain may lead to supra-seasonal drought, where the flows drop below the long-term base flow.

Hydrological droughts are events in which the flows are abnormal low flows or in which water in lentic systems falls to abnormal levels. In trying to define a drought, the available hydrological time series (months, seasons, years) needs to be analyzed to set a threshold below which drought flows occur. Furthermore, there is a need to standardize the observations to describe the spatial extent of the drought (Dracup *et al.*, 1980b). Droughts may then be described in terms of duration, severity (cumulative water deficiency), magnitude or intensity (severity/duration), time of occurrence and spatial extent (Dracup *et al.*, 1980a, b; Bonacci, 1993; Hisdal *et al.*, 2004).

Smakhtin (2001) and Hisdal *et al.* (2004) envisage two ways to characterize droughts in terms of thresholds, namely by constructing either flow duration curves or low flow frequency curves. A flow duration curve can be based on daily, monthly or any arbitrary time interval (Figure 2.4) and usually depicts flow levels and their exceedance frequencies as percentages of time. The median flow or Q_{50} is a measure of average flow, although, as flow data are usually positively skewed, the mean usually exceeds the median. The Q_{90} or Q_{95} may indicate base flow in perennial streams and be set as a threshold of flows, below which drought flows occur (e.g. Tallaksen *et al.*, 1997). In intermittent streams, Q_{80} or Q_{90} may be zero, and thus values such as Q_{50} or much lower may have to be used.

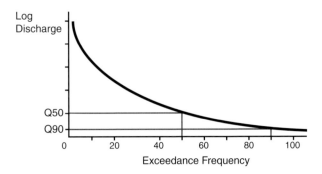

Figure 2.4 A flow duration curve illustrating the Q_{50} and Q_{90} thresholds.

For example, with ephemeral rivers in Nigeria, Woo & Tarhule (1994) set Q values ranging from Q_5 to Q_{20}. For such intermittent systems to arrive at a reasonable Q value, signifying low flow or drought flow, a long record is required. Flows below the set threshold can be measured as a total or cumulative water deficit (Keyantash & Dracup, 2002). Different streams near to each other, with similar precipitation inputs but different catchment geology or catchment land use, may have quite different flow duration curves.

Low flow frequency curves show the proportion of time (months, years) when a low flow is exceeded or 'equivalently the average interval (months, years) that the flow is below a threshold discharge' (Smakhtin, 2001). The curves can be compiled from series of annual flow minima, encompassing time series of annual minima averaged over durations. The curves show the annual minimum n-consecutive day discharges, and for drought detection it is the very low values bounded by zero that are important.

In plotting annual minima series against probability of exceedance, Velz & Gannon (1953) and the Institute of Hydrology UK (1980) detected a break point in the curve at which normal low flows separated from drought flows. They indicated that the break point occurred at an exceedance probability of ≈65 per cent. In south-east Australia, Nathan & McMahon (1990, 1992) suggested that the break point was closer to 80 per cent. For intermittent or ephemeral streams with long periods of zero flow, low flow frequency curves and flow duration curves may not very meaningful. Instead, the durations of zero flows may provide an indication of drought.

Both flow duration curves and low flow frequency curves can be used to set threshold or truncation levels, below which drought sets in. Droughts may occur as single events of long duration, or they may also occur as events interspersed with high flow events, in which case the drought in terms of cumulative water deficit may not be broken. Yevjevich (1967, 1972) applied the theory of runs or general crossing theory to the description of droughts. If a threshold is set, below which the low flows are deemed to be drought flows, there may be periods of sub-threshold flow punctuated by periods of above-threshold flow. Hence, for each deficit flow, the beginning and the end can be set. The low flows may be deemed to characterize the period of drought, so that 'the negative run-length represents the duration of a drought' (Yevjevich, 1967, 1972). Depending on the value of the set threshold, droughts that are dependent may emerge (Hisdal *et al.*, 2004). In this way, predictable seasonal droughts may be separated from the much more serious supra-seasonal droughts, which may themselves consist of dependent droughts as the water deficiency of a prior drought is not eliminated before the next drought sets in (Figure 2.5).

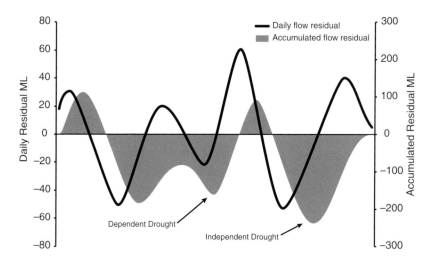

Figure 2.5 The nature of dependent droughts as determined by streamflows, with the cumulative water volumes showing the effect of the first drought leaving a lag in water volume for the second dependent drought. Independent droughts start with no deficit in water volumes.

A further way to characterize drought is the determination of the shape of the water deficit (Yevjevich, 1967). Streamflow deficits with drought can have an 'early steep deficit' and a 'slow decrease' (shape 2 of Yevjevich, 1967), and/or a slow increase in the deficit with a 'late maximum deficit' (shape 3 of Yevjevich, 1967). The total volume of the negative run-length, or the drought deficit volume, measures the severity, while the ratio between the drought deficit volume and the drought duration measures the drought intensity (Yevjevich, 1967; Hisdal et al., 2004).

Applying the theory of runs to characterizing hydrological drought for a stream can produce the total water deficit (Dracup et al., 1980b; Keyantash & Dracup, 2002). The drought is broken when the water deficit set by the selected threshold returns to zero. In the evaluation of hydrological drought indices by Keyantash & Dracup (2002), total water deficit was rated highly.

Groundwater drought usually sets in some time after hydrological drought has been established. It is difficult to characterize groundwater droughts effectively (van Lanen & Peters, 2000; Peters et al., 2005), because it is difficult to assess groundwater storage volumes. Under natural conditions, groundwater droughts may be characterized by a reduction in groundwater recharge, with a consequent lower groundwater head which can be readily measured. As drought sets in, groundwater recharge initially decreases, followed by decreases in groundwater levels and groundwater discharge. If there is human extraction of groundwater, drought may be difficult to characterize, depending on the levels of extraction. Groundwater droughts can have lasting deleterious effects on springs, streamflow, wetlands and riparian condition.

As droughts invariably occur over a large area, characterizing the spatial extent of the drought is an important component to determine. Droughts with large spatial extents occur with long durations, and embedded in the region affected there may be localities with different levels of drought indicators (e.g. severity, magnitude, etc.).

Dracup *et al.* (1980b) suggest three procedures for regionalization of droughts, with the first option being not to do it. The second option suggests regionalization based on the similarity of meteorological characteristics and similarity in geology and physical geography. The assumption is that where there is similarity in the above attributes, the hydrological responses to drought would be similar. The third option involves standardization, in that the grouping of sites depends on the similarity of the statistical properties of drought. In an ideal configuration, there should be a concordance between the climate-geographical regionalization and the standardization.

In relation to the ecology of drought in aquatic ecosystems, it is unfortunate that hydrological drought indicators are very rarely used to characterize the drought under investigation. In many cases, seasonal droughts are treated as if the droughts were legitimate hydrological droughts. Thus, as a note of warning, it must be pointed out that as the droughts studied ecologically are very poorly defined, it is difficult to make firm comparisons between droughts in different localities, or even between different droughts occurring at different times at the same locality.

In assessing hydrologic droughts using streamflow records, it appears that the usual assumption is that the stream systems are unregulated and that water extraction and diversion are negligible. In many, if not most, stream systems with human settlements, agriculture, and irrigation on their catchments, such an assumption is unwarranted. Water extraction and diversion serve to reduce flows and to exacerbate the effects of low flow and drought. In some cases where irrigation water is delivered downstream from dams, released flows may serve to lessen or prevent drought – the 'anti-droughts' of McMahon & Finlayson (2003).

Of the drought indices used, it appears that only some (e.g. PDSI, PHDI and SWSI) allow direct links to be made between precipitation and hydrologic drought. The Drought Monitor for the USA (Svoboda *et al.*, 2002) incorporates many indices, such as the PDSI, Standardized Precipitation Index SPI, Percentage of Normal Precipitation, CPC Soil Moisture Model Percentiles, Daily Streamflow Percentiles and the Satellite Vegetation Health Index, along with some ancillary indicators, to produce a weekly depiction of drought type, spatial extent and severity across the USA (Svoboda *et al.*, 2002). Such a system, giving multiple indicators, is clearly a highly useful and comprehensive way to depict the spatial and temporal dynamics of drought. Hopefully, in time, this could be applied to parts of the world beyond the USA.

3

The perturbation of hydrological drought

For clarity in examining disturbances and their impacts, it is necessary to discriminate between the event of the disturbance and the responses by the abiotic and biotic components of ecosystems to the disturbance (Bender *et al.*, 1984; Glasby & Underwood, 1996; Lake, 2000). Together, the disturbance and the responses to it comprise a **perturbation** (Bender *et al.*, 1984). Different types of disturbances include pulses, which are sharp, rapid events, and presses, which start sharply and then maintain their pressure. Pulse disturbances are common in nature (e.g. hurricanes, bushfires, tsunamis), whereas press disturbances are quite rare in nature and are usually generated by human activities (e.g. pollution, land clearing). The third type of disturbance is a ramp, a disturbance that steadily builds in strength. Droughts, especially supra-seasonal droughts, can be considered to be ramps (Lake, 2000), in that their initiation is difficult to determine precisely and their nature is to build steadily in strength and in spatial extent. Prior to this determination, droughts, rather surprisingly, were regarded as pulse disturbances (e.g. Detenbeck *et al.*, 1992).

As the drought builds and water availability declines, in some systems there may be sharp step-like events (thresholds), such as when a stream ceases to flow. The termination of droughts may be a slow process or, as mentioned earlier, it may be very sharp – marked by a pulse of severe flooding. Indeed, overall, the time taken for a supra-seasonal drought to become established is mostly longer than the time taken for the same drought to be broken. The ecological responses to drought are mainly ramp responses, although, as pointed out by Boulton (2003), there may be step changes in the biota and ecological processes, such as when a stream or a lake water level drops away from its vegetated shoreline.

Droughts are a distinctive type of hazard, as they are a disturbance of deficiency rather than one of excess. They may be characterized by their severity, intensity, duration, spatial extent, frequency (probability of

Drought and Aquatic Ecosystems: Effects and Responses, First Edition. P. Sam Lake.
© 2011 P. Sam Lake. Published 2011 by Blackwell Publishing Ltd.

recurrence) and timing (initiation and termination in calendar time). Ideally, droughts should be described by these characteristics in the same way that other types of disturbance, such as fires and floods, have been. Unfortunately, in many cases, droughts have instead been characterized by their effects. In most studies of hydrological drought and its ecological effects, the drought itself has been poorly characterized, making it difficult to undertake rigorous quantifiable comparisons between different studies and between different droughts occurring at the same place.

Droughts are dynamic events with constant changes in time and spatial extent. An indication of this dynamic nature can be seen in comparing the weekly drought maps in the USA, available on the website of the Drought Monitor (Svoboda *et al.*, 2002). In general, the more severe the drought, the greater will be the spatial extent. Within the extent of a large drought, there can be areas where the drought is most severe.

As mentioned before, in studying the impacts of drought, a distinction must be made between seasonal droughts and supra-seasonal drought that arises from an abnormal deficiency of precipitation and, hence, water availability (Lake, 2003). Supra-seasonal droughts of long duration include megadroughts (droughts lasting longer than a decade), which are regarded as 'large infrequent disturbances' or LIDS (Turner & Dale, 1998). LIDS are defined as disturbances that are not only infrequent but are extreme in terms of duration and severity (Turner & Dale, 1998). Such droughts can have drastic ecological effects with incomplete recovery, such that they may leave a lasting impact on affected ecosystems (Romme *et al.*, 1998). Their effects may be exacerbated by their interactions with other disturbances, including those of anthropogenic origin. In this situation, the combined effects of other attendant disturbances with LIDS, such as drought, may create 'ecological surprises' – ecosystem or community configurations that were not anticipated (Paine *et al.*, 1998).

Many studies do not distinguish between seasonal and supra-seasonal droughts. In some cases, a particularly severe seasonal drought may lead to a supra-seasonal drought. For example, Boulton and Lake (1992a, 1992b) studied two rivers that normally ceased to flow over summer (i.e. seasonal drought). Due to failure of winter rains in 1982–83, the streams suffered from an intense El Niño-driven supra-seasonal drought, which had an average SOI value of −21.7 – the most severe value in the period of 1900–2002 (Barros & Bowden, 2008).

As hydrological drought sets in, there is a reduction in available water; streamflow may go below base flow, shallow wetlands may dry and lake levels may drop. In wetlands, lakes and flowing waters, the water level may recede from the shore, the littoral vegetation and the riparian zone, producing a step function in the ramp trajectory (Boulton & Lake, 2008).

With further drying, longitudinal connectivity in stream channels can be severed, causing the channel to break up into a series of pools and eliminating habitat areas with flowing water, such as riffles and runs – another step in the ramp trajectory. Wetlands may also be broken up into pools, and lakes normally linked by streams may become isolated.

As the drought continues, habitat reduction and compression become severe and are accompanied by changes in water quality. Intra-specific and inter-specific interactions may intensify or even develop *de novo*. In streams, severe and lasting drought may produce a dry streambed, whereas wetlands may become completely dry and lakes greatly diminished in extent. Some time after hydrological drought starts to take its toll, groundwater drought may set in, with a lowering of the water table and a reduction in groundwater springs.

All of these threatening events are marked by changes in both abiotic and biotic ecosystem components. In the case of the biota, the impacts of drought may be countered by two basic and interlinked properties – resistance and resilience (Webster *et al.*, 1983; White & Pickett, 1985; Fisher & Grimm, 1991; Giller, 1996).

- Resistance refers to the capacity of the biota to withstand the stresses of a disturbance. High resistance may be difficult to detect, as it means that with a drought there are few changes in the populations or community structure or ecological processes.
- Resilience refers to the capacity to recover from the disturbance, even though, in some instances, the biota and ecological processes have been greatly diminished.

For most aquatic biota, the response to drought involves both resistance and resilience. As a drought persists, resistance may weaken and resilience strategies of various capacities become increasingly important. Unfortunately, while there is considerable information on the drought-resistant capacities of biota, there is a lack of information on resilience, due to many studies not being long enough to assess recovery – a function of resilience.

As a generalization, it appears that for the biota of stream ecosystems impacted by floods, the resistance is low but the resilience is high, with substantial recovery of macroinvertebrates in many instances being faster than the average generation time (Giller, 1996; Lake, 2007). Unfortunately, we cannot make such an assessment for droughts in freshwater systems, as few studies have evaluated the impacts of drought combined with any subsequent recovery (Lake *et al.*, 2008). Clearly, different biotic components differ substantially in their capacity to deal with drought.

Through the efforts of the Resilience Alliance (Walker & Salt, 2006), the understanding and scope of resilience, especially as applied to ecosystems, has expanded considerably. Ecosystem resilience refers to the capacity of an ecosystem to tolerate disturbance without collapsing into a qualitatively different state that is controlled by a different set of processes (Resilience Alliance, 2008). A resilient ecosystem can withstand shocks and rebuild itself when necessary and the concept of resilience may also be applied to social (human) systems. Resilience as applied to ecosystems, or to integrated systems of people and the natural environment, has three defining characteristics:

- The amount of change the system can undergo and still retain the same controls on function and structure.
- The degree to which the system is capable of self-organization.
- The ability to build and increase the capacity for learning and adaptation.

Two important issues arise from such a definition. First, disturbed ecosystems may not recover to their original state, but may cross thresholds and move into a new state that may or may not be stable. The ecosystem structure and function are no longer reversible (Palumbi *et al.*, 2008) and, even with extended time, recovery may not be possible. The second key component is the realization that most ecosystems are not free of human intervention and are better regarded as socio-ecological systems (SESs), where human actions may have significant and lasting effects that may reduce or strengthen ecosystem resilience (see Chapter 11).

In assessing the effects of drought and other disturbances, it may be insightful to examine some ecosystem properties that relate positively to resistance and resilience. Though the evidence is not conclusive in freshwater systems, diversity and ecological functions can be positively correlated. Disturbances such as drought, in depleting or eliminating species in functional trophic groups of a community, may reduce the efficiency of resource utilization and productivity (Cardinale & Palmer, 2002; Covich *et al.*, 2004; Cardinale *et al.*, 2006). This reduction appears to be dependent on the identity of the species that are lost.

Complementarity between species occurs when species in a community display niche partitioning that allows the processing of similar resources, or when interspecific interactions increases the effectiveness of resource utilization by species (e.g. mutualism, facilitation) when they co-occur (Cardinale *et al.*, 2007). Changes in complementarity may add to the loss in ecosystem processing. In flowing waters, it appears that mutualism is quite rare, whereas there are many examples of facilitation. Thus, in streams, species of shredders, by breaking down intact leaves to leaf

fragments, may facilitate resources for filter feeders and collectors. Therefore, the reduction of species such as shredders, as drought persists, may
weaken complementarity.

Linked with complementarity is the concept of redundancy, which refers
to the situation where, in a community, there are ecologically equivalent
species in terms of function that differ in their responses to environmental
factors (Walker, 1995). Redundancy in a community may strengthen both
resistance and resilience in drought; although species are lost, ecological
function remains, albeit reduced. In the case of drought, that is usually a
large-scale, long-duration disturbance, and both redundancy and complementarity may be spatially distributed across an ecosystem. For example, if
drought is more severe on headwater streams, some streams may be reduced
in ecosystem functions, whereas others feeding into the same river system
may maintain their ecosystem functions through either redundancy
or complementarity.

Aquatic ecosystems marked by high productivity may be able to recover
more effectively from drought than ecosystems that have a low productivity.
Productivity in highly productive systems may return rapidly when the
drought breaks, as there may not be a shortage of nutrients to fuel primary
production and, with drought, there may be a build-up of detritus that
becomes available with the return of water. In the case of some systems, such
as temporary pools, the detritus on the bottom of the pools may, with drying,
increase in nutritional quality (Bärlocher et al., 1978; Colburn, 2004); when
the drought breaks, this detritus may fuel a spell of high production,
relatively free of predation (Lake et al., 1989).

Both resistance and resilience to a drought may be shaped by the legacies
of past events. Ecological memory refers to the capacity of past states and
disturbances to influence contemporary or future states, and has been
applied to populations (Golinski et al., 2008), communities (Padisák,
1992) and places/landscapes (Whillans, 1996; Peterson, 2002). In the
case of a freshwater community, past disturbances (e.g. floods, droughts)
may shape species composition, such that the responses to a current drought
in terms of immediate impacts (resistance) and subsequent recovery (resilience) are strongly influenced by ecological memory – the species pool
shaped by past events. A severe drought, for instance, could exert a strong
influence on the responses to later droughts. Alternatively, a series of short
but frequent droughts could produce a very different memory or legacy to
that produced by a severe drought.

Resilience, or the capacity to recovery after drought, can take a number
of trajectories. Reversibility is regarded as a key component of resilience
(Palumbi et al., 2008), and it may be achieved after a drought, depending
on the time required. It can also, of course, be thwarted by the occurrence

of another drought or another type of disturbance. In many stream ecosystems, with their dynamic communities, the goal for reversibility may be hard to determine. Reversibility may not be achieved after drought, but a functional community or ecosystem does return, with most, but not all, of the original species and with different configurations of relative abundance. This situation is similar to that envisaged in ecosystem restoration, whereby reversibility is thwarted and a different state is achieved in restoration –referred to as the 'Humpty-Dumpty' model (Sarr, 2002; Lake *et al.*, 2007). Conceivably, drought may push a freshwater ecosystem across a threshold so that it becomes an alternative state, stable or otherwise.

As a disturbance, drought could conceivably act as a major force for regulating species diversity, by altering species diversity either directly through disturbance or indirectly by altering the availability of resources, in particular food and nutrients. In streams, it is floods, as a type of natural disturbance, that have been studied in relation to diversity. Thus, at highly disturbed stream localities, species diversity may be reduced, and with decreasing disturbance at different localities within the same region, species diversity may increase. Such a pattern has been found (e.g. Death & Winterbourn, 1995), though it may be also influenced by productivity (Death, 2002).

Alternatively, the intermediate disturbance hypothesis (Connell, 1978) may apply to stream communities. In this theory, a unimodal curve is produced, with low disturbance giving rise to low diversity, due to intense competition. High disturbance also produces a low diversity of hardy opportunistic species and intermediate levels of disturbance, having higher diversity with a blend of species. Evidence for this theory was indicated by Townsend *et al.* (1997) in New Zealand streams and by Reynolds *et al.* (1993) in phytoplankton in lakes.

As floods are rather frequent pulse disturbances, it is feasible to derive evidence for their effects on community structure. However, as droughts are ramp disturbances, usually of long duration and at large spatial extents, it can be difficult to acquire evidence for their effects on community structure, especially if data are needed from a series of droughts. Alternatively, experimental droughts in micro- or mesocosms may be insightful (e.g. Chase, 2007), but their applicability to the large spatial extents which are typical of natural droughts requires considerable caution. Droughts vary in severity and spatial extent, and a very severe drought (intense and of long duration) may be an extreme event that can reduce many populations and species and can drastically alter ecological processes. For example, Thibault and Brown (2008) found that an extreme flood, a 'punctuational perturbation' caused lasting and major changes to a desert rodent 'community'. Evidence that droughts may cause major reconfiguration of aquatic communities is sparse.

3.1 Refuges and drought

To resist the stresses of drought and to strengthen resilience, biota may use refuges. In this book, the Latinized terms 'refugium' and 'refugia' will not be employed, and instead refuge and refuges will be used. This step is taken to distinguish refugia in a palaeoecological and evolutionary sense (e.g., Bennett *et al.*, 1991; Stewart & Lister, 2001) – such as in places where biota survived ice ages – from refuges in a more immediate ecological sense, such as places where biota have survived floods, droughts and wildfires.

Refuges from disturbances were originally defined as 'habitats or environmental factors that convey spatial and temporal resistance and/or resilience to biotic communities impacted by biophysical disturbances' (Sedell *et al.*, 1990), and were initially seen in freshwater ecology in the context of dealing with floods (e.g. Lancaster & Hildrew, 1993; Giller, 1996; Lancaster & Belyea, 1997). The type and availability of refuges are part of an organism's environment and their use is a function of the adaptations of individual species.

Axes of adaptation influencing the use of refuges include: mobility and dispersal capacity; life history type; generation duration; reproductive capacity; and physiological capacity to withstand stress. Lancaster & Belyea (1997) outline four different types of mechanism of refuge use. Two types operate between generations, with one involving movement between similar habitats and the other involving movement between different habitat types (e.g. aquatic/terrestrial). The other two types operate within a generation and either involve movement between different habitats or remaining within a habitat patch, but with changes in behaviour 'habitude' (Lancaster & Belyea, 1997). In terms of dealing with drought, all four types appear to be in operation.

Refuge use differs considerably between biota. For example, fish cannot survive without water, whereas many invertebrates can seek and use refuges without any free water. Stream fish that become confined to pools may differ considerably in their tolerances to high temperatures, low oxygen concentrations and high concentrations of polyphenols (Magoulick & Kobza, 2003; McMaster & Bond, 2008). Fauna moving into refuges to survive during drought may, however, lead to increased densities that in turn could increase intra- and interspecific competition, within-refuge predation and predation by terrestrial predators, notably birds (Magoulick & Kobza, 2003).

Depending on the biota seeking refuges and the setting and type of water body, the variety and abundance of potential refuges can vary considerably. In two temporary streams in south-eastern Australia, for example, Boulton (1989) identified six types of refuge that were used in seasonal droughts and in one severe supra-seasonal drought, whereas in an intermittent stream in Arizona, Boulton *et al.* (1992) identified only three types of drought refuge.

In a review, Robson *et al.* (2008b) define a refuge as a secure place persisting through a disturbance with the critical criterion being that after the disturbance the refuge provides colonists to allow populations to recover. Robson *et al.* (2008b) define five distinct types of refuges, based on the composition of the refuge users and the mode of their formation. In the case of the latter point, they identify 'anthropogenic refuges' (e.g. farm dams, reservoirs, etc.) as a distinct type. Three types of refuges depend on the composition of the occupants: 'ark-type' refuges contain species representative of the surrounding land/waterscape; 'polo-club' type refuges contain species specialized to use them; and 'casino-type' refuges are areas not subject to disturbance and harbouring a biota lucky enough to be there before the disturbance. 'Stepping-stone refuges' are those vital to the dispersal of critical life history stages of certain biota.

As regards drought, Robson *et al.* (2008b) see many of the refuges that biota use as being 'ark-type' refuges, with specialized places such as riparian zones and crayfish burrows being 'polo-club' type refuges. Inter-specific interactions, such as predation and parasitism, may with time, in drought, change an ark-type refuge into one more resembling a 'polo-club' type refuge.

The critical importance of refuge use is the capacity to weather the drought, and for populations to recover after the drought. For invertebrate taxa that withstand drying within a locality, either as tolerant adults or as drought-resistant eggs and cysts, recovery depends on becoming active, feeding and rapid reproduction. For fish that are restricted to refuges with free water, there is the need for connectivity with the breaking of the drought, to allow migration into new localities. In lakes and wetlands, this may not be possible, and in streams, even with water flowing, riffles and shallow sections may act as migration barriers for some species. Man-made barriers such as fords, weirs and dams may reduce connectivity and impair recovery after drought as migration out of refuges is restricted (Magoulick & Kobza, 2003; Matthews & Marsh-Matthews, 2003).

3.2 Traits and adaptations to drought

Linked with refuge use is the range of adaptations or traits that organisms have evolved to deal with drought. Adaptation to predictable seasonal drought may be feasible, as would the development of adaptations to be active when water is available, as shown by the fauna of temporary ponds and streams (Colburn, 2004; Williams, 2006). However, adaptations or traits to contend with supra-seasonal droughts, which are unpredictable, severe, and mostly of long duration, may be more difficult to achieve. Nevertheless, it is clear that many organisms have traits that allow them to resist and/or recover from supra-seasonal drought, and traits to contend with seasonal drought may aid in survival through supra-seasonal drought. However, the above speculations are poorly supported by any evidence, as the analysis of traits in freshwater biota to contend with disturbance has largely concentrated on floods (Lytle & Poff, 2004).

Traits to contend with seasonal drought in the fauna of Mediterranean climate streams may also be suitable for supra-seasonal drought. Bonada *et al.* (2007) analyzed traits of macroinvertebrates in perennial, intermittent and ephemeral streams. They did not find any significant traits for the perennial streams, but they did find significant traits for the fauna of intermittent and ephemeral streams. Fauna at both these latter stream types may possess traits for supra-seasonal drought, such as active aerial dispersal, tegumental and aerial respiration, diapause and dormancy.

It is clear from the literature that there is a wide array of traits/adaptations across an array of taxa, from algae to fish, that assist in resisting and recovering from drought. The array of environments ranges from temporary ponds and streams to large lakes and perennial rivers. While no statistical analysis has been applied to traits for contending with drought, lists of observed traits for both lentic and lotic biota can be compiled from the literature.

3.3   The nature of studies on drought in aquatic ecosystems

Our understanding of the effects of seasonal droughts is far more advanced than our knowledge of supra-seasonal droughts. Those studying seasonal droughts have the advantage that they are predictable and are short-lived (months compared with years). Our knowledge of supra-seasonal droughts is incomplete and patchy, which is not surprising, given the low predictability of their onset and of their termination, along with their large spatial extent. Another hurdle to understanding the effects of drought is that in most studies, the parameters of the drought investigated are rarely given. Unlike many other types of disturbance, such as fire, earthquakes, floods, this lack of characterization of the drought makes comparisons between droughts, either at the same locality or between localities, difficult.

In a survey of 129 papers on drought in freshwater ecosystems, Lake *et al.* (2008) found that there were 51 studies on seasonal droughts, 67 on supra-seasonal droughts and 12 on both types of drought. Of these studies, 29 dealt with standing water bodies, eight focused on floodplain wetland systems and the remaining 92 were on flowing water systems. With the latter, there were two groups: small streams with a stream order of 3 or less, and rivers of order 4 or above. Only 23 of the studies were on rivers, whereas 69 were on the small streams. There were 36 studies on intermittent streams, with 32 dealing with seasonal drought and only four studying both seasonal and supra-seasonal droughts. For the 56 perennial stream studies, 47 were on supra-seasonal droughts, five on seasonal drought and four addressing both types of drought.

Droughts, even seasonal ones, usually are large-scale phenomena. The studies were divided into three spatial categories:

- **site** for studies carried out at one or two proximate sites;
- **local** for studies at multiple sites across a wetland system or within a river system;
- **regional** for studies at multiple sites across large river basins, or regions.

Only 22 studies were at the site level, 27 were at the regional level and most studies (80) were at the local level. Thus, there is a dearth of studies that have been carried out at the large spatial extent, the scale at which droughts are manifest.

Of the biota of concern in drought studies, two studies used microbes, ten used algae, six centred on macrophytes, seven on zooplankton, 59 on macroinvertebrates and 56 on fish. There were only seven studies on ecological processes, principally decomposition. Very few studies dealt with more one biotic group, and there was only one comprehensive study, combining physico-chemical data depicting the drought with the responses of omnivorous, herbivorous and carnivorous infaunal and epibenthic invertebrates and fish in a Florida estuary to a two-year drought (Livingston *et al.*, 1997).

Supra-seasonal droughts can have a long duration, but this is not reflected in the time span of drought studies. Of the 129 drought studies, 110 studies were for three years or less, with 37 being for only one year. There were very few long-term studies, with those of Elliott *et al.* (1997) and Elliott (2006), for 30 and 34 years respectively on drought and trout populations in streams in the English Lake District, being exemplary studies.

After examining studies as to whether data were gathered *before* the drought, *during* the drought or *after* the drought, the studies were divided into six different categories. There were, surprisingly, 83 studies that fitted the before-during-after category, but of these 52 covered three years or less in duration and were thus dealing with short supra-seasonal droughts or seasonal droughts. There were 15 before-during studies, 19 during-after, three before-after, seven during and one after.

All of this indicates that although valuable information has been gathered from all of these studies, very few complete studies, either in duration or in design, have been executed. This is, no doubt, due to the unpredictability of drought (especially supra-seasonal ones), the great difficulty in performing experimental studies on drought, and to the general neglect of long-term studies in ecology, and in freshwater ecology in particular. Whereas we have some understanding of the impacts of drought, the short duration of most studies means that we have a poor understanding of recovery and of the lags (ecological memory) and long-term changes that drought may leave in populations, communities and ecosystems. A particular weakness of the studies of supra-seasonal drought is that the observations of post-drought recovery rarely exceed one year.

4

Droughts of the past: dendrochronology and lake sediments

Droughts, like other disturbances, are singular events. They vary in frequency, intensity, and spatial distribution. While individual drought events directly impact on organisms in natural communities, it is the underlying drought regime that shapes populations, communities and ecosystems.

Characterizing the influence of drought on ecological systems depends on understanding the nature of drought regime and its inherent variability. The relatively short record of direct observations on drought limits inference about this variability and its ecological, economic and social impacts. To understand drought and drought regimes, one must consider longer time frames. This requires examining a variety of biological and geological proxy climate records, such as tree-rings, lake sediments, corals and speleothems. Proxy records of drought have provided unique insights into drought at centennial and millennial timescales and regional and continental spatial scales. However, proxy climate records also have their limitations, which must be accounted for in considering their interpretation. In this chapter, I briefly describe the proxy indicators that are commonly used to reconstruct historical droughts, consider reconstructed drought records over the Holocene ($\approx$10,000 years ago to the present) and the ecological inferences derived from them, and outline their limitations.

Descriptions of the duration and severity of past droughts are valuable for two main reasons:

- First, they provide an understanding of droughts and dry periods in past climates that have influenced past environments (especially aquatic ones) and produced selection pressures that have shaped the contemporary biota and have reduced and eliminated past biota.

Drought and Aquatic Ecosystems: Effects and Responses, First Edition. P. Sam Lake.
© 2011 P. Sam Lake. Published 2011 by Blackwell Publishing Ltd.

- Second, they may serve as a warning to consider in the formulation of current water resource planning, especially as, in the past, changes to drought conditions and drought regimes have been abrupt, not gradual (Overpeck & Cole, 2006). Also, recent droughts may not reflect the magnitudes of past droughts – droughts the like of which may lie in the future.

The focus of this chapter, as with the book as a whole, is on sustained hydrological droughts. These are droughts for which there is evidence of hydrological changes, be they changes in lake levels and volumes, shorelines or streamflows. Different proxy records of drought cover varying time spans and have varying accuracies and resolutions. In most cases, seasonal and short droughts do not leave a discernible signal in the proxy records. As such, many of the droughts detected in the proxy record are relatively long ones – in some cases, megadroughts (>10 years long; Woodhouse & Overpeck, 1998).

The proxy records for drought are geographically biased. In most cases, drought reconstructions are inferred from changes in lake sediments and tree-rings. The most extensive and comprehensive records for these come from North America, although important and steadily increasing records are also available from east Africa, east Asia and South and Central America.

4.1 Indicators of past droughts

Records of past droughts vary in resolution and length. At one extreme, networks of meteorological stations can provide sub-daily measurements of temperature, precipitation, relative humidity and insolation across a variety of spatial scales. These data, which are available for the past ≈100 years, may then be used to assess drought situations at sub-annual conditions.

Beyond the contemporary record, proxy indicators can be used to indicate supra-seasonal droughts over varying time periods and levels of temporal resolution. For example, archaeological records may provide indirect evidence of drought in the past 3,000 years or so. Dendrochronology, the study of tree rings, can provide a sensitive signal of drought (annual), extending back for around 2,000 years at most. Lake sediments provide a less sensitive indication of drought, but they do offer a wide range of signals, from stable isotopes to biotic remains (e.g. diatoms, pollen), that may provide a record of past droughts extending back into the early Holocene and even the Pleistocene (10,000 – 1.8 million years ago) (Cohen, 2003). In analyzing lake sediments, it is usual for a number of indicators to be used; a procedure that increases the capacity to detect droughts. Indicators from lake sediments may, however, only provide aggregate signals of extended droughts and drying periods. Across the range of different proxy records, tree-rings and lake sediments used together are particularly helpful in studying past drought, because they provide a useful compromise between the resolution and length of record.

4.1.1 Dendrochronology

The study of tree-rings (dendrochronology) has been fundamental to the study of past climates because trees are sensitive to changes in soil moisture and evaporative stress and are widely distributed across the terrestrial surface of the earth (Cook *et al.*, 2007). The tree-ring record of drought is best developed for North America (Canada, USA, Mexico), where reliable, large-scale drought records date back as far as 800 AD (Cook *et al.*, 2007, 2010). Dendrochronological drought reconstructions typically use the Palmer Drought Severity Index (PDSI), a metric of soil moisture available to trees, to identify periods (droughts) in which tree-growth may be water-limited. Tree-ring values can be calibrated and verified against PDSI values calculated directly from instrumental data over a common period, using standard statistical modelling techniques (e.g. 'point-by point regression' Cook *et al.*, 1996). The statistical model can then be used to estimate historical PDSI values from the tree-ring values that pre-date the instrumental data, and this reconstructed record can be used to identify past drought events (Cook *et al.*, 1999, 2007, 2010; Stahle *et al.*, 2007).

The most significant dendrochronological reconstruction of drought covering a continent is the Living Blended Drought Atlas (LBDA), a recent extension of the North American Drought Atlas (NADA), which is still being progressively developed (Cook *et al.*, 2010a). The LBDA is based on 5,638 temperature stations and 7,848 precipitation stations from the US, Canada, and Mexico, and it uses 1,845 tree-ring chronologies to reconstruct historical drought over 11,396 $0.5° \times 0.5°$ grid cells. Depending on the region of the grid, the reconstructions are several centuries to more than a millennium, with the majority of the longest records occurring in the West.

Recently and using evidence from tree rings, corals, ice cores, speleothems, ocean sediments and recorded history, it has been possible to reconstruct the severity and spatial extent of megadroughts across Asia back to mid-Mediaeval times (Cook *et al.*, 2010b). The megadroughts resulted from failures of the Asian monsoon – failures which may have been substantially due to changes in tropical sea surface temperatures in the Indo-Pacific Oceanic region (Cook *et al.*, 2010b).

Tree-ring chronologies can be used to reconstruct terrestrial drought conditions that produce hydrological droughts. For example, modern tree-ring

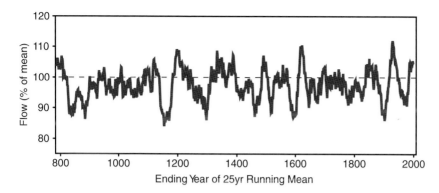

Figure 4.1 Flows of the Colorado River reconstructed from 11 tree-ring chronologies determined in the Colorado River basin. The time-series plot is a 25-year running mean and indicates the severe and lengthy mediaeval megadrought from 1143 to 1155. (Redrawn from Figure 2 of Meko *et al.*, 2007.)

records collected from trees in river catchments can be calibrated and verified with natural river flows or estimated natural flows that take in consideration water storage, extraction rates and diversions. The relationship between the tree-ring chronologies within the catchment and flow of the catchment's river can then be used to derive an estimate of past flow patterns, especially flows during drought. Such investigations have been largely confined to major rivers in the western North America, such as the Colorado (e.g. Tarboton, 1995; Timilsena *et al.*, 2007; Meko *et al.*, 2007; see Figure 4.1), the Sacramento (e.g. Meko *et al.*, 2001; Meko & Woodhouse, 2005) and the Columbia (Gedalof *et al.*, 2004) in the USA and three rivers of the Canadian Prairies (Case & MacDonald, 2003). In a world sense, estimates of past flows during drought in rivers are few, and unfortunately it is difficult to reconstruct the ecological impacts.

4.1.2 Indicators from lakes: tree stumps and sediments

In lakes, water levels may drop considerably in severely sustained droughts. As the shoreline recedes, trees may colonize the newly exposed sites and grow through the extended drought. Subsequent rises in the lake level as the drought breaks may lead to the invading trees being inundated and dying, becoming submerged stumps in the lake. Such stumps can be carbon-14 dated to determine the dates of establishment and mortality, thus providing an indication of both the timing and length of drought periods (Lindström, 1990, 1997; Stine, 1994). For example, submerged stumps in Mono Lake in California's Sierra Nevada mountains provide evidence for two severe megadroughts from ≈AD 892 to ≈1112 and from ≈AD 1209 to ≈1350 (Stine, 1994).

Lake sediments taken in cores from lake bottoms do not have the fine resolution of tree-rings, but they can provide a much longer record. Lakes may be open, in which there are both inflows and outflows, or they may be closed, with endorheic basins which have inflows but no outflows. From the perspective of obtaining relatively interpretable sediment cores, closed lakes are preferable because in open lakes, temporal variability in inflows and outflows may make the estimation of lake levels difficult (Cohen, 2003). However, even in closed lakes, there may be problems in interpretation of cores – for example, if the lake receives groundwater flows (Fritz, 2008).

Lake levels, volumes and areas change with changes in climate (e.g. Mason *et al.*, 1994) and thus, with particularly prolonged droughts, lake levels and volumes will drop and conductivity (salinity) and possibly temperature will rise, inducing chemical and biotic changes that are reflected in the lake sediments. The dating of sediments is usually done using two stable isotopes, carbon-14 (^{14}C) and lead-210 (^{210}Pb). Carbon-14 can be used to date sediments as old as 50,000 year, while lead-210 can be used to date recent sediments from 1 to 150 years old (Cohen, 2003).

Varves are annually formed laminae or layers in sediments derived from seasonal glacier melting, or from seasonal events such as high primary production. Their layering and variations in their contents can be used to estimate past climatic conditions. A notable example is provided by the study of varves from Elk Lake, Minnesota, USA. Recently formed varves consist largely of biogenic material produced within the lake, but material in mid-Holocene varves (8,000–4,000 years ago) is enriched with allochthonous material of aeolian origin – material blown into the lake indicating persistent dry periods (Dean, 1997).

Magnetic susceptibility is a measure of 'how easily a material can be magnetized' (Sandgren & Snowball, 2001) and can be measured in the field. It is often correlated with the level of magnetic minerals, principally magnetite (Cohen, 2003). High susceptibility may be associated with high lake levels and catchment inflows, while low susceptibility may indicate low lake levels and limited inflows, as occurs in extended droughts or drying periods.

The chemicals in lake sediments have been studied for many different purposes (Cohen, 2003), and thus here the concentration will be on chemical signals that may indicate drought. Indicators in drought are proxies which may reflect changes in lake level and volume and which, in turn, through changes in salinity, alter chemical concentrations in materials, both authigenic (produced within lakes) and allogenic (originating from outside lakes) (Boyle, 2001).

Concentrations of total inorganic carbon (TIC) in sediments have been used to indicate changes in lake levels with low lake levels being indicated by high levels of TIC in a sedimentary record (e.g. Benson *et al.*, 2002;

Haberzettl *et al.*, 2003, 2005; see Figure 4.2). The TIC may be produced either by inorganically or biologically-induced precipitation to form carbonates, principally calcium carbonate, which can exist in a variety of forms – principally calcite and aragonite – depending on the chemical conditions in the water at the time of precipitation (Cohen, 2003).

Analysis of elemental ratios in carbonates may provide indications of changes in lake levels. For example, in sediments from an Indonesian lake (Crausbay *et al.*, 2006) and from Lake Edward in central Africa (Russell &

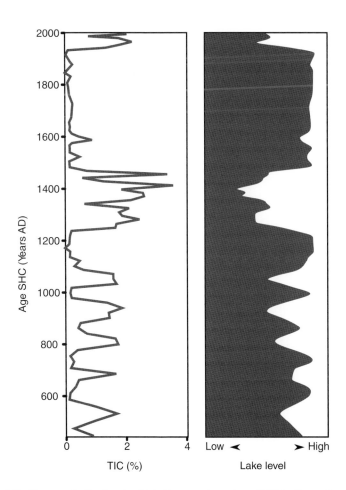

Figure 4.2 Changes in the levels of total inorganic carbon (TIC) in the sediments and changes in depth of Laguna Potrok Aike, southern Argentina. Increased levels of TIC, along with drops in the lake level from 1230 to 1410 AD, may be a drought signal of the Medieval Climate Anomaly. (Redrawn from Figure 7 of Haberzettl *et al.*, 2005.)

Johnson, 2007), the magnesium (Mg) to calcium (Ca) ratio in carbonates was a sensitive indicator of lake levels. Increases in Mg/Ca and in strontium (Sr)/Ca ratios may indicate an increase in salinity (Cohen, 2003). In lakes with sulphate as a major cation, evaporation and consequent drop in water levels may cause the precipitation of gypsum ($CaSO_4$). Lake Chichancanab on the Yucatan Peninsula of Mexico is one such lake. In sediment cores taken from this lake, there were gypsum layers indicating dry conditions, along with organic-rich layers indicating wet conditions (Hodell et al., 2005).

Ostracods are a diverse group of crustaceans with a range of species across the gradient from fresh to highly saline water. Their taxonomy is relatively well established, and the presence of certain species and species groupings in sediments can indicate past conditions. Thus, changes in species in a lake may indicate changes in salinity that, in turn, can reflect changes in lake level. Even in a single lake, different species can live in different habitats, for example littoral habitats versus deep benthic habitats. Changes in species can thus indicate habitat availability linked with changes in lake levels.

In growing to become adults, ostracods moult eight times, shedding the moulted valves and growing new ones. Moulting is rapid, and the valves mineralogically consist of calcite low in magnesium (Chivas, 1986). As in the case of abiotically-formed calcite, the chemicals incorporated into the valves may reflect lake water conditions at the time of moulting (Cohen, 2003). Changes in Mg/Ca, and especially in Sr/Ca, may reflect changes in water chemistry and salinity, and hence lake level (De Deckker & Forester, 1988; Cohen, 2003).

Stable isotopes can be taken up from lake water and incorporated into insoluble inorganic or organic material. Of particular importance for stable isotope analysis are carbonates, either abiotically precipitated as aragonite or calcite, or biologically deposited in the shells or exoskeletons of animals such as crustaceans (especially ostracods) and molluscs – snails and bivalves (Ito, 2001; Leng et al., 2006). Ostracod valves are a very important source of carbonates in lake sediments. Because chemical constituents of ostracods may vary between species, valves from a single species are preferred (Leng et al., 2006).

Mollusc shells may be also used as a source of carbonate for stable isotope analysis (Ito, 2001). A key isotope from carbonates in lake sediments is oxygen-18 (^{18}O). This is expressed in a ratio between ^{18}O and ^{16}O. Evaporation in a lake causes enrichment of ^{18}O relative to ^{16}O, especially in closed lakes (Cohen, 2003, Leng et al., 2006). The key assumption is that the ^{18}O concentration in a shell or valve or abiotically formed carbonate reflects the ^{18}O concentration at the time of deposition or precipitation (Ito, 2001; Yu et al., 2002; Cohen, 2003; Leng et al., 2006). If evaporation such as in a drought increases the salinity of a lake and lowers the volume, this should result in an increase in ^{18}O concentrations. Apart from analyzing ^{18}O from carbonates, for lakes that are acidic and/or dystrophic and hence lack insoluble carbonates, ^{18}O from the silica in diatom frustules may be analyzed (Cohen, 2003).

Particular groups of aquatic biota can leave lasting remains in lake sediments. These remains may be used to identify the taxa to family and even to species. In turn, for many of these groups, there is considerable contemporary knowledge on the habitats and requirements of particular taxa. Such knowledge may involve lacustrine habitats (e.g. benthos, littoral, profundal) and abiotic requirements or tolerances such as salinity, temperature, pH and chemical composition. Thus, past conditions can be ascertained (Cohen, 2003).

Ostracods can be identified by their valves to species, and past conditions that indicate drying and droughts can be estimated (De Deckker and Forester, 1988; Smith, 1993; Curry, 1999). To elucidate the changes in lake levels in the late Holocene from sediments in Lake Tanganyika in east Africa, Alin and Cohen (2003) initially determined from contemporary samples the changes in ostracod species representation with water depth. These data, in the form of ostracod-based lake-level curves, were then used to assess past lake levels and their changes with time, revealing five periods of 'lowstands' (droughts) and three important wet periods over the late Holocene.

Diatoms, with their siliceous frustules, are 'probably the single most valuable group of fossils for palaeolimnological reconstruction' (Cohen, 2003). Species occur abundantly across a wide range of physico-chemical conditions. Their fossil record extends back to the late Cretaceous, but it is their presence in late Quaternary (Holocene) sediments that has been the major focus of research (Cohen, 2003). Diatom species composition can change markedly with changes in salinity, which consequently may indicate changes in precipitation inflows and evaporation of standing waters. Contemporary records from lakes and wetlands of various salinities and their diatom flora allow the construction of diatom-salinity transfer functions that can then be used to determine past salinities and lake levels (Verschuren et al., 2000; Fritz et al., 2000; Laird et al., 2003; Tibby et al., 2007).

In Lake Naivasha, east Africa, Verschuren et al. (2000) examined fossil diatom assemblages to reconstruct conductivity, and thus past lake levels, which revealed that over the period from 900 AD until 1993, the lake had suffered five extended periods of drought. From six lakes on the northern prairies of North America, dramatic shifts in sediment diatom assemblages indicating changes in salinity and lake levels were detected by Laird et al., (2003). The changes occurred between $\approx$500 AD to $\approx$800 AD in the three

Canadian lakes and between ≈1000 AD and ≈1300 AD in the Northern US lakes.

Insect fossils can be preserved in lake sediments, with the head capsules of chironomid midges being of particular interest (Cohen, 2003). From knowledge of the distribution of modern chironomids in relation to conductivity (salinity), changes in conductivity indicating changes in lake level can be deduced from fossil head capsules in lake sediments.

Pollen of both terrestrial and aquatic plants can be preserved in lake sediments, with terrestrial plant pollen being much better preserved than that from aquatic plants. Pollen may be swept into a lake from its catchment and may subsequently be entrained into sediments (Cohen, 2003). Analysis of pollen in sediments can indicate changes in vegetation at the local and regional level. As changes in plants due to changes in drying and wetting occur over a number of years (decades or more), pollen deposits may indicate lengthy periods of drought, but they do not readily allow the discrimination of distinct droughts, in contrast to such indicators as tree-rings, diatoms and ostracods. However, using pollen from particular species may reflect changes in lake level indicative of distinct droughts. In the Great Basin of the western USA, the presence of pollen from plants in the family Chenopodiaceae (its level designated C) indicates dry conditions and a decline in lake levels, whereas pollen from sagebrush (*Artemisia*) (designated A) may indicate wetter conditions (Mensing *et al.*, 2004). Thus, in sediments from Pyramid Lake, Nevada, the ratio A/C reflects changes in lake levels and has proven to be a relatively sensitive indicator of droughts in comparison to such indicators as $\delta^{18}O$ (Benson *et al.*, 2002; Mensing *et al.*, 2004).

As outlined above, there is a range of indicators that may be used to detect droughts of the past. Tree-rings provide a sensitive signal of drought, but dendrochronological records only extend back for around 1,000 years. Lake sediments provide a wide range of signals, from stable isotopes to faunal remains, that may indicate past droughts and indicators in lake sediments that can extend back into the Pleistocene (10,000 – 1.8 million years ago) (Cohen, 2003). In analyzing lake sediments, it is usual for a number of indicators to be used – a procedure that increases the capacity to detect droughts.

Indicators from lakes provide information from droughts that occur across the catchments of the lakes. However, as droughts occur at a large spatial extent, it may be difficult to assess the spatial extent and overall severity of droughts. Where there is a group of lakes, a lake district, from which sediments have been sampled and analysed, the occurrence of lengthy droughts and an indication of their spatial extent can be revealed. Such regions with multiple sampled lakes examined include the northern prairies of North America (e.g. Fritz *et al.*, 2000; Yu *et al.*, 2002; Laird *et al.*, 2003;

Michels *et al.*, 2007), the Yucatan Peninsula in central America (e.g. Hodell *et al.*, 2005, 2007) and tropical east Africa (e.g. Verschuren *et al.*, 2000; Alin & Cohen, 2003; Russell & Johnson, 2005). With tree rings, it is possible, through correlation, to derive Palmer Drought Severity Index values. However, to date, values from lakes – such as salinity and lake level – have not been correlated with PDSI values. This may be because of the lack of contemporary records (lake indicators against meteorological records) to build the necessary correlation, or simply because, in many cases, lake records indicate past droughts of much greater severity (intensity, duration) than contemporary droughts.

4.2 Impacts of past drought on lakes

In severe droughts, shallow lakes may dry out, as described in Chapter 8 in the case of Lake Chilwa in Malawi, which is an endorheic and shallow lake that in 'normal times' is slightly saline. With drought, salinity rises as lake volume and shore levels decline, alkalinity rises and both calcium and magnesium salts are precipitated. As the lake recedes with drought, there are major changes in the phytoplankton (diatoms, cyanobacteria), zooplankton (copepods, cladocerans, ostracods, rotifers), zoobenthos (chironomids) and fish. Dramatic changes occur in shallow lakes, and have presumably occurred in the past in lakes as droughts developed. The changes are reflected in lake sediments (e.g. Verschuren, 2000; Laird *et al.*, 2003; Michels *et al.*, 2007).

In examining sediment cores from shallow prairie lakes, diatom species richness (frustules) has been found to correlate negatively with diatom production, at least over the past 2,000 years (Rusak *et al.*, 2004). This relationship would only have applied when the lakes had fresh water, but would have been disrupted when there were increased salinities, such as occurs in these systems with drought. In brackish salinities (such as in drought), production was maintained even though species richness was reduced. Drought may thus diminish diatom species richness, but not diatom production levels. As mentioned before, as the volumes of shallow lakes decline in drought and water temperatures rise, blooms of cyanobacteria occur, potentially depleting oxygen levels and altering secondary production due to their resistance to herbivores. Thus, there is both fossil and contemporary evidence that drought alters species richness, primary production and trophic structure; however, at this stage there is no integrated diversity-function account of the impacts of and responses to drought in lakes.

In deep lakes, severe droughts may cause water levels to drop and salinity to rise slightly, but the lakes do not dry up. Furthermore, in deep lakes,

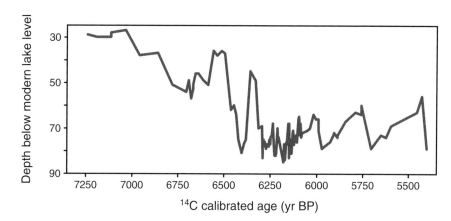

Figure 4.3 Reconstructed lake levels of Lake Titicaca in comparison with contemporary levels. The plot indicates three periods of lengthy megadroughts in the mid-Holocene, with the lowest lake level occurring at ≈6,179 years before present (BP). (Redrawn from Figure 7 of Theissen *et al.*, 2008.)

stratification may be strengthened and may occur for long durations. For example, Lake Titicaca in South America has a 'normal' maximum depth of 107 m. In about 1100 AD, a severe megadrought dropped the water level by an estimated 12–17 m (Binford *et al.*, 1997), while a megadrought in the mid-Holocene dropped the lake level by 29–44 m (Theissen *et al.*, 2008; see Figure 4.3) and very significantly increased salinity and lake stratification. In Pyramid Lake, Nevada, with a natural normal depth of 123 m, mid-Holocene megadrought reduced the lake level to 88 m (Benson *et al.*, 2002).

Even though severe droughts do not cause deep lakes to dry out, they can induce major changes in the lake ecosystem. Changes in conditions in deep lakes due to drought are reflected in reduced littoral production and a strengthening in terms of both gradient and duration of thermal stratification (e.g. Theissen *et al.*, 2008), accompanied by high production in the epilimnetic region of the lake (e.g. Hodell *et al.*, 2007). Changes in water level, and in the availability of habitat at various depths, can be reflected in changes in the ostracod fauna detected in sediments (Alin & Cohen, 2003). If the lake level were to drop to shallow levels, salinity levels would be high and, as envisaged by Benson *et al.*, (2007), the hypolimnion would disappear and destratification would set in during summer, increasing nutrient concentrations in the water column and increasing primary production. As mentioned above, high phytoplankton production and decomposition would increase the risk of periodic deoxygenation, thus eliminating fish and other fully aquatic fauna.

It is rather surprising that, in dealing with past droughts, the emphasis in many papers has been on the effects of past droughts on human populations and civilizations. Studies of past droughts, which invariably focus on lakes, rarely provide much information, let alone speculation, on the effects that the droughts had on the structure and functioning of lake ecosystems.

4.3   Droughts of the Holocene

4.3.1   Early and mid-Holocene droughts

The Holocene, after the terminal Pleistocene deglaciation, was originally viewed as an extended period of relative climatic stability (Cohen, 2003). However, evidence from lake sediments now strongly indicate that the Holocene was a period of abrupt changes in climate (Mayewski *et al.*, 2004; Overpeck & Cole, 2006) and hence in lake levels (e.g. Cohen, 2003; Laird *et al.*, 2003; Michels *et al.*, 2007). Abrupt climate change has been defined as occurring where there is a 'transition in the climate system whose duration is fast relative to the duration of the preceding or subsequent state' (Overpeck & Cole, 2006), with the transitions from wet to dry periods being consistently abrupt.

After analyzing a wide range of palaeoclimate records, Mayewski *et al.*, (2004) concluded that during the Holocene there have been six periods of abrupt climate change in the periods of 9,000–8,000, 6,000–5,000, 4,200–3,800, 3,500–2,500, 1,200–1,000 and 600–150 years ago. These rapid changes may be associated with the abrupt onset of drought and drought regimes, with the development of 'prolonged droughts being almost always abrupt' (Overpeck & Cole, 2006).

In western and central North America, as revealed in lake sediments, during the early Holocene (11,600 to 8,000 years BP) there was a transition from wet to drier conditions (e.g. Laird *et al.*, 1996a; Benson *et al.*, 2002), with periods of cyclic droughts (Stone & Fritz, 2006). In this period, Moon Lake, North Dakota, moved from being a freshwater lake to a shallow saline system, and the surrounding vegetation changed from spruce forest to prairie grassland (Laird *et al.*, 1996a). In Pyramid (Nevada) and Owens (California) Lakes, from 8,000 to 6,500 years BP, highly variable dry conditions prevailed that gave way to severe drought conditions from 6,500 to 3,800 years BP (Benson *et al.*, 2002). Radio-carbon dated tree stumps from previous shores of Lake Tahoe, near to Pyramid Lake, indicate severe drought-dominated conditions between 6,290 to 4,840 years BP (Lindström, 1990).

Black, organic, matter-rich mats are formed around springs in the southern Great Basin of the western USA and harbour a distinctive fauna (Quade *et al.*, 1998). Carbon-14 dating of these mats indicate that they were present from 11,800 to 6,300 years BP and from 2,300 years BP until now, but that they were absent in the intervening period, indicating dry conditions over this time. The severe drought-dominated conditions of the mid-Holocene in central North America are also reflected in evidence of strong activity in dune fields over the period from 7,500 to 5,000 years BP (Forman *et al.*, 2001).

In South America, in Lake Titicaca from 7,200 BP to 6,200 years BP, due to prolonged dry conditions, the level of the lake dropped substantially (29–44 m) (Theissen *et al.*, 2008; see Figure 4.3), whereas in Lake Edward in east equatorial Africa, mid-Holocene aridity began at ≈5,200 years BP and the lake dropped 14 metres to reach its lowest level between 4,000 to 2,000 years BP (Russell *et al.*, 2003). Evidence from lake sediments of the dry conditions of the middle Holocene are indicative of a widespread global phenomenon – a drought-dominated period of severe drying that 'was probably not synchronous' (Cohen, 2003).

A dramatic large-scale change from a wet climate to a dry climate appears to have occurred across the Sahara region around 5,500 years BP. From being a region with lakes, wetlands and widespread vegetation cover, the Sahara became stricken with drying and droughts and rapidly turned to desert (Foley *et al.*, 2003). The causes for this dramatic shift from a 'green Sahara' to a 'desert Sahara' (Foley *et al.*, 2003) appear to be twofold: slowly-acting changes in solar radiation due to changes in the Earth's orbit (Milankovitch variations), combined with fluctuating feedback interactions between vegetation cover and water availability (Claussen, 1998). This dramatic shift has been posited as a prime example of an environmental system having two alternative stable states – the wet and the dry Sahara (Foley *et al.*, 2003; Lenton *et al.*, 2008).

Severe and extended droughts appear to have greatly affected very early civilizations (deMenocal, 2001). Two ancient civilizations, the Akkadian in the Middle East and the Indus civilization in west Asia, appear to have declined, if not collapsed, due to abrupt climate change with extended megadroughts (deMenocal, 2001; Weiss & Bradley, 2001; Staubwasser & Weiss, 2006). The Akkadian civilization occupied the fertile plain between the Tigris and Euphrates rivers from ≈4,350 to ≈4,170 years BP and was based on irrigation agriculture. Its main city, Akkad, later became Babylon. Abrupt climate change due to the onset of an extended mega-drought was a major force in the collapse of this civilization (Kerr, 1998; deMenocal, 2001). The Indus civilization flourished between 5,500 to 4,200 BP and then rapidly declined. Its waning appears to have been caused by a climate change event that caused decline of the south-west monsoon and the onset of severe extended drought (Staubwasser & Weiss, 2006).

4.3.2   Late Holocene droughts

Following the warmer and dry conditions of the middle Holocene, in the late Holocene, at around 3,000 years BP, modern wetter conditions – albeit with rather unpredictable droughts – started to prevail in central North America (Laird et al., 1996a; Benson et al., 2002; Overpeck & Cole, 2006). In east Africa, the transition to milder and wetter conditions in the late Holocene appears to differ in time at a regional level. Sediments from Lake Tanganyika (Alin & Cohen, 2003) indicate that arid conditions with extended droughts persisted until recently, whereas in Lake Edward, to the north, although the late Holocene began with arid conditions, wetter conditions caused the lake level to reach its contemporary level by 1,700 years BP (Russell et al., 2003). In South America, the level of Lake Titicaca rose at around 3,900 years BP (Mourguiart et al., 1998) or 3,500 years BP (Abbott et al., 1997), to be followed by four significant drought-dominated periods (Abbott et al., 1997). Thus, the late Holocene was a period with milder conditions than those that prevailed in the middle Holocene, but severe and extended droughts unpredictably occurred.

As for ancient civilizations, the past 2,000 years have seen civilizations being forced to retreat due to the ravages of severe droughts. The Mayan civilization of Central America was a complex and creative society, with remarkable buildings in populous cities, intricate agricultural schemes and a written language (Diamond, 2005, 2009). Maize production with irrigation was very important as the supplier of the staple diet. The 'Early Classic Period' of Mayan civilization began about 250 AD, with major increases in population, in buildings and monuments. From about 770 AD, an extensive dry period commenced, with severe droughts in $\approx$810, 860 and 910 AD (Haug et al., 2003), with intervening wetter periods (Hodell et al., 2005). The droughts in this period, the 'Terminal Classic Period', appear to have played a major role in the decline of the Mayan civilization by greatly reducing water availability and hence agricultural production (deMenocal, 2001; Diamond, 2005; Hodell et al., 2007).

Evidence for climatic conditions during the Mayan civilization has come substantially from the analysis of sediment cores from two lakes on the northern Yucatan Peninsula – Lakes Chichancanab and Punta Laguna (Hodell et al., 2005, 2007) as well as from anoxic marine sediments off Venezuela (Haug et al., 2003). While other factors associated with resource depletion, such as warfare, were significant in the collapse, the vulnerability of the complex Mayan society to drought appears to have been high (Diamond, 2005; Haug et al., 2003). The cause of the severe droughts may have resulted from variation in solar radiation (Hodell et al., 2001; Brenner et al., 2002), though this suggestion has been questioned (Me-Bar & Valdez, 2003; Hunt & Elliott, 2005).

Lake Titicaca at the altitude of 3,812 m in the Altiplano basin, straddles the border between Peru and Bolivia. It is an endorheic lake, with an average depth of 107 m, and which is fed by rainfall, glacial meltwater, five major rivers and 20 smaller rivers. The lake is made up of two basins – the Lago Grande and Lago Huiñaimarca – joined by the Strait of Tiquina. Analysis of sediment cores from the lake indicate that, after a wet early Holocene period (Mourguiart et al., 1998), during the mid Holocene from about 7,200 years BP, the lake level started to drop, reaching its lower level, ≈44 m below the modern level, at ≈ 6,179 years BP, with three periods of severe mega-droughts from 7,000 years BP to 6,200 years BP (Theissen et al., 2008; see Figure 4.3). From this low level, and with fluctuations till 3,350 years BP, the lake level rose (Abbott et al., 1997). This was followed by four late Holocene periods of drying, with extended droughts from 2,900–2,800, 2,400–2,200, 2,000–1,700 and 900–500 years BP (Abbott et al., 1997).

Beginning around 300 BC, human settlements with agriculture became established around the lake (Binford et al., 1997), which led to the flowering of the Tiwanaku civilization. The agricultural production was marked by the construction and maintenance of raised fields in wetlands verging the lake and in river valleys. These raised fields served to foster high production, to protect the crops from winter frosts, to retain nutrients and to control salinization (Binford et al., 1997; deMenocal, 2001). The golden age of the Tiwanaku was from 600 to 1,100 AD, with extensive construction activity and increases in population supported by the raised field agriculture. However, as revealed in lake sediments (Binford et al., 1997; Abbott et al., 1997), severe droughts reducing water availability appear to have halted the use of the raised fields, leading to sharply declining agricultural production and the collapse of the Tiwanaku civilization (Binford et al., 1997; de Menocal, 2001).

To gain an understanding of past droughts that may relate to contemporary conditions, climate conditions for the last 2,000 years are relevant. For this late period of the late Holocene, there are numerous records from sediments, and more precise records from tree-rings. With tree-rings, in particular, a clear depiction of the spatial extent and of severity of droughts can be produced. Such information can also be used to gain an understanding of the weather conditions that produced past droughts.

In North America, especially in central and western regions, over the past 2,000 years droughts of considerable severity and large spatial extent have occurred. Records from lake sediments of Pyramid Lake indicate that hydrologic droughts over the last 2,000 years have occurred, with intervals between them from 80 to 230 years (Benson et al., 2002). Tree-ring records can go back 2,000 years or so, though records of this age are very few. For example, in reconstructing the past droughts of North America in the North

American Drought Atlas (NADA), Cook *et al.* (2007) had a few records stretching back for about 2,000 years, but to gain statistical reliability and adequate spatial coverage, the reconstruction started at 951 AD. Tree-ring records have three advantages over lake sediment records in that they allow more precise dating of the initiation and termination of droughts, they can provide a measure of the severity of droughts and they allow a reconstruction of the spatial extents of droughts.

During the period from the 9th to the 14th century, there is comprehensive evidence of a widespread increase in temperature (Lamb, 1977, 1995) – a period called the Mediaeval Warm Period or Mediaeval Climate Anomaly (MCA) (Stine, 1994). The warming was not tightly synchronous across localities, with considerable differences in the times of warming and subsequent cooling. It also does not appear to have been global, with some regions showing no or very little warming (Hughes & Diaz, 1994). Thus, for example, there was warming in this period in northern Europe, China, south-east Asia, western North America and even Tasmania (Cook *et al.*, 1992), but not in southern Europe or southeastern North America (Hughes & Diaz, 1994).

The increase in temperatures in affected regions was accompanied by droughts (Bradley *et al.*, 2003). Over central and western North America, this period was marked by the occurrence of droughts of long duration (20–40 years – megadroughts), as detected in tree-rings (e.g. Cook *et al.*, 2004, 2007, 2010; Herweijer *et al.*, 2007), lake sediments (e.g. Laird *et al.*, 1996b, 2003; Laird and Cumming, 1998; Benson *et al.*, 2002) and from submerged tree stumps (e.g. Stine, 1994). Herweijer *et al.* 2007) depict the PDSI values and spatial extent of four of these droughts (1021–1051, 1130–1170, 1240–1265 and 1360–1382 AD), emphasizing the point that, while the PDSI and the spatial extent are similar to modern droughts, the durations were much longer. These droughts would have had major hydrological consequences, such as marked drops in lake levels and in streamflows.

In reconstructing the streamflows of the Sacramento River, Meko *et al.* (2001) identified severe 20-year low flows (droughts) from 1139 to 1158 AD and from 1291 to 1311 AD for the Upper Colorado River, and also an 'epic' drought from 1130 to 1154 which was the most extreme drought between 762 AD and 2005 AD (Meko *et al.*, 2007). These mediaeval droughts appear to have been widespread over western North America, as dendrochronological evidence indicates low flow conditions between 900 and 1300 AD in rivers of the Canadian Prairies (Case & MacDonald, 2003).

In the Four Corners region of south-western USA, between 850 to 1300 AD, the Anasazi people developed complex settlements based on maize agriculture that was supported by extensive and intricate systems of water capture and storage (Diamond, 2005; Benson *et al.*, 2007). The development of the settlements was accompanied by deforestation of the surrounding hinterlands. The viability of the settlements was very dependent on sufficient rainfall to grow maize. However, two megadroughts in the middle 12th and late 13th centuries greatly reduced maize yields, leading to abandonment of the settlements (Axtell *et al.*, 2002; Diamond, 2005; Benson *et al.*, 2006, 2007), with the megadrought from 1276–1297 AD being particularly severe (Cook *et al.*, 2007). In short, the Anasazi society was highly vulnerable to the hazard of intense, prolonged droughts and was not sufficiently resilient to endure such droughts. The Anasazi population was greatly diminished, with the survivors abandoning the region.

The Cahokian culture in the mid-Mississippi region flourished from ≈900 to 1300 AD and depended on horticulture, hunting and gathering (Benson *et al.*, 2007; Cook *et al.*, 2007). At Cahokia, the centre of the culture, over 120 large pyramidal mounts were constructed. However, it appears that two severe and prolonged droughts in the 14th century (1344–1353 AD, 1379–1388 AD) overwhelmed the Cahokian culture causing it to collapse (Benson *et al.*, 2007; Cook *et al.*, 2007).

In southern South America, prolonged droughts associated with the Mediaeval Climate Anomaly (MCA) occurred in Patagonia, Argentina, at the same time as those occurring in California (Stine, 1994). The evidence comes from drowned stumps in two lakes and a river gorge in California, and also from drowned stumps of *Nothofagus* around two lakes in Patagonia. In Patagonia, analysis of sediment cores from Laguna Potrok Aike, a deep lake in a cold semi-desert, indicates an extended period of lowered lake level from 1230 to 1410 AD, with droughts of 'varying durations and intensities' (Haberzettl *et al.*, 2005) occurring late in the MCA.

In east Africa, data from the analysis of sediment cores from five lakes – Lake Turkana (Halfman *et al.*, 1994), Lake Naivasha (Verschuren *et al.*, 2000), Lake Tanganyika (Alin & Cohen, 2003), Lake Victoria (Stager *et al.*, 2003) and Lake Edward (Russell & Johnson, 2005) – indicate that major droughts occurred between 900 and 1400 AD. It thus appears that in east Africa, the MCA with droughts commenced at least 100 years before its onset in Europe and North America (Russell & Johnson, 2005; see Figure 4.4). Sediment cores from Ranu Lamongan, a lake in Java, Indonesia, indicate some evidence of the MCA, with two severe dry periods with severe droughts from 1275–1325 AD and from 1450–1650 AD, with the severity of the latter period occurring in the 'Little Ice Age' and possibly correlated with a change in the nature of the ENSO system (Crausbay *et al.*, 2006).

Evidence from tree rings (a 759-year record) indicate that in southeast Asia, notably in southwestern Thailand, Cambodia and Vietnam, severe droughts occurred in the 14th and 15th century (late MCA), with an El Niño-linked weakening of the summer Asian monsoon (Buckley *et al.*, 2010). These droughts appear to have led to the demise of the Angkor society, which was dependent on a vast and complex water distribution system and was thus highly vulnerable to extended periods of drought coupled with short intervening periods of heavy flooding (Buckley *et al.*, 2010).

The 'Little Ice Age' was a period of intense cooling, advancement of glaciers and severe winters, stretching from the 16th to the 19th century. There were three periods of extreme low temperatures in about 1650, 1770 and 1850. In some regions, this was also a period in which prolonged and severe droughts occurred. The most severe megadrought in North America in the past 500 years lasted 24 years, from 1559 to 1582 (Stahle *et al.*, 2000, 2007), and occurred in an extended wet period from ≈1500 to 1800 AD (Herweijer *et al.*, 2007). This megadrought contained within it shorter, regional droughts that ranged across North America from northern Mexico to Canada and from the southwest to the southeast of USA (Stahle *et al.*, 2000, 2007; Woodhouse, 2004).

Reconstructions of streamflow in the upper Colorado river (above Lees Ferry) reveal severe drought over the period 1580–1604 AD (Timilsena *et al.*, 2007), with the ten-year moving average reconstructed flow at Lees Ferry being extremely low (Woodhouse *et al.*, 2006). This drought elevated salinities (lowered shore levels) in lakes of the Northern Plains, albeit with some variability between lakes (Fritz *et al.*, 2000; Yu *et al.*, 2002; Shapley *et al.*, 2005), and increased the levels of wind-borne dust (Dean, 1997).

Cores from Chesapeake Bay in the eastern USA were analyzed for changes in salinity using benthic foraminiferans and ostracods (Cronin *et al.*, 2000). Elevated salinities indicated 'sustained droughts during the middle to late sixteenth and early seventeenth century' that were considerably more severe than 20th century droughts (Cronin *et al.*, 2000). These severe droughts have been linked to the failure of the English colony of Roanoke Island (North Carolina, 1587–1589) and the near demise of the colony at Jamestown, Virginia (1606–1612) (Stahle *et al.*, 1998a).

In east Africa, during the Little Ice Age, droughts were not uniformly spread across the region. High Mg/Ca ratios in authigenic calcite showed that low lake levels and drought conditions occurred in Lake Edward from 1400 to 1750 AD (Russell and Johnson, 2007) (see Figure 4.4). Similarly low lake levels indicating drought occurred in Lake Tanganyika from ≈1550 to 1850 (Alin & Cohen, 2003) and in Lake Malawi from 1400 to 1780 AD (Brown & Johnson, 2005). However, in Lake Naivasha, to the east of these lakes, diatoms and chironomid fossils suggest a wet Little Ice Age with only three short dry periods (Verschuren *et al.*, 2000). The reasons for

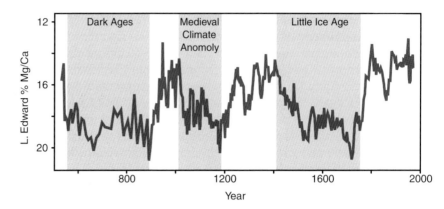

Figure 4.4   Ratios of magnesium to calcium in calcite from the sediments covering the past 1,400 years of Lake Edward, Uganda. High values of the magnesium to calcium ratios are reliable indicators of drought conditions, and thus they indicate persistent droughts from ≈540 to 890 AD, droughts in the Medieval Climate Anomaly from 1000–1200 and further persistent droughts in the Little Ice Age from 1400–1750 AD. (Redrawn from Figure 3A of Russell & Johnson, 2007.)

this discrepancy in the proxy climate records from the region remain unclear.

Similarly, in other parts of the world, the evidence for droughts in the Little Ice Age differs. For example, in Lake Titicaca and Patagonia in South America, there is no evidence for droughts over this period (Abbott *et al.*, 1997; Haberzettl *et al.*, 2005), while in Indonesia, Crausbay *et al.* (2006) found evidence of prolonged drought during the Little Ice Age. Thus, whereas droughts during the MCA appear to have been a global phenomenon, droughts associated with the Little Ice Age were spatially variable.

After the Little Ice Age, severe and persistent droughts occurred in the Upper Colorado from ≈1644 to 1681 AD and from ≈1877 to 1909 AD (Timilsena *et al.*, 2007). In examining correlations in flow between the Sacramento and the Blue River in the upper Colorado River basin, Meko & Woodhouse (2005) noted that during drought periods, the flow correlations between two systems were stronger than at other times in the 538-year reconstruction period. A reconstruction of the Columbia River back to 1750 by Gedalof *et al.* (2004) revealed that the worst drought period was in the 1840s, a time of severe drought in the Sacramento River (Meko *et al.*, 2001). The next most severe period of drought was the 'Dustbowl' drought of the 1930s.

For many years, there has been little available scientific evidence of major droughts in Asia, although there have been historical accounts, such as the vivid account of the Victorian Great Drought by Davis (2001). The summer

Asian monsoon system is crucial to the agricultural economies of Asia, and failures of the monsoon can result in crop failures, civil unrest and famine. Recently, using tree-ring data from sites across monsoonal Asia, Cook *et al.* (2010b) have produced a gridded reconstruction of wet and dry events and periods (Palmer Drought Severity Index (PDSI) and Drought Area Index (DAI)) over Asia for the past millennium, which is called the Monsoon Asia Drought Atlas (MADA).

Four major droughts in the past 400 years have been reconstructed in terms of their severity and spatial extent. The 'Ming Dynasty Drought' (1638–1641) occurred in the Little Ice Age and was at its greatest intensity in north-eastern China; the resulting food shortages and civil unrest may have contributed to the fall of the Ming Dynasty. A major drought (1756–1768) called the 'Strange Parallels Drought' was focused on India and south-eastern Asia, and the short but very severe 'East India Drought' (1792–1796) focused on northern and central Asia with outliers in southern India and Burma. The late Victorian 'Great Drought' was both severe and widespread, occurring across southeastern Asia, northern and central Asia and India. It 'ranks as the worst of the four historical droughts' (Cook *et al.*, 2010b), and as described by Davis (2001), it resulted in millions of people dying. As more tree data are accumulated, the MADA will grow in precision and coverage, and will undoubtedly serve to elucidate the climatic patterns which produce severe, widespread droughts.

Over the past 150 years, there have been four prolonged North American droughts (1855–1865, 1889–1896, 1931–1940, 1950–1957), which have spanned the continent longitudinally, extending from Canada to Central America. While the Dustbowl drought is perhaps the best known of these, all were economically and ecologically damaging. However, none was as severe and prolonged as the megadroughts from 1000 to 1400 AD (Laird *et al.*, 1996b; Herweijer *et al.*, 2007).

Since record-keeping began in Australia, there have been six prolonged droughts: 1880–1886, 1895–1903, 1911–1916, 1939–1945, 1958–1968 and the recent drought of 1997–2010.

From North America, with the recent supra-seasonal droughts, there have been many valuable ecological studies, but corresponding studies have been few in Australia or Europe. From Europe there has been a considerable number of studies on seasonal droughts or dry seasons, and some regional studies of past supra-seasonal droughts (e.g. Masson-Delmotte *et al.*, 2005; Planchon *et al.*, 2008), but somewhat surprisingly there appear to be very few reconstructions of past droughts covering Europe as a whole. In an exception, Lloyd-Hughes & Saunders (2002), using precipitation records across Europe for the period 1901–1999, showed that some regions, especially Mediterranean ones, were more prone to have droughts than regions in the northern temperate zone, but that there was no significant change in the proportional area of Europe affected by drought.

The south-western USA is currently in a severe drought that has lasted for more than 10 years and is placing great strains on water supplies within the region (Seager *et al.*, 2007). The LBDA demonstrates that the current drought is not exceptional – two droughts of similar severity occurred in the 20th Century. More worryingly, the LBDA shows that, over the past millennium, there have been numerous droughts of significantly greater duration and intensity. Several of these droughts occurred during a multi-century period of elevated aridity in the Medieval Climate Anomaly (MCA).

The LBDA (Living Blended Drought Atlas) provides an excellent representation of the temporal and spatial variability of regional droughts. For the western USA, the region with the longest reconstruction, the LBDA shows dramatic high-frequency (i.e. year-to-year) variability in area subjected to drought, with some periods showing interannual changes in drought occurrence of as much as 60 per cent of the region. There is also significant medium-frequency (i.e. decadal) variability, with most centuries experiencing at least one or two periods of alternating low and high drought conditions. Finally, as noted above with respect to the MCA, there is also low-frequency (i.e. century-scale) variability, in which multiple century runs of above- or below-average drought conditions dominate within the region. Understanding the low-frequency context of high- and medium-frequency events is critical to understanding the regional ecological systems that are shaped by drought. For example, the severe droughts of the 20th Century both occurred during the 500+ year period of relatively low aridity that followed the MCA.

The LBDA demonstrates how droughts vary spatially across North America (Cook *et al.*, 2010a). The most obvious spatial feature of the droughts represented in the LBDA is that each severe drought has a unique spatial signature. While there may be some overlap in core areas, the shape of the drought-affected area is highly irregular over time. As a consequence, the long-term frequency of droughts in any given region may vary considerably from point to point. Another important spatial feature of the LBDA is that areas which are not typically associated with arid conditions and drought, such as the south-eastern USA, have experienced intense, long-lasting droughts over the past 800 years.

The drought atlases (e.g. NADA, LBDA, MADA), though difficult to compile, do produce highly interesting spatio-temporal patterns of droughts, which in themselves may provide valuable clues as to the meteorological/climatic forces that created and maintained severe droughts and mega-droughts. For example, the extended droughts of the MCA in western North America may have been induced by cool La Niña-like conditions, while the megadroughts of the MCA in the Mississippi region may have been linked with the North Atlantic Oscillation (NAO) (Cook *et al.*, 2000a). This development of an understanding of the operation of drought-creating forces may greatly help in making projections of the large-scale effects of global climate change (Cook *et al.*, 2010a).

5

Water bodies, catchments and the abiotic effects of drought

This chapter is concerned with the types of water bodies exposed to drought and the abiotic effects that drought exerts on water bodies, their riparian zones and their catchments. The effects occur as drought tightens its grip, during the drought and when it breaks. Effects generated by drought may produce lag effects with long-term consequences. In many cases, the effects of natural drought may be mixed with, or even greatly heightened, by human interventions such as river regulation, water extraction and catchment clearance.

In this and the following chapters, we are dealing with supra-seasonal droughts rather than seasonal droughts. The latter are clearly normal in many places, but supra-seasonal droughts are extreme events that can cause serious ecological effects. In quite a few studies, it is difficult to determine the nature of the drought and, in some cases, the drought may be a seasonal drought of great severity and/or of abnormal duration.

5.1 Water body types

Drought can affect both standing (lentic) and flowing (lotic) water body types. These water bodies may be either temporary or perennial/permanent. Temporary systems cover those that are:

- ephemeral – ones which receive water for a short period very occasionally and highly unpredictably;
- episodic – those that fill occasionally and which may last months even years; and
- intermittent – which receive water quite frequently either unpredictably or predictably (Boulton & Brock, 1999; Williams, 2006).

Drought and Aquatic Ecosystems: Effects and Responses, First Edition. P. Sam Lake.
© 2011 P. Sam Lake. Published 2011 by Blackwell Publishing Ltd.

Although perennial systems always contain water, such systems may dry out in severe droughts, with major consequences for the biota (Wellborn *et al.*, 1996). Chase (2003) called perennial systems which may occasionally dry up 'semi-permanent'. In ephemeral and episodic, systems the occurrence of hydrological drought as an abnormal period of drying may not be readily detectable.

Lentic systems cover a great array of different water body types. They range in existence from being permanent to ephemeral and may range in size from being very small water bodies such as pockets of water in trees (phytotelmata) to large permanent lakes such as the African rift lakes and the North American Great Lakes. Over this size range, there are gnamas (rock pools), small permanent and temporary pools, puddles and ponds, swamps, bogs, mires, marshes, and immense floodplain wetlands with lagoons. Lentic water bodies made by humans include reservoirs, farm ponds, moats, ditches, cisterns, bomb craters and water-filled car tyres.

Water bodies vary in their extent of openness to water inputs and outputs. Many systems are *endorheic* in that they receive water from their catchments but, apart, from water loss by evaporation, transpiration or to groundwater, there is no flow out of the water body to other surface water systems. Endorheic systems can become highly saline due to salt accumulating over long periods of time in their basins without being exported. Open, or *exorheic* systems, receive water from their catchments and export water out of them to go ultimately into the sea (Wetzel, 2001).

During floods, floodplain ponds and lagoons become part of the flowing water system of their rivers and are then exorheic systems. As the floods recede, they become standing waters and endorheic systems. The effects of drought on these water bodies mainly occur when they are isolated from the river channel, and the major way that drought occurs is through the failure of normal floods.

Flowing waters cover an immense array of water body types and include systems that are either perennial or temporary. Lotic systems range from small springs to large rivers with immense floodplains. Natural flowing waters include springs, bournes, runnels, brooks, burns, becks, creeks, rivulets, streams and rivers. Humans, in implementing various forms of water use, can create flowing water systems such as canals, drainage channels, sloughs and diversions. Humans have also reduced natural connectivity with the construction of many barriers, principally dams with reservoirs – creating in-stream lentic ecosystems. Connectivity may be also created by humans through the construction of pipelines, canals and channels which can create avenues of inter-basin transfer.

Most rivers flow into the sea and, as estuaries are formed and maintained by rivers, drought due to reducing freshwater inputs may greatly change estuarine ecosystems. Streams whose channels are horizontally connected to their water table, or in which the channel is impervious (e.g. rock, clay), are called gaining or effluent streams. Streams in which the channel water is above the prevailing water table may lose water (depending on the permeability of the channel) to the water table, in which case they are called losing or influent streams (Allan & Castillo, 2007). In any one stream there may be gaining and losing sections, and this pattern may become obvious during drought.

5.2   Aquatic ecosystems, their catchments and drought

Both lotic and lentic systems derive water, sediment and many chemicals, including nutrients and organic matter, from their catchments or drainage basins. Streams receive their water from a variety of paths. A small amount (throughfall) may enter directly as precipitation, while the major paths are surface runoff and sub-surface flow, as interflow which flows through the soil but above the water table, and as groundwater linked with the water table. The harvesting of water by a stream is usually carried out in the upper parts of the catchment – in the headwaters.

The permeability of the catchment surface largely controls the paths by which a stream receives its water. If the surface is pervious, groundwater is the major contributor to streamflow, whereas if the surface is relatively impervious, surface runoff may be a major contributor, producing a stream with a 'flashy' hydrograph and little water storage in the catchment. Lentic systems may receive water from throughfall, which can be an important source for small systems, such as phytotelmata. However, by far the major contribution for most standing waters comes from their catchments as surface runoff that includes inflowing streams. Some may also be derived from groundwater. In the case of many lentic systems, the filling phase may be relatively brief, such as occurs in the flooding of floodplain lagoons – a brief period which is followed by a long period of low water inputs.

Within a catchment, water is lost to the atmosphere by evaporation and by transpiration from the vegetation. In forested catchments, evapotranspiration can exert a strong influence on streamflow. However, in urbanized catchments, transpiration, and even evaporation, may not be important in influencing streamflow, since rapid drainage and short residence time of flow are the usual management goals. Water from the catchment may also go into deep groundwater and be lost, in the short term, to streamflow. This is a notable aspect of streams in karstic catchments (e.g. Meyer et al., 2003).

Drought effects on freshwater ecosystems are largely mediated via their catchments, through decreases and changes in surface runoff, interflow and groundwater flows which affect linked aquatic systems. As drought persists, it exerts direct effects on the catchment, including decline in soil moisture, the lowering of the water table and the loss of vegetation. The drying of land surfaces may precipitate major biogeochemical changes in soils of such substances as organic compounds, nutrients and heavy metals. With the breaking of the drought, chemical compounds (e.g. sulphate, nitrate) produced in the soil or added to it by human activities can enter the downslope aquatic system.

In regions with cold winters, droughts may be due to failure for the snowpack to reach normal levels. Drought arises from the lack of normal precipitation – rain and/or snow – falling on the catchments of streams. The lack of precipitation in the snow/rainy seasons, combined with high levels of evapotranspiration in the dry seasons (late spring and summer), serves to lower water levels and water tables and create or maintain hydrological drought. If this condition persists after a considerable time, in most cases, groundwater drought sets in.

A drought usually encompasses lengthy periods of high temperatures, often accompanied by winds and, in many places, dust storms and wild-fires. At first, evapotranspiration levels are high, but they may drop as surface water volumes and soil moisture levels are depleted and high soil moisture deficits are created. Streamflow is reduced and water quality may be substantially altered, due to processes both in the catchment and in the water bodies. Changes in water quality in both standing and flowing systems may stress the biota.

5.3   Drought and effects on catchments

With drought, catchment plant cover may be steadily reduced and litter may accumulate on the ground. Inputs of water, ions, nutrients and dissolved organic carbon to streams and standing waters decline. As soils dry, their surfaces may crack, especially in peaty and clay soils, and air (and oxygen) can penetrate much deeper than normal, increasing the aerobic decomposition of detritus and the oxidation of chemicals such as nitrogen compounds. Nitrates from natural sources, such as from plant decomposition and nitrification, can accumulate in sediments. Furthermore, nitrates from human sources, such as atmospheric pollution (generated by automobile exhausts) and from fertilizers, may also accumulate in soils, especially those of wetlands (Reynolds & Edwards, 1995; Watmough et al., 2004).

Sulphur dioxide generated by smelters and the burning of low grade coal forms acid rain that delivers sulphur to catchments and wetlands downwind of the polluters. Normally, in the high water tables and oxygen-deficient reducing conditions of wetlands, sulphur is stored as sulphides (Dillon *et al.*, 1997; Schindler, 1998). However, under drought conditions, with the lowering of the water table, the reduced sulphur is oxidized to sulphates (e.g. Hughes *et al.*, 1997; Dillon *et al.*, 1997; Warren *et al.*, 2001) and, with the outflow from the wetland restricted or stopped, these sulphates may accumulate, with subsequent formation of sulphuric acid. The water in the wetlands thus becomes acidic (pH $\approx$4) (Adamson *et al.*, 2001; Clark *et al.*, 2006), and this may serve to increase the concentration of dissolved metals, in particular aluminium, zinc, lead, copper and nickel (Tipping *et al.*, 2003). Similarly, as described later in this book in the case of floodplain wetlands (see Chapter 7), the drastic lowering of water levels in floodplain lagoons during drought can give rise to the oxidation of sulphides and the production of highly acidic conditions when water levels rise.

In catchment soils during drought, especially those with large amounts of particulate organic carbon, such as peat bogs, dissolved organic carbon (DOC) can build up. Under normal climate conditions the DOC is delivered to downstream water bodies, but in drought this delivery can be reduced. The accumulation favoured by aeration, the decline of the water table and increased soil temperatures is largely the result of increased phenol oxidase activity (Freeman *et al.*, 2001a, b). Consequently, in the reduced volume of water in wetlands, the concentration of DOC can rise during drought (Worrall *et al.*, 2006; Eimers *et al.*, 2008). This increase may be tempered in wetlands with high sulphur concentrations that are oxidized to sulphates during drought. Under these circumstances, the sulphates, in increasing the acidity by forming sulphuric acid, may serve to inhibit the production of DOC (Clark *et al.*, 2006). However, as Eimers *et al.* (2008) have suggested, the strong inverse relationship between sulphates and DOC concentrations may be more a function of hydrology, with low drought flows reducing DOC export from wetlands.

While most readers will be familiar with the importance of primary production in water bodies, it is increasingly being recognized that DOC, both allochthonous (from the catchment) and autochthonous (produced in the water body) can be a major driver of aquatic metabolism in both lentic and lotic ecosystems:

- Lentic systems can contain such amounts of DOC that it is their most significant carbon component (Wetzel, 2001).
- It has come to be accepted that DOC is the 'major modulator of the structure and function of lake ecosystems' (Sobek *et al.*, 2007).
- Its humic compounds impart colour to lakes, and thus DOC increases heat absorption and influences photosynthesis by absorbing PAR (photosynthetically available radiation).

- The radiation-absorptive capacity of DOC can also moderate inputs of damaging UV radiation (Scully & Lean, 1994).
- DOC contributes to the food web of lakes as it is metabolized by bacteria, which in turn may be consumed by protozoans (Wetzel, 2001).
- It is also of metabolic importance in lotic systems and even estuaries (Battin *et al.*, 2008).
- In streams, DOC is a major fuel for organisms, predominantly bacteria, living in biofilms on surfaces.

Changes caused by drought in catchments and their soils, such as those outlined above, constitute a prelude to the abiotic events that occur when drought breaks, water tables rise and downslope flows return. However, these changes illustrate the widespread effects of drought across ecosystems and indicate that drought can serve to generate other disturbances besides those encompassed in the drying of aquatic ecosystems.

5.4   Riparian zones and drought

Between the terrestrial hinterland of a catchment and the water body lies the riparian zone. This zone is found around standing water bodies, but it has not received anywhere near the same scientific attention as the riparian zone of lotic systems. It is usually identified by its distinctive vegetation, and it serves as a critical ecotone or boundary ecosystem (Naiman & Décamps, 1997). In steep erosional streams, riparian zones can be narrow, but along large meandering lowland rivers, riparian zones comprise the floodplain (Mac Nally *et al.*, 2008). The large scale of the floodplain as the riparian zone may worry some, but it is the floodplain that is periodically flooded and which interacts with the river channel.

The riparian zone is crucially important as a mediator in the transactions of energy, sediment, nutrients and biota between the river and its catchment. From the catchment, in both periodic surface runoff and in groundwater, riparian zones receive inputs of soluble chemicals –inorganic ions and nutrients and organic dissolved organic matter (DOM) or carbon (DOC) (Naiman & Décamps, 1997; Fisher *et al.*, 2004). Riparian zones also receive sediments from their catchments. These sediments are captured and chemicals may be arrested, transformed (e.g. denitrification) or taken up by the vegetation (Naiman & Décamps, 1997; Naiman *et al.*, 2005; Jacobs *et al.*, 2007). Depending on the strength of inputs (e.g. sudden heavy storms), matter from the catchments may move through the riparian zone largely unaltered, and this particularly occurs when the riparian zone has been degraded.

The riparian zone vegetation provides both fine (FPOM) and coarse (CPOM) particulate organic matter to streams, which may be the basic feedstock for the trophic web of streams (e.g. Reid *et al.*, 2008). In addition, it provides wood in the form of logs and branches to water bodies, which then becomes valuable habitat. Invertebrates of the riparian zone may form an important prey subsidy for the fauna, notably fish, of the water body, and conversely hatching insects may be a prey subsidy for riparian invertebrates (e.g. spiders) and birds (Ballinger & Lake, 2006).

In drought, the inputs of water, sediments and chemicals to the riparian zone are greatly reduced (e.g. Jacobs *et al.*, 2007). With the loss of water and drying of the soil, decomposition may slow, and there is a reduction in the rates of nutrient processing (although, as in the case of nitrogen, dry periods may allow the breakdown of organic nitrogen complexes (Pinay *et al.*, 2002)).

The water table of a riparian zone may be maintained from three sources: water from the adjoining hinterland, water from the stream channel and water from aquifers. The effects of drought on the zone may take time to become evident, as declines in groundwater levels are likely to set in long after hydrological drought affecting surface waters has become firmly established. Water from the hinterland, and from the channel in particular, may be greatly reduced, resulting in a lowering of the water table which, in long droughts, may take it below the level of the root zone of riparian trees. For example, in the 1997–2010 drought in south-eastern Australia, the water table of a riparian floodplain of the Murray River dropped to more than 15 m – much deeper than the effective root zone (9 m) of the dominant tree, river red gum *Eucalyptus camadulensis* (Horner *et al.*, 2009). In this situation, with time, tree death occurs.

Most of the available information on the impacts of drought and other forces which reduce streamflow and lower groundwater tables on riparian vegetation comes from studies carried out in western North America, especially the south-west (e.g. Scott *et al.*, 1999; Lite & Stromberg, 2005; Stromberg *et al.*, 1996, 2005), with cottonwoods (*Populus* spp.) being a major focus. For these trees, there is a considerable amount of information on physiological and morphological drought-stress responses (Rood *et al.*, 2003), induced mostly by river regulation and water extraction and not by natural droughts. Physiological responses to drought include stomatal closure, reduction in transpiration and photosynthesis and xylem cavitation, while morphological responses include reduced shoot growth, branch sacrifice and crown die-back (Rood *et al.*, 2003).

Cottonwoods in alluvial sand riparian zones were subjected to manipulated water table declines similar to those produced by drought (e.g. Scott *et al.*, 1999; Amlin & Rood, 2003). Such declines produced extensive leaf senescence, leaf desiccation, branch dieback and abscission (Scott *et al.*, 1999; Amlin & Rood, 2003) in the short term, and decreases in live crown volume and stem growth, with high (88 per cent) mortality in three years (Scott *et al.*, 1999).

In a model of riparian cottonwood forest condition and hydrological flow regimes developed by Lytle & Merritt (2004), the flood frequency was critical to recruitment and stand development, and 'multiple drought years' could greatly reduce, if not eliminate, cottonwood populations and thus create vacant habitat. Into such vacant habitat, the drought-tolerant saltcedar *Tamarix ramosissima* may invade, especially along rivers with flow regulation and riparian zones occupied by the Fremont cottonwood (*Populus fremonti*) and Goodding's willow (*Salix gooddingii*) (Stromberg, 1998; Lite & Stromberg, 2005). *Tamarix* is well adapted to deal with droughts (Nippert *et al.*, 2010) and the high salinities which may occur in the groundwater of streams in semi-arid and arid environments (Vandersande *et al.*, 2001).

The effects of drought on riparian trees is to reduce productivity, transpiration and photosynthesis, shoot growth and leaf area, and to induce branch sacrifice and crown die-back, all of which reduces both the quality and quantity of food resources available to consumers (both terrestrial and aquatic). Thinning of the canopy may reduce shading and expose the water body to greater light and heat. Seedling recruitment of many riparian trees is dependent on the provision of floods, and there is likely to be high mortality if there is a drought in their first few years, (e.g. Lytle & Merritt, 2004).

Drought not only affects the dominant riparian trees, but also the ground cover of herbs and grasses. In extended drought, the riparian zone may be invaded by drought-tolerant plants from the surrounding hinterland (Jacobs *et al.*, 2007). In riparian zones with normal perennial flow, there may be a high diversity of hydric perennials and annuals (Katz *et al.*, 2009), whereas in riparian zones with intermittent flow (Stromberg *et al.*, 2005), and when exposed to a supra-seasonal drought, the cover is lower in species richness and is dominated by mesic annuals. During drought, the riparian zone in many places is increasingly occupied by large wild or domestic herbivores (e.g. Jacobs *et al.*, 2007; Strauch *et al.*, 2009) which can destroy vegetation, cause erosion and increase nutrients by their excretion.

In drought, along lowland streams in pastoral areas in south-eastern Australia, there is a steady increase in both litter and in bare ground (Lake (personal observation)). These changes lower the retentive capacity of the riparian zone when surface runoff returns with the breaking of the drought. This degradation serves to increase erosion, sediment entrainment and nutrient loads in streams, and it is a clear example of how the effects of drought can be greatly exacerbated by other disturbances, such as overgrazing by large herbivores.

5.5 Sequence of changes in water bodies with drying

In all types of water bodies, drought reduces the volume of water (Figures 5.1 and 5.2). In standing water bodies, the morphology of the basin strongly determines the spatial pattern of drying.

The rate of water loss with evaporation is related to both surface area and volume of exposed water bodies. In any one region, small bodies of water, such as phytotelmata and shallow pools, usually dry out long before nearby lakes. In many water bodies, such as shallow lakes and ponds, volumes can decrease more slowly than surface areas. For example, in large moorland pools subjected to drought, a 1 m drop in depth resulted in losses of 3 per cent in volume and of ≈ 15 per cent in surface area (Van Dam, 1988). In ponds and shallow lakes, with their small volumes, relatively larger surface areas and increased turbidity (from suspended sediments and phytoplankton), temperatures, especially in surface layers, can rise (Williams, 2006). If the water body is clear, heat may be absorbed in the

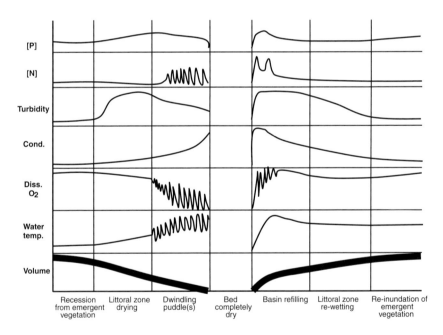

Phases during drying and re-wetting of standing waters

Figure 5.1 General changes in selected abiotic variables (water temperature, dissolved oxygen concentration, conductivity, turbidity, nitrogen and phosphorus) with declines in volume in lentic systems through drought and with re-wetting and re-filling in drought recovery. The trends may differ considerably between different lentic systems.

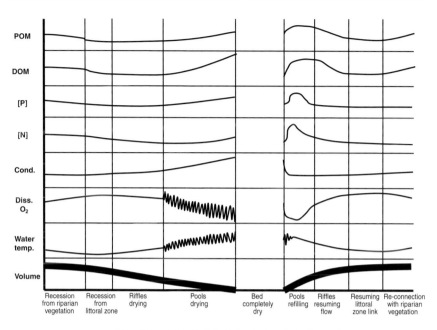

Phases during drying and re-wetting of flowing water

Figure 5.2 General changes in selected abiotic variables (water temperature, dissolved oxygen concentration, conductivity, turbidity, nitrogen and phosphorus) with declines in volume in flowing waters through drought and with the recovery to normal flow volumes. These trends are a guide and may be quite different under different conditions in various streams.

bottom sediments, which also raises water temperatures. The relatively large surface area to volume ratios in ponds and shallow lakes means that temperatures may be high in the day and can drop rapidly at night, further taxing the biota. The increase in temperature, especially at the surface, may serve to create stratification, albeit transiently in small water bodies (e.g. Eriksen, 1966).

With the loss in water volume in both lentic and lotic systems, in terms of biota the initial phase is dominated by the loss of habitat, for example the loss of riparian and littoral habitats (see Figures 5.1 and 5.2). This not only reduces biodiversity but also may produce severe declines in both primary and secondary production. In the second phase of drying, the decline in water quality becomes a key force challenging the biota, while habitat availability continues to shrink.

With the drawdown of natural lentic systems, the highly productive and habitat-rich littoral zone becomes disconnected from the waters and exposed, with damaging effects on the aquatic biota – especially the sedentary components (Figure 5.1). The morphology of lakes will have a strong influence on the effects of this drawdown. Those with extensive shallow littoral zones and wetlands (e.g. Lake Chad – see Carmouze & Lemoalle, 1983; Dumont, 1992) may, with a small drop in depth, lose a large amount of important littoral habitat and surface area. In contrast, deep lakes which have steep shores (e.g. Lake Baikal) may lose their narrow littoral zone but only a little of their surface area (Wantzen *et al.*, 2008). Many lakes have shallow and extensive littoral zones and a deep basin, and thus drought may expose extensive areas of the shallows, but not greatly affect the volume of the deep basin(s) (e.g. Lake Constance and Lake Titicaca).

Droughts are often associated with an increase in the duration of windy periods, further increasing evaporative water loss and increasing the erosive power of waves and turbidity around their shores. With the shoreline of shallow lakes retreating away from the littoral zone, fine sediments can be exposed to the wave zone and be re-suspended (Lévêque *et al.*, 1983; Søndergaard *et al.*, 1992; Hofmann *et al.*, 2008), increasing turbidity, releasing nutrients, reducing light availability and potentially increasing water temperature (Figure 5.1).

As the water level declines along lake shores, groundwater seeps may become very evident (Wantzen *et al.*, 2008). Such seepages can enrich the adjoining lake water, creating patches of high algal and bacterial growth. Similarly, nutrients from groundwater seeping into drying streams may stimulate autotrophic production (Dahm *et al.*, 2003).

Lentic water bodies may suffer from droughts that begin to become evident in summer or the dry season. The drought may be due to the failure of winter or wet season rains. However, droughts can occur in winter due to previous failures in normal precipitation, such as may occur for lakes of the North American Great Plains region (McGowan *et al.*, 2005) and for lakes in central Europe (e.g. Lake Constance – see Werner & Rothhaupt, 2008; Baumgärtner *et al.*, 2008). In this case, with antecedent low precipitation, lake levels drop, exposing the littoral zones and their biota to freezing and ice scouring. Such freezing may kill littoral benthos such as mussels (e.g. Werner & Rothhaupt, 2008) and, with ice completely covering the lake, oxygen levels may so decline as to induce fish kills (Gaboury & Patalas, 1984).

As a lake is drawn down in drought, then, depending on the morphology of the basin, a hitherto continuous volume of water may become divided into isolated pools or even basins. For example, in Lake Chad in central Africa, a long drought started in 1973. As the drought developed, the lake started to contract (Carmouze & Lemoalle, 1983), separating the northern basin from the southern (Figure 5.3). In the northern basin, as drying set in, turbidity rose sharply and oxygen concentrations varied greatly, 'with frequent periods of anoxia' (Carmouze *et al.*, 1983). Later, the northern basin dried

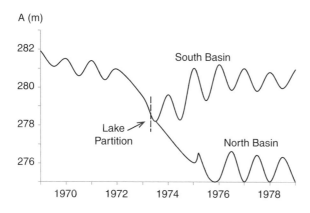

Figure 5.3 The parting of the basins (northern and southern) of Lake Chad in August 1973 as droughts set in. The southern basin maintained water all year, but water was only present in the northern basin during each annual wet season. (Redrawn from Figure 10, page 329 of Carmouze *et al.*, 1983a.)

out, while the southern basin retained water, albeit with a great drop in surface area (Carmouze *et al.*, 1983).

In shallow lakes, and to a much lesser extent in deep lakes, significant changes in abiotic variables occur with drying (Figure 5.1). As volume declines and the lake shallows, major changes can occur in water temperature, dissolved oxygen, conductivity (salinity), turbidity and major nutrients (e.g. nitrogen and phosphorus). The re-wetting and filling of shallow lakes tends to be more rapid than the drying. In the initial stages of re-wetting, many abiotic variables undergo major changes as indicated by sharp increases. Thus, dissolved oxygen concentration, conductivity, turbidity and macro-nutrient concentrations may reach peak values, from which, as water volume increases, there are progressive declines to the normal long-term values (Figure 5.1).

As a stream enters drought, water volume and depth drop and areas of high velocities decrease in area. (Figure 5.2). Normal shallow sections (e.g. riffles) can be so reduced in depth that they become barriers to the movement of fauna, especially fish (e.g. Schaefer, 2001). This step may occur shortly after the stream level drops away from the littoral edge or the toiche zone (Everard, 1996; Figure 5.2). As volume declines further, breaks can start to occur in longitudinal connectivity as shallow patches such as riffles and runs dry up, leaving pools. Some pools, especially deep ones, may be maintained by groundwater flows, whereas others steadily shrink through evaporation.

With the creation of lentic conditions, water quality may steadily or even rapidly deteriorate. Pools persisting in drought and unshaded may gain heat and become too hot for much of the fauna (e.g. Mundahl, 1990). Due to decomposition of organic matter and respiration by the pool inhabitants, combined with high temperatures, pools may reach very low oxygen concentrations that can kill most aquatic animals (e.g. Tramer, 1977). Frequently, hypoxia and hyperthermia may occur together. Pools in streams and on floodplains can contain large amounts of particulate organic matter, the decomposition of which can create 'blackwater' conditions, with high DOC levels accompanied by low oxygen and high carbon dioxide concentrations (e.g. Slack, 1955; Paloumpis, 1957; Larimore *et al.*, 1959; McMaster & Bond, 2008). Finally, pools dry, leaving moist sediments that, with time, become dry sediments.

The changes that occur in stream pools with drought are similar to those that occur in pools and ponds of lentic systems. Although drought is a ramp disturbance, the eco-hydrological response occurs as a series of steps or thresholds are crossed in both lentic and lotic systems, isolating or removing habitats with their associated biota (Boulton, 2003; Boulton & Lake, 2008).

Again, as in lakes, in streams with drying there are distinct trajectories in the levels of abiotic variables. In the early stages of drying, habitat loss is paramount, while in the later stages, habitat space continues to decline but water quality emerges as a decisive force (Figure 5.2). Major changes occur in water temperature and in the levels of dissolved oxygen when flow ceases and pools form. In flowing waters, with pool formation, the levels of particulate organic matter (POM), and especially dissolved organic matter (DOM) can rise substantially. Indeed, as described above, the combination of increased water temperatures, high levels of dissolved organic and low oxygen levels can be toxic.

The breaking of droughts in streams tends to be far more rapid than the onset of drought. In many cases, floods may break the drought, especially in streams with relatively impervious catchments. With channel re-wetting, and in the initial stages of recovery, there may be early peaks in both POM and DOM, and in nitrogen and phosphorus, when pools start to be filled (Figure 5.2). Conductivity may rapidly drop from a high value to return to normal expected levels. Dissolved oxygen levels may be lowered by the sharp increase in metabolism as pools are filled, but may then stabilize as flow returns.

In supra-seasonal drought in floodplain rivers, water in the channel may recede from the littoral edge. More seriously, lateral connectivity with the river's floodplain may be cut for years. If this happens, floodplain wetlands may dry up, but in those wetlands that persist, severe water quality problems can occur, such as stratification, high temperatures and deoxygenation.

Streams, particularly low-order ones, can dry in four basic spatial patterns. The most common form occurs when the headwaters of a stream dry and flow continues downstream. This pattern is common and means that the biota of headwater streams may be adapted to periods of intermittent flow, to which downstream biota are never exposed. In some cases, the downstream part of a stream dries, but the upper sections may continue to hold water due to springs, or to the fact that the stream flows out of a relatively well-watered section to a lower section on pervious substrate, such as streams flowing out from mountains into deserts.

In some unusual cases, a stream may become dry in its middle sections. For example, creeks in central Victoria, Australia, have their middle sections filled with sand slugs created by severe erosion in their headwaters (Davis & Finlayson, 1999). In drought, the upstream and the downstream sections hold water, but the middle sections consist of channels full of dry sand (Bond & Lake, 2005). In some streams, especially those of low gradient and with intricate channels, drought may leave pools with water interspersed with dry stretches. For example, in a low gradient, lowland creek, drought gave rise to a series of pools with variable durations (Perry & Bond, 2009). The likelihood of a river drying into sections and longitudinal connectivity being severed declines as the flow volume of a river increases. Even so, in severe drought, sections of major rivers can become dry, such as the River Murray in the severe 1914–15 drought (Sinclair, 2001). The pattern of drying along a stream channel will largely determine the availability of refuges for water-dependent animals such as fish. Overall, it appears from the literature that the most common form of stream drying is that where the low order headwater streams dry and the higher order downstream stream sections have water or, in extreme cases, dry out after the headwaters.

5.6   Changes in water quality with drought in lentic systems

The changes in water quality with drought are more dramatic in both lentic and lotic systems that normally have small volumes of water. Without drought, in small water bodies such as ponds, oxygen concentrations can fluctuate diurnally, being high in the day through photosynthesis and low at night through respiration. With drought serving to increase the levels and diel range of water temperatures, dissolved oxygen may be depleted, especially in systems with relatively high levels of phytoplankton and bacteria.

Along with a decrease in oxygen, carbon dioxide levels may be altered by the loss of water. Depletion of carbon dioxide by phytoplankton photosynthesis may lower concentrations, while temperature-augmented benthic decomposition can increase dissolved carbon dioxide concentrations.

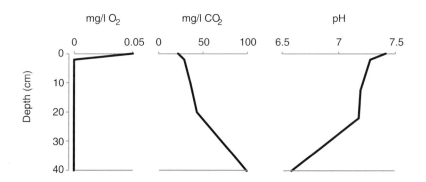

Figure 5.4 Hostile conditions in dissolved oxygen concentrations, carbon dioxide concentrations and pH in a pond in Big Cypress Swamp, Florida, that resulted in a fish kill. (Redrawn from Figure 1 of Kushlan (1974a).)

During a severe seasonal drought, in a pond in the Big Cypress Swamp, Florida, Kushlan (1974a) found that dissolved oxygen was only available in the thin upper surface layer, and carbon dioxide concentrations increased from 16.8 mg per litre at the surface to 98 mg per litre at the bottom (depth 40 cm) (Figure 5.4). The increase in carbon dioxide concentration with depth was accompanied by an increase in acidity. Coincident with these conditions was a major fish kill, with only 6 of 22 species and 0.6 per cent of total fish populations surviving (Kushlan, 1974a).

With evaporation causing water loss, salinity (conductivity) can rise (Figure 5.1). If the water body is isolated from the surrounding water table, salinity may rise, but if the drying of the water body is by the retraction of the water table along with evaporation, salinity may only rise slightly. The effects of drought on salinity (conductivity) are dramatically exemplified by the changes in two African lakes, Lake Chad and Lake Chilwa. In the northern basin of Lake Chad, as drought set in from 1973 to 1978 the salinity (already high) rose from $1,000\,\mathrm{mg\,l^{-1}}$ to $3,000\,\mathrm{mg\,l^{-1}}$ (Carmouze et al., 1983). In Lake Chilwa, in the 1967–68 drought, salinity rose from $870\,\mathrm{mg\,l^{-1}}$ to $11,000\,\mathrm{mg\,l^{-1}}$ (McLachlan, 1979a; Figure 5.5) On the Pongolo floodplain in Botswana in the 1983 drought, some floodplain lagoons (pans) became saline due to the input of saline groundwater seepage as water levels dropped, whereas others became saline due to evaporation and drying reaching maximum salinities in the range of 18–$30\,\mathrm{g\,l^{-1}}$ (White et al., 1984).

Drought increases dissolved salt concentrations in systems with relatively high calcium and free carbon dioxide/bicarbonate/carbonate concentrations, such as in hard water lakes, causing calcium carbonate to be precipitated as calcite (Cole, 1968; Wetzel, 2001; Cohen, 2003). Thus, as

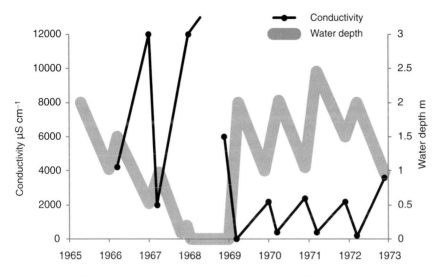

Figure 5.5 Changes in conductivity (dissolved salts) and depth in Lake Chilwa, Malawi, before, during and after the severe drought in 1967–1969. (Redrawn from Figure 4.2, page 67, in Kalk *et al.*, 1979.)

water levels dropped in Lake Chilwa, turbidity, alkalinity and pH ($\approx$10.8) increased and calcite was precipitated (McLachlan, 1979a). Such loss of calcium from the water column causes a proportional increase in concentrations of magnesium and sodium. Because of this, changes in the Mg/Ca ratio in deposited fossil carbonate may indicate past droughts (Cohen, 2003, Russell & Johnson, 2007). With increasing ion concentrations in some lakes, such as Lake Chichancanab Mexico, calcium sulphate can also be precipitated with water loss (Hodell *et al.*, 2005).

In standing waters, when drought lowers water volumes and depths, changes in nutrient dynamics in the water column can be expected, but surprisingly this area remains relatively unstudied. For example, in a long-term study (1964–2001) of Lake Vortsjärv, a small, shallow lake, in Estonia, Nõges *et al.* (2003) found that low lake levels in the 1976–77 European drought did not cause total nitrogen in the water column to change significantly, whereas total phosphorus concentrations doubled to $\approx$80 mg m^{-3} and the TN/TP mass ratio declined.

In Turkey, in the shallow and eutrophic Lake Eymir, after two years of bio-manipulation, a severe drought set in and there were marked increases in salinity and conductivity as a function of the sharp rise in hydraulic residence time (i.e., the lake volume divided by the volume of water flowing into the lake per unit time) (Beklioğlu, 2007; Beklioğlu & Tan, 2008). With drought reducing inputs to the lake, nutrient processes became less dependent on external loading and more controlled by internal forces. Thus, there were increases in the water column of total phosphorus due to resuspension of bottom sediments – a process found in shallow lakes and reservoirs with decreases in depth and the breakdown of stratification (e.g. Søndergaard *et al.*, 1992; Naselli-Flores, 2003; Boqiang *et al.*, 2004; Baldwin *et al.*, 2008). As the drought progressed, anoxic conditions became established due to an increase in phytoplankton (cyanobacteria) production and benthic decomposition and there were fish kills (Beklioğlu, 2007). Under the anoxic conditions, concentrations of ammonium salts increased, and this ammonification greatly reduced nitrification and denitrification (Beklioğlu, 2007; Beklioğlu & Tan, 2008). With the breaking of the drought, both total phosphorus and soluble reactive phosphorus concentrations declined and concentrations of dissolved inorganic nitrogen rose (Beklioğlu & Tan, 2008).

Thus, in eutrophic shallow lakes, as clearly exemplified by the changes in Lake Eymir, the reduction in water levels with drought can induce processes that increase nutrient concentrations, which in turn stimulates high phytoplankton production, especially of cyanobacteria, and may even lead to anoxia. The anoxic conditions, in turn, create dramatic changes in nitrogen processing (ammonification), which further lowers water quality. Nutrient concentrations can be also expected to rise in shallow systems due to increased bioturbation of the bottom sediments by the increased densities of such fauna as fish and crayfish (Covich *et al.*, 1999).

Although data are scarce, it appears that in oligotrophic lakes, provided there are not dramatic changes in depth and volume, drought does not lower water quality to levels that would tax the biota, though slight changes may favour particular biota, e.g. phytoplankton. For example, in an alpine lake exposed to a severe drought (2002, the driest in 110 years), summer surface water temperatures and hydraulic residence times increased (Flanagan *et al.*, 2009). Conductivity, acid-neutralizing capacity and concentrations of calcium, potassium, chloride and sulphate all rose, but not markedly. In contrast, during the drought, silica concentrations declined quite significantly. This change was undoubtedly due to its uptake by high densities of a species of hitherto rare diatom, *Synedra* sp. (Flanagan *et al.*, 2009).

In the 1980s, droughts occurred on the Boreal Shield of western Ontario, specifically in the region of the Experimental Lakes Area (Schindler *et al.*, 1990; Findlay *et al.*, 2001). These lakes are dimictic (stratifying twice a year) and oligotrophic. Their responses to drought, both abiotic and biotic, were temporally coherent (Magnuson *et al.*, 2004). During drought, the length of the ice-free season, the depth of the thermocline, Secchi depth and water residence depth increased as precipitation and direct runoff declined (Schindler *et al.*, 1990). With the decline in direct runoff, the DOC concentrations in the lakes declined and so did light attenuation, resulting in an increase in the depth and volume of the euphotic zone (Findlay *et al.*, 2001). Even with these changes, the concentrations of nitrogen and phosphorus were low and did not change significantly with drought.

Lakes in the same locality may vary greatly in response to drought and not be temporally coherent. James (1991) studied ten shallow acidic lakes in Florida, with five being clear oligotrophic systems and the other five being dark dystrophic systems. The lakes were sampled during and after a severe summer drought. In the clear lakes, both during and after the drought, there were only small changes in acidity and water temperature, but significant decreases in sulphate and nitrate. No changes occurred in total bacteria, DOC, oxygen and chlorophyll-*a* concentrations. However, in the dystrophic lakes during drought, total bacteria and chlorophyll-*a* were greatly reduced, while DOC increased almost fourfold (from 7.4 to 28.4 mg l^{-1}), acidity more than doubled and pH dropped from 4.7 to 4.3. The main effect of drought in the dystrophic lakes was to increase DOC levels and acidity, which may have inhibited the growth of bacteria and phytoplankton. James (1991) examined the relationships between DOC and bacteria and chlorophyll-*a*, which suggested that at high concentrations of DOC, both bacteria and chlorophyll-*a* (phytoplankton) were inhibited. Whether the increased levels of DOC that can occur in pools serves to inhibit phytoplankton production is uncertain.

Although this study showed that DOC concentrations in lakes can rise with the advent of drought to levels that inhibit bacterial growth, in some lakes, drought may cause DOC concentrations to decline (Schindler *et al.*, 1997). This may be due to either a drop in DOC inputs because of low runoff volumes, or to increased breakdown of DOC (possibly by increased UV radiation in drought) in the lake – or perhaps both (Schindler *et al.*, 1997; Sobek *et al.*, 2007).

In summary, by lowering depths, volumes and surface areas, drought in standing waters can cause major changes in basic morphology, temperature regimes, turbidity, suspended solids, water chemistry and nutrient concentrations (Figure 5.1). The initial conditions of shallow lakes can exert a strong influence on the nature of changes due to drought, and variation in initial conditions between water bodies can generate different responses in any one region. For deep lakes, there is a dearth of reports on how they are affected by drought, but effects would probably not be nearly as substantial as those induced in shallow lake systems.

5.7 Drought in connected lakes

As drying occurs in the catchments of lakes, water inputs from streams are greatly reduced. Lakes in a catchment may be linked by both surface and groundwater flows. During drought, the surface water links may be greatly reduced, while the groundwater connections remain. The importance of inter-lake connections is exemplified in the long-term study of a series of lakes – a 'lake district' in northern Wisconsin (Soranno *et al.*, 1999). The lakes are spatially separated, but they are hydrologically linked by surface or groundwater flow paths. Important in the influence of flow paths on the chemical properties of such lakes is landscape or hydrological position (Webster *et al.*, 1996; Kratz *et al.*, 1997; Soranno *et al.*, 1999). Lakes at high elevation in the landscape are strongly dependent on water and chemical inputs from precipitation rather than groundwater flows, whereas lakes at lower elevation in the same landscape receive water and chemical inputs from both precipitation and groundwater.

Interestingly, the nature of these elevational differences became evident in a four-year drought (Webster *et al.*, 1996; Kratz *et al.*, 1997; see Figure 5.6). During the drought, because of a deficit in precipitation, lakes high in the landscape (the 'hydraulically mounded lakes') received reduced inputs of water and still lost water to evaporation and groundwater flow. Lakes lower in the landscape (the 'groundwater flowthrough lakes' received similar reductions in precipitation, but groundwater inputs were maintained, whereas the lowest set of lakes ('the drainage lakes') received both surface runoff and groundwater inflows from the 'groundwater flowthrough' lakes. A significant outcome of this elevational gradient was that calcium and magnesium masses were diminished in the high-elevation, precipitation-dominated lakes and increased in the low-elevation, groundwater-dominated lakes and drainage lakes (Webster *et al.*, 1996; Kratz *et al.*, 1997). Thus, in this system, extended droughts may so reduce calcium and magnesium masses that calcium concentrations can drop below the thresholds for calcium-dependent fauna such as snails and crayfish (Kratz *et al.*, 1997).

Lakes in a landscape responding to the stress of supra-seasonal drought may show uniform temporal coherence (high inter-lake correlation) (Magnuson *et al.*, 2004) in variables such as temperature and chemical concentrations, and there may be structured coherence between the lakes, depending on the nature of the hydrological linkages (Magnuson *et al.*, 2004). This case stresses the points that even when drought has caused surface water movements to cease, groundwater connections may still be operating in a landscape, and that connections between water bodies in drought may heighten variability in essential components (e.g. ions) between water bodies in a catchment. With the breaking of drought in these lakes, the lower drainage lakes may recover faster than the groundwater flowthrough lakes, due to the increased precipitation affecting surface drainage long before groundwater flows. Recovery from drought and cation

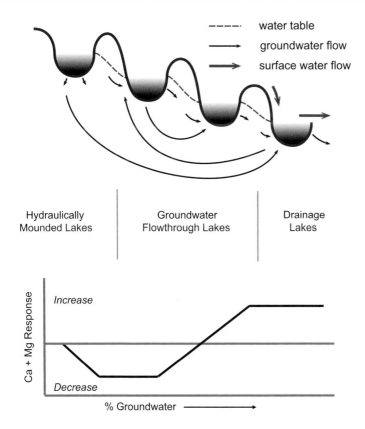

Figure 5.6 Drought across connected lakes in a catchment linked by groundwater pathways, showing the progressive movement of calcium and magnesium from elevated 'hydraulically modified lakes' via 'groundwater flowthrough lakes' to 'drainage lakes' which receive both groundwater and surface runoff. (Redrawn from Figure 1, page 978, in Webster *et al.*, 1996.)

depletion in the high 'hydraulically mounded lakes' would probably be much slower than in the lower lakes, due to cation inputs being very strongly dependent on precipitation.

5.8   Drought and water quality in flowing waters

Drought can change the water quality of streams in three ways: through changes in the water from the catchment moving into streams; or by changes occurring within the stream channel; or by both of the above changes interacting together. As drought progresses, the two latter sources of change become more important than the former. Changes in chemistry in catchments due to drought can result in major, if not dramatic, changes in water quality when the drought breaks and aquatic links between the catchment and the stream are restored.

In drought, as flow is reduced and air temperatures rise, stream water temperatures can rise above normal levels (Figure 5.2) and extend beyond normal durations (e.g. Ladle & Bass, 1981; Boulton & Lake, 1990; Caruso, 2002; Sprague, 2005), which in turn can stress fish and invertebrates (e.g. Quinn *et al.*, 1994; Elliott, 2000) and may result in fish kills (e.g. Huntsman, 1942; Brooker *et al.*, 1977). The increase in temperature can lower oxygen concentrations by reducing its solubility and stimulating decomposition. Low oxygen concentrations can create hypoxic conditions, which, combined with high temperatures, may severely stress and kill fauna, especially fish (Brooker *et al.*, 1977; Smale & Rabeni, 1995; Elliott, 2000; see Figure 5.7).

A consistent characteristic of drought in streams is the increase in conductivity (salinity) (e.g. Foster & Walling, 1978; Boulton & Lake, 1990; Schindler, 1997; Sprague, 2005; Zielinski *et al.*, 2009) as a result of decreased dilution, increased evaporation and increased residence time of subsurface water that discharges as baseflow (Caruso, 2002). Fluctuations

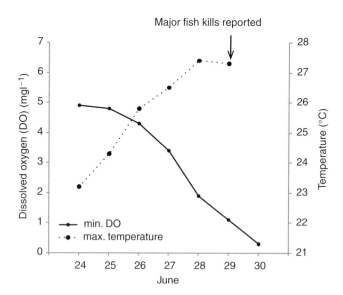

Figure 5.7 Water temperatures and dissolved oxygen concentrations in the River Wye at Kerne Bridge, Wales, UK in the summer of the 1976 drought, in which the low oxygen concentrations and high temperatures resulted in major fish kills of Atlantic salmon. (Redrawn from Fig. 4, page 414 in Brooker *et al.*, 1977.)

can occur even on a daily basis. Kobayashi *et al.* (1990) found that due to evapotranspiration during drought, daytime streamflow and conductivity were reduced in a small, forested stream in Japan. This effect appeared to be due to ion-poorer groundwater flow being the major source of streamflow in the day, while at night, the ion- richer throughflow (interflow) exerted an influence.

Depending on the nature of the catchment, changes in conductivity may be accompanied by changes in pH. For example, with forested streams in catchments with limestone and sandstone in Colorado during the 2002 drought, Sprague (2005) recorded more alkaline pHs, due possibly to decreased dilution of the groundwater.

Concentrations of particular nutrients and ions show varying response to drought. During drought in an English chalk stream, concentrations of potassium and soluble phosphates increased, while nitrate concentrations did not change (Ladle & Bass, 1981). Similarly, in the 1976 drought in Devon, England, there was no significant change in nitrate concentrations in several small streams (Foster & Walling, 1978; Walling & Foster, 1978; Burt *et al.*, 1988). In streams in New Zealand with a severe drought (1998–1999), Caruso (2002) found that both total phosphate and total nitrogen concentrations decreased, even though the streams were in agricultural landscapes. The declines were attributed to greatly decreased non-point source runoff.

A similar situation occurred during drought in Georgia, USA (Golladay & Battle, 2002). Concentrations of particulate organic matter (POC), particulate inorganic matter (PIM), DOC, dissolved inorganic nitrogen (DIN) and soluble reactive phosphorus (SRP) all declined in lowland streams. This decline was attributed to the severing of aquatic links between the creeks and their sources in forests (Golladay & Battle, 2002). In Polish streams in drought (2000), calcium and bicarbonate concentrations rose, while DOC concentrations greatly declined due to the weakening of linkages between streams and their catchments (Zielinski *et al.*, 2009).

The weather conditions of drought, in drying the vegetation and producing high temperatures and winds, increase the risk of wildfires. Drought creates effects in stream chemistry which can be accentuated by wildfire. For example, in streams in north-western Ontario, Bayley *et al.* (1992) found that a wildfire increased acidity (pHs moved from 5.15 to 4.76) and concentrations of sulphates and base cations (calcium, magnesium), causing a decline in acid-neutralizing capacity. These strong effects persisted for several years after wildfire. Drought produced similar effects, but not of the same strength (Bayley *et al.*, 1992). Williams & Melack (1997), studying montane Californian streams, found that fire increased stream concentrations of sulphate, chloride and nitrate and, to a lesser extent, base cations.

Drought increased concentrations of calcium, sodium, sulphate and acid, neutralizing capacity, possibly as a result of reduced dilution by stream water. Silicate concentrations dropped during the drought, possibly due to much less weathering. Wildfires during drought may exacerbate the effects of drought, compromise the recovery capacity of catchments and limit the recovery of stream biota from the combined disturbances of drought and wildfire.

During drought, DOC concentrations in soil, especially peat, may increase, but flux and concentrations in streams can decrease due to low water levels to mobilize the DOC. In an experiment simulating summer drought in a peatland stream, Freeman et al. (1994) observed a decline in DOC concentrations, increases in inorganic nutrient concentrations (nitrate, ammonium, Ca, Mg, K) and a lowering of the ratio of organic to inorganic nutrients. These changes appeared to have diminished heterotrophic production and stimulated autotrophic production. Similarly, in a montane New Mexico stream with drought, both DOC concentrations and nitrate concentrations declined, while phosphate concentrations remained low. The low nitrogen concentrations may have been due to active uptake by algae, increasing autotrophic production, whereas declining concentrations of DOC served to limit heterotrophic production (Dahm et al., 2003). Thus, during drought in streams which normally have significant heterotrophic production, DOC inputs may decrease, favouring an increase in autotrophic production – provided, of course, that sufficient light is available.

In general, during drought in flowing waters, the concentrations of suspended particles and turbidity decline (e.g. Golladay & Battle, 2002; Caruso, 2002) and fine sediments may be deposited in and on the stream bed, generating low flow depositional habitats (e.g. Wood & Petts, 1999; McKenzie-Smith, 2006). During periods of low flow and drought, particulate organic matter can accumulate in the stream channel (e.g. Boulton & Lake, 1992c; Maamri et al., 1997a; McKenzie-Smith, 2006). This is especially noticeable in Australian streams which have evergreen catchment vegetation that sheds leaves all year with a peak in summer. This material in pools can create serious water quality problems, principally 'blackwater events' with low dissolved oxygen levels and high DOC concentrations (Towns, 1985; McMaster & Bond, 2008).

During drought, the volumes of water delivered to estuaries decreases, and consequently saline estuarine water can move upriver, dramatically changing water quality. Thus, in the 1961–1966 drought in the eastern USA, the 'salinity invasion' in the Delaware estuary extended upriver some 32 km more than normal (Anderson & McCall, 1968). In eastern England in the 1976–77 drought, severe saline intrusions occurred in lowland rivers, resulting in high salinities that made the river water unusable for irrigation and stock (Davies, 1978). High levels of salinity may also occur in drought in streams in regions affected by salinization. For example, pools in the lower Wimmera River, Victoria, Australia during the severe 1997–2000 drought, salinity increased from $\approx$2.1 to $\approx$21.2 or 60 per cent, a level that few freshwater animals can tolerate (Lind *et al.*, 2006).

All of the above applies to fairly natural streams. However, in areas with high human populations generating water-borne wastes, drought in lowering water volumes can create water quality problems by reducing the potential dilution of wastes. For example, in the 1975–76 drought in Europe (e.g. Slack, 1977; Brochet, 1977; Davies, 1978; Van Vliet & Zwolsman, 2008), the 2003 drought in Europe (Van Vliet & Zwolsman, 2008, Wilbers *et al.*, 2009), the 1961–1966 drought in north-eastern USA (Anderson & McCall, 1968) and the 1982–83 drought in eastern Australia (Chessman & Robinson, 1987), water quality in rivers declined during and immediately after the drought. The declines were largely due to decreased dilution of sewage and wastewater discharges.

Sewage can contain large amounts of dissolved solids, both organic and inorganic, and consequently total dissolved solids (conductivity) and ions, such as chloride and sulphate, can rise considerably during drought as a result of pollution (Slack, 1977; Davies, 1978; Chessman & Robinson, 1987). With large amounts of readily metabolizable organic matter entering rivers from sewage, dissolved oxygen concentrations can decline (Anderson & McCall, 1968; Chessman & Robinson, 1987), with a corresponding rise in biochemical oxygen demand (BOD) values (Anderson & McCall, 1968).

Nitrogen, as ammonia, nitrites and nitrates, along with phosphates, can be discharged into rivers from point and non-point sources (Slack, 1977; Davies, 1978; Van Vliet & Zwolsman, 2008). These inputs, combined with the low flow and high temperatures of drought-stricken rivers, may create blooms of algal and cyanobacterial growth – eutrophication. Eutrophic conditions create great fluctuations in oxygen concentration ranging from supersaturation (e.g. 410 per cent saturation (Davies, 1978)) in the day to extremely low concentrations late in the night (Davies, 1978; Van Vliet & Zwolsman, 2008). Accordingly, such conditions can result in fish kills (Davies, 1978).

Cyanobacteria blooms during drought have occurred in inland rivers of south-eastern Australia (Whittington, 1999). Whether the blooms were induced by nutrients from human-created sources is uncertain. In the 1991 drought in the Darling River, there was a cyanobacteria bloom that extended for a thousand kilometres (Sinclair, 2001). In summer (2006–2007) during the recent Australian drought, in a large reservoir (Hume Dam) at very low capacity (9–3 per cent), a cyanobacterial bloom was generated that stretched at least 150 kilometres long down the Murray

River from the dam to Corowa (Baldwin *et al.*, 2010). As in the case of Hume Dam, drought may serve to trigger cyanobacteria blooms in reservoirs, posing a serious problem for management. For example, a survey of 39 reservoirs during the 1998 El Niño drought in north-east Brazil found blooms of the cyanobacterium *Cylindrospermiopsis* in 27 of the reservoirs (Bouvy *et al.*, 2000), indicative of temporal coherence and the widespread, but similar, effects of drought.

5.9   Drought and benthic sediments

As water levels drop in drought, bottom sediments are exposed to the air, especially in shallow lakes, where a slight drop in depth can expose large areas of sediment. As drying takes place, the sediments, especially those with a high clay content, may crack and develop deep fissures, whereas sand sediments may only develop minor cracking on the surface due to drying organic matter. Drying of sediments strongly changes their chemistry, mineralogy and microbiology (De Groot & Van Wijck, 1993; Qiu & Mc Comb, 1995, 1996; Mitchell & Baldwin, 1999; Baldwin & Mitchell, 2000). For example, with drying, iron, as ferrous sulphate (FeS) may be oxidized to ferric oxyhydroxides, which may then bind to any available phosphate (De Groot & Van Wijck, 1993).

In the early stages of drying, the sediments may be two-layered: an oxic layer above an anoxic one. Nitrogen metabolism is largely a function of bacterial activity; thus, in the early stages of drying in the top oxic layer, mineralization of organic nitrogen proceeds, along with nitrification. In the anoxic layer, ammonification occurs, along with denitrification (Baldwin & Mitchell, 2000). As the oxic layer deepens, the microbiota (bacteria, fungi and protozoans) may proliferate and break down organic matter, releasing carbon dioxide to the atmosphere. As the anoxic layer retreats, there is a shutdown of anaerobic microbial activity, reducing denitrification and increasing phosphorus retention (De Groot & Fabre, 1993; Baldwin & Mitchell, 2000). Further desiccation greatly reduces microbial biomass, with microbial mortality contributing to the nitrogen and phosphorus concentrations in the dry sediments (De Groot & Van Wijck, 1993; Qiu & McComb, 1995; Mitchell & Baldwin, 1998).

Sediments of lakes and pools may contain significant amounts of sulphur, principally stored in anoxic sediments as FeS. As drying proceeds in drought, such sediments may be exposed and the sediment sulphur re-oxidized to become sulphates. Subsequently, with re-wetting, sulphuric acid forms and may be flushed into lakes (Van Dam, 1988; Yan *et al.*, 1996). The consequent acidification of lakes can result in a decline in DOC concentrations and increased penetration of UV-B radiation (Yan *et al.*, 1996; Schindler and Curtis, 1997).

In two moorland pools in the Netherlands, Van Dam (1988) recorded that drying of the bottom during drought caused a decrease in pH and an increase in sulphate concentration. In Lake Jandalup, a shallow lake in Western Australia, over a period of intense summer droughts, Sommer & Horwitz (2001) observed that as the lake level dropped and bottom sediments dried, the lake acidified, with the pH dropping from a range of 6–8 to a range of 4–5, accompanied by a marked increase in sulphate, iron and ammonium concentrations. Such a shift induced major changes in the aquatic fauna.

In south-eastern Australia, floodplain wetlands have been greatly altered by river regulation, irrigation and salinization. Furthermore, droughts lower the levels of floodplain lagoons and expose their sediments. Some wetlands have sediments with significant levels of sulphides that, upon exposure to air, re-oxidize to sulphates (Hall *et al.*, 2006; Lamontagne *et al.*, 2006; Baldwin *et al.*, 2007). Acidification of these wetlands may occur with rain events, even if the rains are not sufficient to break the drought.

5.10   The breaking of drought – re-wetting and the return of flows

Dry sediments on the bottom of lakes, lagoons and flowing waters may be re-wetted by rain events that do not persist, so that the sediments again dry out. In the breaking of a drought, with the re-wetting of dry sediments there may be an initial release of nitrogen and phosphorus produced by microbial mortality in the drying (Qiu & McComb, 1995, 1996; Baldwin & Mitchell, 2000). Similar to the release of nitrogen from dried sediments in standing waters, upon re-wetting, stream sediments may release nitrogen compounds.

In an upland stream that dried with a supra-seasonal drought, Baldwin *et al.* (2005) observed that, upon re-wetting, dried sediments released nitrogen mainly as ammonia and nitrates, along with some urea. Such a flush of nutrients with sediment re-wetting may stimulate high levels of microbial activity that, among other processes, leads to an increase in nitrification (Stanley & Boulton, 1995; Baldwin & Mitchell, 2000). Accompanying the nutrient release, there can be increased algal activity, which may be followed by a burst of macrophyte growth (Baldwin & Mitchell, 2000).

Floodplain soils are often rich in organic matter from desiccated macrophytes and algae, as well as from leaf litter from floodplain vegetation. The flooding of floodplain soils after a drought may induce the nutrient processes outlined above, as well as the leaching of soluble carbon compounds (DOC) from the soil and litter (O'Connell *et al.*, 2000). This mixture of nutrients and available carbon (DOC) can create intense microbial activity that, with high water temperatures and no water movement, may create anoxic conditions, These conditions can result in the very damaging blackwater events (Howitt *et al.*, 2007).

During drought, soils and sediments in the catchments of water bodies may accumulate chemicals through atmospheric inputs or via metabolic processes, such as decomposition, that continue during the drought. With the lack of available free water, many of these chemicals have limited mobility. With the breaking of the drought and the wetting of soils and water body sediments, concentrations of a range of chemicals can rise and, in some cases, persist.

Soils and wetlands may, in drought, accumulate nitrogen as nitrates. This, with the breaking of drought, produces a pulse of nitrates into catchment wetlands and streams (Reynolds & Edwards, 1995; Watmough *et al.*, 2004). For example, with the breaking of the 1976–77 drought in England, Slack (1977), Foster & Walling (1978), Walling and Foster (1978) and Davies (1978) recorded dramatic increases in nitrate concentrations and in nitrate/nitrogen loads (Burt *et al.*, 1988; see Figure 5.8). In the case of catchments in Devon, the increase in nitrate concentration was 45–50-fold (Foster & Walling, 1978; Walling & Foster, 1978). Multiplying concentrations by volumes reveals the 'enormous nitrate loads' (Burt *et al.*, 1988) that flowed through the waterways. This pulse of nitrates is no doubt accentuated by human activities in the catchment, such as the extensive spreading of nitrogenous fertilizers (Burt *et al.*, 1988).

Decomposition occurs in soils during drought, especially in organically-rich soils such as peat, and DOC can accumulate. With the drought breaking and the water table rising, DOC may be mobilized and DOC fluxes in streams

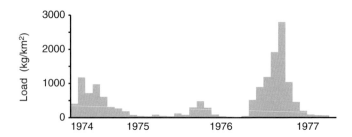

Figure 5.8 Nitrate loadings in streamflows from a small catchment (Slapton Wood) in Devon, UK, before, during and after the severe 1976 drought, illustrating the high nitrate loading that occurred with the breaking of the drought. (Redrawn from Figure 6, page 280 in Burt *et al.*, 1988.)

increase (e.g. Worrall *et al.*, 2006). In Britain, there is evidence that DOC fluxes from peatlands and into rivers have increased in recent times (Worrall & Burt, 2007), though the influence of droughts on this increase may be minor in comparison with large-scale hydrological changes such as increased streamflows (Worrall & Burt, 2007, Worrall *et al.*, 2008).

In streams where the channel dries and where CPOM (leaf litter) accumulates during drought, with the first flows as a drought breaks, there can be a pulse of high concentrations of dissolved organic carbon (Romani *et al.*, 2006) – DOC that may be readily incorporated into the trophic structure of the recovering stream community (Dahm *et al.*, 2003; Acuña *et al.*, 2005). This pulse may be accompanied by increased concentrations of DOC, POM and nutrients, such as nitrate and phosphorus, which are conceivably essential ingredients fuelling the recovery of stream ecosystems. However, this stimulus may be greatly dampened if there is increased acidity, excessive sulphate loads and mobilized heavy metals. At present, apart from the preliminary work with seasonal droughts of Romani *et al.* (2006), the biogeochemistry of stream ecosystem recovery after drought remains unexplored.

As drying proceeds during drought, sulphur-containing sediments may be exposed and the sediment sulphur then oxidized to sulphates. With re-wetting, the sulphates form sulphuric acid, which consequently acidifies the water body (Van Dam, 1988; Yan *et al.*, 1996). This is a particular peril for lakes and lagoons along the lower Murray River in south-east Australia, when droughts break (Hall *et al.*, 2006; Simpson *et al.*, 2010).

In particular regions downwind of heavy industrial activity, such as northern-eastern North America and Scandinavia, fossil-fuel combustion, the operation of metal smelters and the prevailing westerly winds have produced acid rain with sulphuric acid from the emitted sulphur dioxide. In the downwind region's wetlands with high water tables, DOC and reducing conditions, the sulphur from acid rain and locally-derived sulphur is stored as sulphides (Schindler, 1998). In times of drought, however, the water levels in the wetlands drop and the stored sulphides are oxidized, producing sulphuric acid (Warren *et al.*, 2001). With the breaking of the droughts and increased runoff, sulphates are delivered in pulses to downslope lakes, creating acidic conditions (LaZerte, 1993; Yan *et al.*, 1996; Dillon *et al.*, 1997; Devito *et al.*, 1999; Eimers *et al.*, 2007, 2008). In streams with these high sulphate concentrations, there is a drop in pH and a decline in acid neutralization capacity (ANC) (e.g. Laudon & Bishop, 2002; Laudon *et al.*, 2004; Eimers *et al.*, 2007, 2008). High sulphate concentrations and elevated flux rates in streams which started to flow after El Niño droughts produced pulses of acid into downstream lakes (Dillon *et al.*, 1997). Such drought-induced export of sulphates from catchment wetlands is likely to continue for a considerable time (years to decades) (Dillon *et al.*, 2003; Eimers *et al.*, 2007, Laudon, 2008).

Wetlands that undergo acidification by sulphur oxidation during drought can also accumulate high concentrations of mobilized metals (e.g. cadmium, zinc, aluminium), which may be flushed out with the breaking of droughts and have damaging effects on downstream ecosystems (Lucassen *et al.*, 2002; Tipping *et al.*, 2003; Adkinson *et al.*, 2008).

Increases in acidity in streams and lakes as a drought breaks is a major setback for any recovery from drought. Conversely, the frequency of droughts with post-drought acidification events may regulate the recovery of streams and lakes affected by acidification (Laudon, 2008). In Ontario, Canada, the macroinvertebrate communities of streams subject to acidification were compared with streams with uncontaminated catchments (Bowman *et al.*, 2006). While there had been some recovery in invertebrate communities in the streams affected by acidification, it was found that recovery was set back by both acid and toxic metal catchment inputs with the breaking of droughts. The interactions between low water levels due to drought, creating an accumulation of sulphates, metals and nitrates in soils, which are then flushed downslope as the drought breaks, is a startling example of the synergy between two forms of disturbance – the ramp of drought and the pulse of acidification. Both of these disturbances derive from changes in the temporal pattern of precipitation, moving from extended drought to normal conditions.

It appears that there has been little concern from researchers about the way that droughts can break. Crucial in this regard are the state of the catchment, the groundwater levels and the nature of the precipitation. After a drought, the catchment of a water body may be bereft of ground cover, either from desiccation and the breakdown of plant cover, or from heavy grazing, or as a result of wildfire. The riparian zone may have lost its ground cover and, with trampling by grazing animals, the banks may be unstable.

If a drought is broken by a prolonged spell of steady precipitation, then streams may gradually start to flow and standing waters steadily fill. Even if the catchment is in poor condition, damage may be low, although there will be a movement out of the catchment of soluble chemicals, including nutrients. If, on the other hand, a drought breaks with heavy downpours, then erosion may be severe and extensive, and large amounts of sediments, along with nutrients, may be exported rapidly out of the catchment. In south-east Australia, El Niño-induced droughts may be broken quite rapidly by La Niña-induced floods. Care of catchments could reduce the damage incurred when droughts break, but this, unfortunately, is not widely practised. The aftermath of droughts, in terms of how they are broken, remains a major gap in ecological research.

5.11   Concluding remarks

In terms of the abiotic changes in water bodies as a result of drought, there are some obvious patterns, the nature of which are governed by the intensity and duration of the drought. In droughts in lotic systems, headwater streams may cease to flow, while the lowland mainstem rivers, provided they are not regulated and exploited, may be relatively unaffected. However, severe and long droughts can cause mainstem river channels to be depleted of water, and even cease to flow. In terms of lentic waters, small water bodies are particularly vulnerable to drought, as are shallow lakes in contrast to deep, large lakes.

Drought in both lentic and lotic systems serves to break connectivity. A critical avenue of connectivity that may be severed, or at least greatly reduced, is that between a water body and its catchment. Droughts reduce catchment inputs of water, nutrients, dissolved organic carbon and particulate organic and inorganic matter to downslope water bodies. Lakes and streams are withdrawn from their normal shorelines and as groundwater inputs retract. Links between standing waters, such as lakes in a region, may be broken or at least greatly reduced. In flowing waters, longitudinal connectivity may be severed, such as headwater streams becoming disconnected from their mainstem river. Lateral connectivity may be broken, with streams being disconnected from their riparian zones and floodplain river channels being disconnected with their floodplains. Overall, in drought, movements of biota, chemicals, sediments and nutrients with a lack of water become greatly restricted.

In both lentic and lotic systems the initial steps that occur as drought sets in involve reduction in habitat availability and diversity As volumes dwindle further habitat loss continues, but changes in water quality, mostly stressing to biota, become the dominating force (Figures 5.1 and 5.2).

Drought is always a disturbance of large spatial extent and, as a disturbance, it creates heterogeneity. For example, streams may be broken up to a series of pools, all with water quality conditions generated by local and not catchment-wide conditions. Similarly, floodplain lagoons may be converted as they dry into a complex landscape made up lagoons, each with differing water levels, water quality and biota. Within isolated and receding pools and lagoons, as exposed sediments dry, major changes in the forms and concentrations of chemicals and nutrients take place. Some of these changes can exert strong effects when a drought breaks.

By reducing the volume of water bodies, droughts concentrate their chemicals, and this process, combined with high temperatures and low oxygen concentrations, may greatly reduce water quality. In a stream, the deterioration of water quality is accelerated when and if flow ceases and pools form.

An important ecosystem component in the changes in water bodies during drought is the microbial biota and their processes. This applies both to the benthic sediments and the water column, and yet this fascinating area remains relatively unexplored. When droughts break, nutrients and soluble carbon are released from dried sediments and from catchments. These releases may produce a pulse of great magnitude, but there does not appear to be an understanding of how such a pulse is processed by the biota (although there are some hints, from work on lentic systems, that there may be a burst of production). Unfortunately, when droughts break, the pulse from the catchment may contain damaging chemicals such as acids, excess sulphates and heavy metals, which can seriously harm freshwater biota and ecosystem processes. The effects of these harmful inputs are better understood than those that may be beneficial to ecological functioning.

It seems that the abiotic changes to water bodies caused by drought are much better known than the abiotic changes that occur when a drought breaks and recovery set in. In many cases, the breaking of a drought is much more rapid than the development of a drought. Rapid changes occur in abiotic variables with the return of water, especially in the early re-wetting stage (Figures 5.1 and 5.2). With recovery, abiotic variables may return eventually to pre-drought levels, though in some cases there may be persistent changes. Whether such changes have marked effects on the future structure and functioning of water bodies remains uncertain. Needless to say, such information is crucial for the effective management of water bodies and water resources, especially with the increasing threats arising from global climate change.

5.12   The next chapters

The next six chapters will deal with the effects of drought in four major types of freshwater environments. This approach has been adopted because it appears that the biotic effects of drought are strongly influenced by the type of aquatic system. In temporary waters, one would expect the biota of such water bodies to be well adapted to the stresses of drought.

Thus, Chapter 6 deals with drought effects on both temporary lentic and lotic systems, which are normally prone to drying and which may be ephemeral, episodic or intermittent. In some cases, in a land/waterscape, drought may create a mixture of persistent and temporary water bodies; thus, one can compare drought effects along the perennial to temporary axis.

Chapter 7 looks specifically at large wetland systems that have a regular wet/dry season climate, with flooding in the wet season, a flood pulse and then drying in the dry season. These large wetlands may be on the flood-plains of large rivers or an immense wetland complex, such as the Florida Everglades, through which large amounts of water slowly move. In both cases, it is a challenge for the biota to survive the dry season. Most of the literature on these systems is concerned with fish dynamics and their use of refuges, both seasonally and in drought. Thus, given the regularity of wet/dry seasons, this chapter will examine the effects of drought that arises from failed wet seasons and the lack of floods.

In Chapter 8, the effects of drought on the biota (excluding fish) of perennial water bodies will be covered. These water bodies are both lentic and lotic, and they range from permanent lagoons to large lakes, and from brooks to large rivers. In terms of the studies of drought on freshwater ecosystems, perennial systems, which in severe droughts may dry up, have attracted a lot of attention.

As drought and its effects on fish in running waters have been the focus of many studies, this area will be dealt with in a separate chapter – Chapter 9. Of particular interest in this regard are studies of fish population dynamics and on changes in assemblage structure that may occur during and after droughts.

Most freshwater systems eventually go into the sea. Droughts drastically reduce the volumes of water going into estuaries, as well as the supply of soluble and insoluble constituents that may be important to the ecosystem dynamics of estuaries. In recent years, there has been a welcome increase in work on the effects of drought on estuaries. This will be covered in Chapter 10.

Finally, there is a critical need to consider how human activities on catchments and in water bodies have served to exacerbate the damaging effects of drought. Thus, Chapter 11 addresses this important and greatly neglected problem, which, with the current and projected effects of global climate change, can be expected to become more critical. The need for proactive rather than reactive measures to contend with drought strength-ened by climate change is espoused.

6

Drought and temporary waters

The term 'temporary waters' can cover both standing and running water bodies. The temporary nature of these bodies encompasses three forms – ephemeral, episodic and intermittent (Boulton & Brock, 1999; Williams, 2006):

- Ephemeral waters are those which receive water for a short period very occasionally and unpredictably;
- Episodic waters are those that fill occasionally and which may hold water for months even years;
- Intermittent water bodies receive water quite frequently and, in most cases, predictably.

In both ephemeral and episodic waters, given the unpredictable way that they receive water, it may be difficult to detect hydrological droughts. Rather, as many of these water bodies are in semi-arid and arid regions, drought may be detected as a meteorological phenomenon based on long-term data. In the case of intermittent waters, precipitation usually falls in a predictable wet season and water is lost in the dry season, producing a regular seasonal drought. Such aquatic systems occur in regions with a Mediterranean climate or in tropical/sub-tropical regions that have a distinct wet/dry season climate. Supra-seasonal drought in these intermittent systems usually occurs because wet seasons fail.

For standing waters, temporary water bodies range from systems as small as phytotelmata, through vernal pools and on to large ponds. For flowing systems, temporary waters range from mere trickles to rivers which only flow occasionally. Humans have created many temporary water bodies, ranging from water in tin cans and car tyres to large storages that may become seasonally empty.

Drought and Aquatic Ecosystems: Effects and Responses, First Edition. P. Sam Lake.
© 2011 P. Sam Lake. Published 2011 by Blackwell Publishing Ltd.

This chapter will not be covering large floodplain wetlands, where flooding is a regular event and where fish are an important ecosystem component (see Chapters 7 and 9). Fish are absent from most lentic temporary water bodies, but they may be present in lotic temporary systems, especially if there are refuges of persistent water in the stream system. The lack of fish in lentic temporary systems, and thus the absence of a major force of predation, has important consequences for community composition and structure in these systems (Wellborn *et al.*, 1996).

Supra-seasonal droughts in temporary waters are events that have abnormal dry periods in duration and severity (see Chapter 2). Exactly what effects supra-seasonal drought has on temporary systems is difficult to determine; many systems have regular seasonal droughts and consequently, in most studies, seasonal drought has been the focus of attention. A key question in this chapter is whether the biota of temporary waters are pre-adapted to survive the strong challenge of supra-seasonal drought. Such adaptations would be incorporated in life history strategies and in a range of desiccation-resistant mechanisms, which are all presumably moulded by the prevailing duration, variability, and predictability of the hydroperiod that normally prevails (Brock *et al.*, 2003; Williams, 2006).

Many temporary waters have small catchments, and the abiotic effects of drought on the catchments may be limited compared to the effects on the water bodies themselves. Temporary streams, on the other hand, may have large catchments, especially in arid regions, and as drought builds, hydrological links with the catchment are steadily lost. However, when a drought breaks, especially with storms, there can be rapid inputs of chemical, sedimentary and organic matter. As such pulses are likely to be the way that droughts break in arid areas, the degree of retention of the inputs in stream channels is crucial as such retention may control the nature of recovery. This question is relatively unexplored. Indeed, it is probably a feature of many temporary waters that, while drying may take quite some time, the refilling of basins or the flooding of streams can be very rapid.

6.1 Drought and the biota of temporary waters

6.1.1 Algae

As drought sets in, the low flows, increased water clarity and increased nutrients may all promote algal growth (Freeman *et al.*, 1994; Dahm *et al.*, 2003). However, at the same time, wetted habitat is being reduced and desiccation of exposed attached algae and biofilms occurs. When flow ceases, pools form and may persist through the drought. In such pools, water quality typically deteriorates and, if large amounts of particulate organic matter are present, 'blackwater' conditions may eventuate. Blackwater events may limit algal production by limiting light availability and by limiting nutrient resources due to enhanced microbial activity. This area also remains unexplored. If the pools persist through the drought, they may serve as refuges for algae and as recolonization foci when flow resumes (Robson & Matthews, 2004).

Algae in the biofilms of exposed surfaces may resist the stress of desiccation. Rapid drying in experiments (e.g. Mosisch, 2001; Ledger et al., 2008), or with human flow adjustments (e.g. Benenati et al., 1998; Ryder, 2004), or under natural conditions, can kill algal cells. In drought with low flows, nutrient concentrations may promote algal growth (Freeman et al., 1994; Dahm et al., 2003). As drying continues in both ponds and stream pools, algae may employ various mechanisms to resist drying, such as producing desiccation-resistant cysts and zygotes and developing extracellular mucilage layers to retain water (Stanley et al., 2004). Cyanobacteria such as Nostoc can withstand lengthy periods of desiccation (Dodds et al., 1995), and in the algal communities of some Canadian intermittent ponds which fill after drying, cyanobacteria dominate the cell counts, while Chlorophyta and Bacillariophyta dominate the biovolume (Williams et al., 2005). The responses of algae to re-wetting at the end of a drought depend on a number of factors, including whether the drought broke with a flood, how long the drought lasted, the availability of nutrients and the presence of algae to colonize the newly inundated surfaces. Studies on algal recovery in streams after drought are few.

Recovery from seasonal drought was rapid on rocks in a Kansas prairie stream, with the return to pre-drought biomass levels within eight days after re-wetting (Dodds et al., 1996). Interestingly, in this experiment, algal recolonization was mediated by the stream drift rather than through the revival of algae of the biofilm (Dodds et al., 1996), though, in a further study in a Kansas prairie stream, initial recovery from drought was heralded by the revival of attached desiccation-resistant filamentous algae (Murdock et al., 2010).

After a seasonal drought in Spain, the stromatolitic biofilm composed of the cyanobacteria Rivularia and Schizothrix rapidly recovered (Romani & Sabater, 1997); within three hours of inundation, it reached metabolic levels higher than pre-drought values. In an English acid stream, the epilithon, dominated by coccoid green algae and diatoms, remained intact when the channel dried up in a severe drought and post-drought recovery was rapid – within three days (Ledger & Hildrew, 2001).

Robson (2000) compared algal recovery in two Australian creeks, one having rocks with residual cyanobacteria-dominated biofilm and the other having rocks with little biofilm. In the creek with residual biofilms on the rocks, recovery after five weeks after flow returned was strongly influenced by the composition of the residual biofilm, whereas in the other creek, where the rocks lacked dried biofilms, recovery was slower and dependent on the colonization of algae from other sources (Robson & Matthews, 2004).

Further studies developed a model of algal recovery that depended on revival of algae from residual biofilm and colonization by drift from algae surviving in refuges, with persistent pools being particularly important (Robson *et al.*, 2008a). Such refuges, as regards algae, appear to be the 'Ark' type of Robson *et al.* (2008b). The recovery of algal biofilms after drought is critical to the recovery of the fauna but, as yet, the nature and dynamics of this link remain unstudied.

As pointed out by Stanley *et al.* (2004), our knowledge of drought impacts on primary production by algae (and other processes, e.g. decomposition) in streams is substantially based on investigations carried out at individual sites. Within sites, drought may lower water levels and expose pieces of habitat, such as stones and wood, and thus generate a mosaic of productive patches with different responses to drought. The concentration on sites greatly limits our understanding of drought and ecosystem processes at the appropriately large spatial extent which is relevant to drought.

From small headwater streams to large lowland rivers, drought creates different responses that vary in strength (Stanley *et al.*, 1997, 2004). If, for example, a stream dries from upstream down, effects are exerted sooner and probably more severely in the headwater streams, whereas in larger, high-order streams, the stream may continue to flow throughout the drought. On the other hand, the effects of re-wetting after drought may be stronger in the headwater streams. To date, no such large catchment-based investigation of drought and aquatic primary production has been carried out.

Thus, in temporary waters – at least in temporary streams and ponds – the algae which dwell in them have quite strong resistance (e.g. desiccation-resistant mechanisms), which allows for rapid and strong recovery. Such recovery may actually peak well before the arrival of the first herbivores.

6.1.2 *Vascular plants*

A considerable variety of vascular plants are to be found in temporary water bodies. These plants vary from annuals to perennials and from littoral semi-aquatic, through emergent aquatic, to fully submerged species. Their resistance and resilience to the stresses of drought varies considerably and, in many situations, the survival of plants is critical for the recovery of faunal assemblages and for the development of structural habitat and trophic resources.

Plants of water bodies subject to drought may survive drought as intact plants by employing two different survival strategies: drought avoidance (e.g., propagule production) and drought tolerance. Drought tolerance involves the use of mechanisms not only to reduce water loss, such as decreasing stomatal number and changing leaf orientation, but also to maintain cell turgor through changes in osmotic physiology. In an experiment with five species of wetland herbs, and simulating a one in 20 year drought, Touchette *et al.* (2007a) found that four of the species used drought avoidance physiological mechanisms, whilst one species, *Peltandra virginica*, showed drought tolerance. Similarly, Romanello *et al.* (2008) found that the aquatic macrophyte *Acorus americanus* displayed drought avoidance mechanisms such as reducing surface and below-ground biomass. However, for the vast majority of wetland species, the relative strengths of drought avoidance versus drought tolerance are unknown.

Wetland plants in many situations are likely to be stressed by too much water (flooding) and too little (drought), and tolerance of these two extremes may thus be an important axis of selection. Results from a study with three wetland plants strongly suggests that tolerance of droughts and tolerance of flooding involves a trade-off (Luo *et al.*, 2008), at least for two sedges, *Carex lasiocarpa* and *Carex limosa*, and the grass *Deyeuxia angustifolia*. The order of tolerance to flooding from high to low was *C. lasiocarpa* > *C. limosa* > *D. angustifolia*, whereas the order of tolerance to drought was the reverse (Luo *et al.*, 2008). It remains to be seen whether this trade-off in tolerance is a general phenomenon for wetland plants, though one would expect that the trade-off largely governs plant distribution in wetlands, especially those with great fluctuations in water availability.

Many wetland plants as mature plants cannot survive drought. Thus, the ultimate form of drought avoidance for many wetland plants is to invest in resistant propagules, such as seed and vegetative fragments above ground, and in below-ground vegetative structures. Below-ground structures that may survive drought include rhizomes (e.g. sedges in the genus *Cyperus* and *Cladium*), tubers (e.g. *Hydrilla verticillata* (Parsons & Cuthbertson, 2001)) and turions (protected buds found in aquatic plants such as *Utricularia* and *Potamogeton* (Philbrick & Les, 1996)). For short but severe droughts, vegetative fragments, such as from *Elodea* species, may allow post-drought recolonization (Barrat-Segretain & Cellot, 2007).

Many aquatic plants produce seeds that can remain viable for long periods of time and which may be stored in sediments as drought sets in (Brock, 1998; Brock *et al.*, 2003). In some plants, as exemplified by some charophytes, falling water levels may serve to accelerate oospore production and maturation (Casanova, 1994). Thus, in many water bodies subject to seasonal drying and supra-seasonal drought, seed and egg banks occur in the sediments. These banks may consist of seeds and oospores (*Chara*) from plants, cysts from phytoplankton, and cysts and eggs from animals (Leck, 1989; Bonis & Grillas, 2002; Williams, 2006). As such, they constitute an important drought refuge, vital to maintaining the resilience of aquatic communities.

Brock *et al.* (2003) proposed that the maintenance of species in seed banks requires a number of sequential steps to be carried out successfully. Thus, initially, when water is present, the plants need to grow and produce seed that is viable and well protected. Such seeds need to lodge in the sediment and be capable of undergoing dormancy for considerable lengths of time. Dormancy involves surviving with extremely low metabolic levels until cues for hatching/germination, such as re-wetting and/or abrupt changes in temperature and in oxygen concentrations, occur. The way that different species may respond to re-wetting depends on an array of different environmental conditions, such as seed depth in the sediment, temperature and water salinity (Brock *et al.*, 2003). Upon germinating, the next step is establishment, which usually involves rapid growth if favourable conditions, such as enhanced nutrient levels, are present. The final step is reproduction to produce new seeds, some of which may germinate directly into new plants, while others may go to the seed bank.

In many cases, not all the viable seeds may hatch with re-wetting. For example, in an experiment with sediment from an Australian temporary wetland, of the seeds of 50-plus plant species, viable seeds of 20 species were still present after eight years, with each year having a wetting event (Leck & Brock, 2000). Many wetland plant species display 'bet hedging'. In this, not all of the seeds will germinate on only one or two occasions; consequently, even though some seeds may die in the sediment, others may survive to germinate later, possibly in more favourable, occasions (Williams, 2006).

Plant communities of temporary water bodies can be altered by drought, with the aquatic species being reduced and terrestrial species expanding in the water body. In moorland pools in the Netherlands affected by a short but severe drought in 1975–76, Sykora (1979) observed the decline of wetland plants, such as *Sphagnum crassicladium*, and an expansion of more terrestrial plants such as the grasses *Molinia caerulea* and *Eriophorum angustifolium*, which persisted after the drought broke. Holland & Jain (1984), in sampling vernal pools in years of normal precipitation and in a drought year (1975–76), found that in the drought, terrestrial species (e.g. *Hypochaeris glabra*, *Erodium botrys*, *Bromus mollis*) occurred in vernal pools, and that some semi-aquatic plants that normally occur were missing. Similarly, Panter and May (1997) found that in a pond in Epping Forest, UK, during normal years the vegetation was dominated by the aquatic plant *Glyceria fluitans*. However, with a drought in 1995–96, the pond became colonized by terrestrial grasses; creeping bentgrass *Agrostis stolonifera* became dominant, and this dominance persisted.

Both field observation and experimentation on the seed bank of a prairie marsh in Iowa, USA, revealed that were three different types of seed bank (Van der Valk & Davis, 1978). One consisted of emergent species that germinated on mudflats or in very shallow water, while another consisted of submerged and floating species that germinated when there was standing water. The third group comprised terrestrial species that had seeds which germinated on dry mud when there was no standing water due to drought. Thus, drought as a disturbance in these normally fluctuating systems may create an opportunity for more terrestrial species to flourish briefly. Droughts, especially those of long duration, can lead to temporary, and possibly even long-term 'terrestrialization' of the flora of lentic systems. Similarly, channels of temporary streams may be invaded by terrestrial plants during drought and, combined with deposited sediment and terrestrial litter, these plants, in long droughts, may change the stream channel morphology.

Droughts can produce major changes in the aquatic plant communities of particular temporary wetlands. In the Okefenokee Swamp, Georgia, there is a complex of marsh wetlands which have different hydroperiods and which are affected differentially by drought (Greening & Gerritsen, 1987). In persistent and deep marshes during a drought (1980–81), there were no major changes in the aquatic vegetation, nor in the vegetation of a shallow wetland subjected to regular and predictable seasonal drought. However, in a normally inundated deep marsh that dried out in the drought, with subsequent inundation a new 'fugitive' species, the beak sedge *Rhynchospora inundata*, became dominant (Greening & Gerritsen, 1987). Following its boom in 1982–83, the beak sedge declined, while the 'resident' group of species continued to increase to normal levels.

In wetlands, plant assemblages can be arranged along gradients of water availability which correspond to relative water levels in wetlands. Thus, in Californian subalpine wetlands, Rejmánková *et al.* (1999) delineated four distinct vegetation zones that, from wet to dry and from high to low water levels, were dominated respectively by *Nuphar polysepalum*, *Scirpus acutus*, *Carex rostrata* and *Juncus balticus*.

In response to extended hydrological disturbances – floods and droughts – Van der Valk (1994) proposed a 'migration model' for plants in wetlands. This model maintains that, in times of drought, plant assemblages may shift along an axis from dry to wet. The Californian wetlands studied by Rejmánková *et al.* (1999) were exposed to a long and severe drought from 1988 to 1994, with wetland water levels dropping and an overall drop in plant biomass. Both *Carex* and *Juncus* species maintained their dominance, whereas the *Scirpus*-dominated assemblage was eliminated from the wetland and did not return after the drought broke (Rejmánková *et al.*, 1999). Both the *Scirpus* and *Nuphar* assemblages were invaded and came to be dominated by two 'stress tolerators', *Hippuris vulgaris* and *Polygonum amphibium*.

Fortunately, the study was able to gather post-drought data. Both the *Carex* and *Juncus* assemblages recovered rapidly after the drought and may be regarded as having both high resistance and high resilience. The *Nuphar* assemblage had a low resistance and a moderate resilience, whereas the *Scirpus* assemblage appears to have had neither resistance nor resilience. Overall, the results suggest that the 'migration' model was not applicable to this system and that individual attributes or traits of species allowed them to contend with the drought. It is suggested by Rejmánková *et al.* (1999) that the two 'stress tolerators' can tolerate a wide range of wet-dry conditions, but that their success depends on gaps being created by disturbance in competitive communities.

In summary, plants have a wide range of adaptations, allowing them to persist through drought and to successfully recover after it. Drought can substantially alter the plant community structure of wetlands by eliminating or reducing particular species or assemblages and by creating gaps into which highly dispersive and opportunistic species may invade and establish. In some cases, this allows the invasion of other aquatic plants; in others, it allows terrestrial plants to invade. Such changes may be transitory or enduring, and can have strong knock-on effects in influencing the development of post-drought community structure.

6.2   Fauna of temporary standing waters and drought

6.2.1   Fish of temporary lentic waters

A very low diversity of fish occurs in temporary lentic systems. As might be expected, temporary ponds and wetlands are very difficult environments for fish to survive drought in. As drought builds and water levels drop, wetlands become fragmented and some parts persist, whereas others dry up (e.g. Snodgrass *et al.*, 1996; Baber *et al.*, 2002). For fish in such wetlands, there is pressure either to have drought-resistant adaptations or to emigrate at the right time from vulnerable wetlands to more secure ones (refuges).

There is, however, a small group of fish species that can survive as adults in the temporary pools of wetlands that dry out in drought. For example, in swamps of south-western Western Australia, there is the salamander fish, *Lepidogalaxias salamandroides*, which dwells in pools in acidic peat swamps exposed to predictable summer droughts (Pusey & Edward, 1990). In these droughts, *L. salamandoides* aestivates by burrowing into the mud and it survives on lipid reserves (Pusey, 1990). In New Zealand there are three endemic mudfish species (*Neochanna* spp.) which dwell in swamps. Two species, *N. burrowsius* (Eldon, 1979a, 1979b) and *N. diversus* (McPhail, 1999) can aestivate by moving into shaded plant material. The very hardy oriental weather loach (*Misgurnus anguillicaudatus*) (Ip *et al.*, 2004) can successfully dwell in temporary swamps and in drought it aestivates by burrowing into mud and wet soil (Anonymous, 2008).

An alternative strategy is to evolve a life history that, instead of having adult aestivation, has desiccation-resistant eggs. Species of the family Cyprinodontidae live in temporary waters with short life histories and a high level of reproduction. Faced with receding water levels in Floridan marshes, one cyprinodontid, the marsh killifish (*Fundulus confluentus*), mates and lays eggs (Harrington, 1959; Kushlan, 1973). The eggs are left stranded around the pond shores and hatch when water returns (Harrington, 1959). Indeed, many cyprinodontid fish can survive drought in temporary ponds and marshes. Key adaptations to do this are the capacity to lay diapausing desiccation-resistant eggs (Wourma, 1972), which can hatch as soon as water is present, producing fish that mature rapidly and lay many eggs (Lévêque, 1997; Hrbek & Larson, 1999). In Africa, species in the genera *Nothobranchius*, *Fundulopanchax* and *Aphyosemion* mature quickly and lay diapausing, drought-resistant eggs, whereas in South America, species in the genera *Rivulus*, *Cynolebias*, *Leptolebias* and *Plesiolebias*, among others, are similar (Hrbek & Larson, 1999). The ability to produce diapausing eggs seems to have evolved independently a number of times (Hrbek & Larson, 1999).

In temporary waters, rather than having mechanisms to resist drought, another strategy is to leave. This involves exploiting the resources of temporary wetlands and then retreating to permanent waters as drought set in. Such migration has to be precisely timed. Cucherousset *et al.* (2007) investigated the timing of migration from a temporary wetland in the Grande Brière Mottière Marsh in France during a drought in 2004. Fish species migrated in sequence, with pike (*Esox lucius*) being the first, closely followed by pumpkinseed (*Lepomis gibbosus*) and then the European eel (*Anguilla anguilla*). After the eels came rudd (*Scardinius erythrophthalmus*), followed by black bullhead (*Ameiurus melas*), and finally mosquitofish (*Gambusia holbrooki*). The pumpkinseed, black bullhead and mosquitofish are all exotics. The timing scored as the emigration moment (the number of days required to catch 50 per cent of the total number of fish caught) was positively correlated with three variables or indices, the 'physiological tolerance index', a 'coefficient of water quality flexibility' and the 'temperature of upper avoidance' (Cucherousset *et al.*, 2007). These results strongly suggest that perception of deteriorating water quality provides a species-specific cue to emigration.

6.2.2   Invertebrates

In contrast to fish, many invertebrates living in temporary waters have adaptive mechanisms to stay and contend with the dry period. Such mechanisms include desiccation-resistant eggs, larvae and adults, cysts, and avoidance strategies such as burrowing and leaving for other systems with water (Williams, 2006). Relatively immobile animals, such as oligochaetes, may undergo encystment, though this may be confined to only a few species. Cysts of *Tubifex tubifex* survived a five-month drought in a dry pool on the Rhine river floodplain (Anlauf, 1990). In floodplain wetlands along the Paraná River, Argentina, after a dry period of 2–4 weeks, Montalto & Marchese (2005) found that of 26 species of aquatic oligochaetes, only two species – *Dero multibranchiata* (Tubificidae) and *Trieminentia corderoi* (Opistocystidae) – underwent encystment.

An alternative strategy for many species, especially insects, is to move to refuges. Insects generally have good dispersal capabilities and can flee temporary waters as they dry and survive in permanent water bodies. Williams (2006) developed a model of the adaptive traits that insects needed to have in order to exist successfully in three basic types of temporary waters:

- In water bodies with relatively long and predictable hydroperiods, insects would have relatively long-lived adult stages, good powers of dispersal, long development times, staggered egg hatching and 'diapause capability in several life cycle stages'.
- In intermediate types of water bodies, the generation time of the insects is about the same as the length of the hydroperiod, and the adults are semelparous (reproduce only once), good dispersers, and lay short-diapause eggs that produce obvious cohorts (Williams, 2006).
- At the other extreme, of water bodies with short and unpredictable hydroperiods, insects may be 'short-lived highly vagile adults with high fecundity' that produce diapause eggs with staggered hatching, and have immature stages that can survive drying.

Studying the fauna of temporary ponds, Wiggins *et al.* (1980) devised a scheme of four groups – a scheme that also fitted the faunal groups found in Australian vernal ponds (Lake *et al.*, 1989):

Upon the pond filling, the initial fauna (Group 1 of Wiggins *et al.*) consists of species, predominantly crustaceans, that hatch from eggs, as well as some species of insects and molluscs that survived the drying as adults. This group has a very poor dispersal capability. The pond may then receive insects that either breed in the pond and then disperse (Group 2), or lay their eggs in the bottom of the pond when it is dry, and which hatch when water is present (Group 3). The fourth group consists almost entirely of predators that fly in from permanent water bodies and which retreat to permanent water when drying occurs. For example, adult water beetles (Dytiscidae) have been observed to fly out of desert ponds as they dried and move to persisting water bodies (Lytle, 2008).

A critical component of the ecology of the biota of temporary waters is the need to have mechanisms to survive the dry periods in the waterless water body or to migrate at the appropriate time to persisting water bodies – a general 'stay or go' strategy.

With long dry periods, the viability of stored propagules may decline, weakening the response to water when it comes. For example, in experiments on hatching micro-invertebrates eggs in sediments of flood-plain wetlands, Jenkins & Boulton (2007) found that dry periods (natural and human-created) lasting longer than six years caused a very significant decline in the densities of microinvertebrates that hatched with the addition of water. Their model, which may be applicable to many temporary aquatic habitats, suggests that if droughts are long (> 6–10 years), the egg bank of microinvertebrates may be significantly depleted and, consequently, the post-drought hatching of microinvertebrates may be too low to fuel the effective recovery of higher trophic levels. However, studies of the fauna that live in temporary pools after a supra-seasonal drought are rare, and thus we do not know whether severe droughts have any effect on the aquatic fauna of temporary ponds. As temporary ponds are often isolated, the depletion of the egg bank, plus limited migration, may lead to low densities of some species and even, possibly, to local extinction.

Small water bodies, such as car tyres and tree holes, may be very susceptible to drought, and one would expect drought to act as a powerful force determining the biota of such systems (Bradshaw & Holzapfel, 1988; Aspbury & Juliano, 1998). Tree holes – a form of phytotelmata – are a very suitable habitat for mosquito populations, and drought is clearly a hazard. In Florida, mosquito species have differing tolerances to drought, with three drought-intolerant species coexisting in tree holes that are large enough to survive drought, and one drought-tolerant species, *Aedes triseriatus*, breeding in tree holes that are subject to drought (Bradshaw & Holzapfel, 1988).

In the drought-intolerant group, there is a predator (*Oxorhynchites rutilus*) with which the other two drought-intolerant species can co-exist. However, this predator can eliminate *A. triseriatus* when it happens to co-occur, and thus *A. triseriatus* survives by using a hazardous habitat that the predator cannot tolerate (Bradshaw & Holzapfel, 1988).

With the drought-tolerant species of mosquitoes, Aspbury & Juliano (1998) found that in the post-drought filling phase, both resource condition (leaf litter) and intraspecific competition strongly influenced population increase. The drought-tolerant *A. triseriatus* lays its eggs during a drought above the water line. The eggs can accumulate to such levels that upon hatching after re-filling, intraspecific competition of the 'scramble' type (e.g. Nicholson, 1954) occurs and limits population recruitment.

In a survey of treehole mosquitoes, Srivastava (2005) found that drought at a local level was a major contributor to variation in species richness, but not at a regional level, stressing the point that drought, by exerting strong small-scale effects, can strongly influence species composition and richness. These studies reveal how drought may favour some species and not others, and thus exert strong local effects on species composition and richness. This is an important finding in that it reveals that species – even closely related species – vary tremendously in their tolerance of drought, and that these differences allow species to coexist if there is variability in habitat persistence when drought occurs.

Other dipterans in temporary waters are affected by drought. In a temporary pool in France, Delettre (1989) sampled emerging chironomids in a drought year (1980) and two subsequent normal years (1981, 1982). In the drought year, only 11 species were recorded, while in the subsequent normal years, 15 species were recorded. Of the eight most common species normally found, one species was absent in the drought and another was only recorded once. Three other species emerged from the pond before the drought but did not emerge in the year after the drought broke. All eight of the common species emerged in the years after the drought.

Thus, as expected, drought can disrupt the emergence of aquatic insects and influence local species richness. If a drought occurs before recovery from previous droughts has been completed, it is likely that species richness will be reduced for a considerable duration thereafter. This illustrates the important point that a series of droughts can exert cumulative effects not shown in a single isolated drought. In other words, in dealing with drought, history matters. The importance of past history of drought in shaping the effects of a contemporary drought is an example of ecological memory (*sensu* Padisák, 1992).

6.2.3   Invertebrates in regional standing water bodies of differing hydroperiods

In many landscapes, permanent and temporary water bodies together form a mosaic that alters in pattern as drought lowers water availability at a regional level. Drought may be a major force determining community structure in such landscapes, depending on the relative proportions of water bodies with different hydroperiods, their respective biota, the drought history and the spatial patterns of hydrological and (the more general) ecological connectivity between the water bodies.

In a complex with temporary and permanent ponds, Jeffries (1994) sampled macroinvertebrates from 29 ponds in a Scottish wetland in 1986 and 1987 before a drought, and in 1992 immediately after a four-year drought. The study thus assessed drought impacts rather than recovery. The series of small ponds was created by the removal of anti-tank barriers and was close to Marl Loch, a permanent pond, which survived the drought and was a source of aquatic fauna. The drought impacts varied according to pond final condition, with six ponds having some water, 12 'still wet' and 11 being dry in 1992. The ponds that did not dry out accumulated taxa, as did the ponds that were still wet, with the latter group acquiring species more typical of temporary ponds. The ponds that dried lost taxa, the loss being most severe in those ponds that were temporary before the drought. The drought extinction rates were high in both temporary and permanent ponds, and permanent water taxa were poor colonists compared with the taxa from temporary ponds (Jeffries, 1994). Overall, the results illustrate the highly variable effects of drought at a landscape scale, and how the previous history of water bodies strongly influences the faunal responses to drought – a further example of ecological memory.

With the restoration of beaver populations in North America, beaver dams now occur in many wetlands. Depending on their hydrology, the impoundments range from temporary to permanent. Wissinger & Gallagher (1999) sampled the macroinvertebrate fauna of two permanent beaver pond wetlands and two semi-permanent impoundments that filled in autumn a year (1994) before a drought. The fauna of the two types of wetlands were both rich and similar. In 1995, there was a short but severe drought in which three wetlands dried completely, and one of the permanent wetlands was greatly reduced in volume. Samples were again taken in late 1995 and in 1996.

During the drought, sediment samples were taken from the basins of the three wetlands that dried out. These sediments were used in a rehydration experiment to find those taxa that had drought-resistant propagules in the sediments. Four major groups of taxa were found: terrestrial arthropods coming into the dried basins, diapausing crustacean zooplankton, 'flightless invertebrates and wetland insects' and desiccation-resistant aquatic insects. Combining the experimental findings with the samples from the wetlands indicated that the two autumnal wetlands had a greater proportion of fauna recolonizing from drought-resistant stages (71 per cent and 63 per cent of species) than the permanent wetland (38 per cent). All three wetlands had similar proportions of recolonization by immigrating adult insects (22–27 per cent of species). Recovery of the fauna within one year differed between the wetlands with the two autumnal, temporary wetlands recovering about 90 per cent of their pre-drought fauna, whereas recovery in the one permanent wetland was 77 per cent of the species of the initial fauna. Furthermore, in the temporary wetlands, only a small number of species in the post-drought period were new taxa, compared with 16 new taxa in the 'permanent' wetland. This suggests that in the permanent wetland, drought as a disturbance created an opportunity for new taxa to invade, but in the temporary wetlands, the fauna, being pre-adapted to drought, recovered rapidly and drought did not provide opportunities for invasion.

Wissinger & Gallagher (1999) speculated that the recovery of the fauna of wetlands after drought is strongly dependent on the nature and distribution of neighbouring wetlands that serve as sources from which colonists can disperse into the drought-affected areas. The above results strongly suggest that adapting to the taxing conditions of intermittent systems produces a tolerant and robust fauna which readily bounces back from drought, whereas adapting to permanent conditions is less challenging and produces a fauna that is likely to be changed considerably by drought. This endorses the view that the past evolutionary and ecological history of the fauna in wetlands with different hydrological regimes strongly governs the impacts and patterns of recovery after drought.

Prairie wetlands in Minnesota, both permanent and temporary, were subjected to a severe drought from 1987 to 1990 (Hershey et al., 1999), with some wetlands becoming dry and others being greatly reduced in depth. Samples of those wetlands with some water in the drought, compared with samples after the drought, revealed that densities of molluscs (gastropods, sphareiid bivalves) increased, but densities of insects overall decreased. In the insects, the Chironomidae and Ceratopogonidae were most depleted, while the Stratiomyidae, Tipulidae and Coleoptera were not affected. Species richness declined in the drought, with most of the decline being due to the loss of chironomid taxa.

After the drought broke, there was a one-year lag in the recovery of chironomids and ceratopogonids but, three years after the drought, the insect abundance was three to five times greater than in the drought years (a post-drought boom). The causes of this boom are uncertain; drought may have reduced predation or may have served to lift primary production by increasing nutrient availability. Mollusc densities dropped significantly after the drought – perhaps reflecting not mortality but a habitat concentration effect during the drought. In the zooplankton, both rotifers and cladocerans had more taxa and higher abundances in the post-drought years than during the drought, while copepods and ostracods appear not to have been affected by the drought at all.

Considering these major faunal changes due to drought in the Minnesota wetlands, Hershey *et al.* (1999) devised a scheme for the changes in community structure between normal wet periods and drought with greatly lowered water volume. In times of drought, the non-insect macroinvertebrates (gastropods, sphaeriids, annelids) are abundant and, with the drought breaking, insects increase in abundance (with chironomids, somewhat surprisingly, having a lag in their recovery). In the zooplankton during drought, copepods and ostracods dominate; with the drought breaking, there is a rapid recovery by copepods, rotifers and cladocerans.

The lag recovery in the insects is held to be due to the need for dispersal from more permanent wetlands (refuges), while the rapid recovery of the zooplankton is driven by the rapid hatching from the egg bank in the wetland sediments (Hershey *et al.*, 1999). The results suggest that drought has major and predictable effects on assemblage structure, and that drought is a major force governing diversity, particularly beta diversity (between water bodies).

In many parts of the cold temperate zone of the Northern Hemisphere, snowmelt ponds are a common form of temporary pond. If the snowpack is low, drought may occur in the subsequent spring-summer period. The hydroperiods of the ponds normally range in duration from a few days to up to ten months or so. In a series of such ponds, Schneider & Frost (1996) found that species richness increases with pond duration, as does the number of predatory taxa, and that the proportion of taxa with desiccation-resistant life history stages (predominantly eggs) declines as pond duration increases. Ponds with short duration tend to harbour a fauna adapted for fast growth and reproduction, with mechanisms to survive the dry period. In ponds of long hydroperiod, biotic interactions such as competition, and especially predation, may exert a powerful influence on community structure (Lake *et al.*, 1989; Schneider & Frost, 1996).

In 1987–1989, there was a severe drought in the upper Midwest of the USA that caused the ponds studied by Schneider (1999) either to dry out or to have only short hydroperiods. Extensive faunal samples across seven ponds were taken in 1985, prior to the drought, and in 1989, immediately after the drought. Drought decreased the species richness. In two ponds that usually had long hydroperiods, gastropods and caddis flies were reduced, if not eliminated. At the same time, in one of these ponds, the hitherto very rare fairy shrimp *Eubranchipus*, along with *Aedes* mosquito larvae, underwent massive population increases. In ponds usually with long hydroperiods, drought eliminated predators with relatively long life cycles, such as dragonfly larvae and notonectids. This reduction in predation pressure provided an opportunity for rapid population growth of the fairy shrimps and mosquito larvae (Schneider, 1997, 1999). In normal wet years, fairy shrimps were eliminated by predators but had unhatched eggs in the sediment. These eggs hatched after dry years and allowed the production of many eggs that may hatch when another drought once more eliminates the predators.

With drought and the consequent loss of taxa, the number of links, the link number per taxon and connectivity in the resulting food webs all declined (Schneider, 1997). The drought favoured the existence of taxa with desiccation-resistant stages and depleted the fauna without these adaptations. Consequently, the impacts of drying, with changes in community structure and food webs, were more marked in those ponds that normally had long hydroperiods. Again, as indicated in Wissinger & Gallagher's (1999) study, drought can temporarily eliminate taxa, notably predators, and hence provide a window of opportunity for fast growing and predator-intolerant taxa – a further example that history matters. Not only do these effects alter biodiversity patterns across wetlands, but they also alter food webs by simplifying them.

As shown above, drought may favour some species while inhibiting others. Working in Pennsylvania, Chase & Knight (2003) classified a series of wetlands into three types:

- permanent, which always contain water;
- temporary, which regularly fill and dry out;
- semi-permanent, which only dry out with severe drought.

The wetlands were sampled for mosquitoes, mosquito predators and competitors from 1998 to 2001. A major drought occurred in 1999. Both predators and competitors were drastically reduced by drought in the semi-permanent wetlands, but not in the permanent or temporary wetlands. Without pressure from predators and competitors, mosquito populations in the semi-permanent wetlands boomed the year after the drought (Chase & Knight, 2003). This result was supported in experimental mesocosm experiments, using drought as the disturbance, that showed that the semi-permanent treatment had the highest abundance of emerging adult mosquitoes and the lowest biomasses of competitors and predators (Figure 6.1). For predator-prone and poor competitive taxa such as mosquitoes, drought

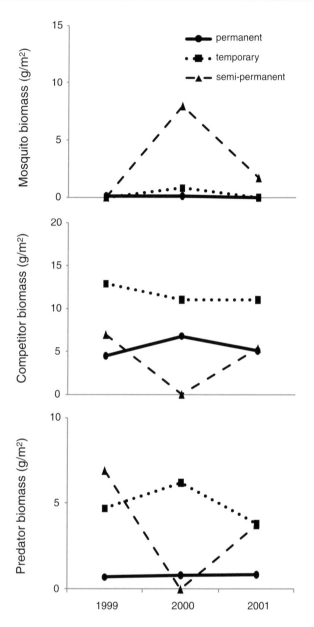

Figure 6.1 The abundance of adult mosquitoes emerging from three types of meso-cosms (temporary, permanent and semi-permanent) (top graph) after an experimental drought, along with the biomass of competitors (middle graph) and biomass of predators (bottom graph). (Redrawn from Chase & Knight, 2003.)

may be a strong force in reducing both predation and competition and providing a chance for rapid recruitment of such opportunistic species.

In another survey, Chase (2003) sampled ponds for macroinvertebrates, amphibians and small fish across the permanent, semi-permanent and temporary gradient. Semi-permanent ponds only had occasional droughts, while temporary ponds had regular and predictable dry spells. Drought as a disturbance had more severe effects on semi-permanent ponds than on permanent ponds and on temporary ponds that dry regularly. For ponds of the same level of primary productivity, Chase (2003) found that, at a regional level, permanent ponds were the most dissimilar to each other in community composition, temporary ponds were the most similar and semi-permanent ones were intermediate in similarity. Local species richness was highest for the semi-permanent ponds, and both permanent and semi-permanent ponds had similar levels of regional species richness (Chase, 2003).

At the local level, the results support the intermediate disturbance hypothesis (Connell, 1978; Chase, 2003) but, at the scale of regional richness, dispersal and patterns of connectivity may override local patch diversity. In terms of the effects of drought, this may mean that drought-intolerant species may seek refuge during drought in permanent ponds, and that drought-tolerant species that move into semi-permanent ponds with drought and may persist regionally in temporary ponds.

As drought is invariably a large-scale disturbance operating at the regional and larger spatial scales, the mosaic of ponds with different states of permanency may provide resilience to drought at the regional level. At a regional level, disturbances such as drought can generate a lag signal, especially if dispersal is low, that may produce a range of dissimilar communities spatially and with time.

6.3   Insights from experimental studies of drought in temporary waters

Field observations on small bodies of water, such as phytotelmata and ponds, have provided insights on the ecological affects of drought on the biota of small, temporary lentic systems. Indeed, overall, experimental studies of drought have almost exclusively used standing waters, and they may mimic successfully the conditions found in small natural and temporary water bodies. While the experiments have proven to be insightful, there is a problem in scaling up the short-temporal-scale and limited spatial extents of the experiments to the long-term, large-scale phenomenon of natural drought. Drought may be only one type of disturbance among others that can act on natural systems, and small-scale experiments may provide opportunities to examine the interaction of drought with other disturbances, both natural and human-generated.

Disturbance history can be a major force shaping communities. Using water-filled bamboo stumps, a form of phytotelmata, Fukami (2001) assessed the effects of two different types of disturbance on populations of protozoans and small metazoans (e.g. rotifers, nematodes). The two disturbances were drought and the introduction of a microorganism-consuming predator – mosquito larvae. The two disturbances were delivered in four different sequences. On its own, drought had a much greater effect than predation, but recovery of species richness was faster and more complete after drought than after predation in the short term. Different sequences of disturbance induced different successional pathways after the final disturbance, and there was considerable variation in community structure and in resilience to the disturbance sequences. In terms of droughts, this work suggests that the interactions between drought and other disturbances may produce unpredictable responses. As drought lowers water availability and, depending on the local conditions, other disturbances such as excessive turbidity, high water temperatures and hypoxia may arise in distinct sequences, producing context-dependent outcomes and giving rise to different recovery pathways.

Using microcosms containing populations of protozoans and rotifers from artificial phytotelmata, Kneitel & Chase (2004) assessed the effects of three different treatments applied fully factorially. The treatments were: disturbance by drought; different resource levels (dried leaves); and predation by mosquito larvae (*Aedes albopictus*). Drought disturbance and predation, as individual factors, decreased species richness and total abundance. An increase in drought severity made community composition between the various treatments more similar, suggesting strong winnowing of drought-intolerant species. Drought and predation interacted significantly, with drought having a stronger impact when there was no predation, and predation being more effective in the absence of drought. Drought and availability of resources appeared not to interact. Community composition became more similar as disturbance increased (experimental drought frequency).

Thus, predation and drought may interact strongly, but not necessarily in a straightforward additive way. Nevertheless, the outcome of disturbance is strongly moderated by interspecific interactions such as predation, and both can very significantly change community composition.

In aquatic ecosystems, especially small systems, the disturbance of drought is clearly a strong environmental force that can regulate community structure, especially after a drought has occurred. Communities in a system may be rebuilt after drought, through recruitment by survivors that tolerated the drought and/or by recruitment by dispersal and recolonization from external water bodies. The level of habitat isolation in a region may determine the extent to which dispersal influences recolonization of habitats after drought.

Using a similar experimental set-up as that used by Kneitel & Chase (2004), Östman *et al.* (2006) tested the effects of disturbance – drought – on regional species richness of multiple microcosms. Local sets of microcosms comprised a region, and local microcosms in a region were either connected or not. With drought, regional species richness was lower in regions with isolated microcosms than in those that were connected to each other. The overall regional species richness consisted of drought-intolerant and drought-tolerant species and thus, in the connected microcosms, dispersal between microcosms partly offset the effects of drought. Hence, as suggested above, the interaction between dispersal and habitat isolation may be critical to the emergence of community structure after disturbances such as drought.

The role of dispersal in assembling communities through influencing colonization is stressed in the Neutral Theory of Hubbell (2001). As colonization may be a stochastic process, there may be considerable differences between local communities within a region after a drought (Chase, 2007). An alternative, more deterministic, view is that environmental filters regulate the recruitment of species to local communities from the regional pool and, if this filter is strong, uniformity is imposed on the composition of local communities. Such a process requires successful species to have the appropriate traits to conform to criteria of the filter – a process of 'niche selection' (Chase, 2007) regulating community structure.

Chase (2007) set up an experiment with artificial ponds that developed communities for two years, after which one half of the ponds were subjected to a short and severe drought. The invertebrates of the communities were sampled two years later. Drought reduced both local species richness (α diversity) and regional diversity (γ diversity) significantly. As depicted in multivariate ordination (Figure 6.2), there was a high similarity among the drought-affected communities and a wide dissimilarity between each of the control permanent pond communities, along with a significant difference between the drought and the non-drought communities. The experiment clearly showed that drought was acting as a strong environmental filter, reducing the three forms of spatial diversity (α, β, and γ diversity). Given that there was only one drought and that it was, in the scale of droughts, a fairly mild event, the possibility is strengthened that drought may be a powerful environmental filter to determine the composition and diversity of many aquatic communities.

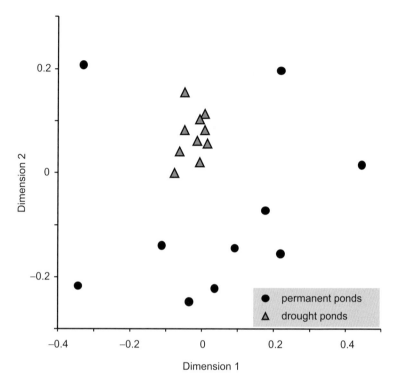

Figure 6.2 Non-metric multidimensional ordination of the biota of experimental undisturbed permanent and drought-disturbed ponds. The ordination clearly indicates the strong homogeneity of the biota created by the disturbance of drought. (Redrawn from Figure 2, Page 17431 in Chase, 2007.)

6.4   The biota of temporary streams and drought

In this section, the focus will be on macroinvertebrates. This reflects the amount of investigation that has been directed at the effects of drought on this group rather than on microinvertebrates. Fish will be mentioned, although treatment of the effects of drought on fish dwelling in mostly permanent, but some temporary, waters are covered in separate chapters (7, 9 & 10).

Drought can affect a range of surface water streams, from those that have highly variable flow regimes (such as intermittent streams) to streams and rivers with stable and highly predictable flow regimes. For the biota of temporary streams, drying is a normal environmental event that has shaped their evolution (viz. Williams, 2006). In temporary streams, supra-seasonal drought can occur and exert significant effects.

6.4.1   Drying in desert streams

Desert streams, as exemplified by the well-studied Sycamore Creek in Arizona, have widely fluctuating flows, from large, pulsing floods (Fisher *et al.*, 1982) to long periods of desiccation (Stanley *et al.*, 2004), and so they may provide some insights on the effects of stream drying. Drying of the channel in desert streams can be severe, with the channel being completely dry for kilometres. Sections with water may persist as isolated pools, which may be linked by subsurface flows through sand (Stanley *et al.*, 1997). Drying not only influences the distribution of biota, but also fragments and reduces production (Stanley *et al.*, 1997, 2004).

In desert streams, seasonal drought, and especially supra-seasonal drought, causes considerable mortality of invertebrates, as they may be stranded in pools that dry out and stressed as pools decline in water quality (Boulton *et al.*, 1992b; Stanley *et al.*, 1994). The hyporheos is affected by drying with losses in the shallow zone (<50 cm), but not necessarily in the deeper phreatic zone (>50 cm) (Boulton & Stanley, 1995). Macroinvertebrates do not appear to use the hyporheic zone as a refuge, with the exception of *Probezzia*, a ceratopogonid (Stanley *et al.*, 1994).

As drying progresses, macroinvertebrates may migrate to reaches with persistent water (Stanley *et al.*, 1994). For example, Lytle *et al.* (2008) observed that, as drought strengthened in a desert stream, there were mass migrations of the water beetle *Postelichus immsi* (Dryopidae) (Figure 6.3) and nymphs of the dragonfly *Progomphus borealis* (Gomphidae) to upstream persistent water. Adult insects can avoid drought by flying to persistent water (Stanley *et al.*, 1994). Recolonization after drought is by immigration from flying adults and by in-stream migration from persistent pools. In Sycamore Creek, and perhaps in desert streams generally, droughts have more enduring impacts than floods (Boulton *et al.*, 1992b; Stanley *et al.*, 1994; Lake, 2007).

6.4.2   Mediterranean streams

Mediterranean climate streams have a predictable flow regime, with high flow in winter and very low or no flow in summer and autumn (Gasith & Resh, 1999; Bonada *et al.*, 2006). Most studies of these systems have been short-term and have described seasonal drought, but there are some valuable studies that encompass both seasonal and supra-seasonal droughts.

In Mediterranean climate streams, the more complex and species-rich macroinvertebrate assemblages occur in spring to early summer, and the

Figure 6.3 Bands of adult dryopid beetles (*Postelichus immsi*) plodding upstream to persistent water as drought develops in the Santa Maria River, Arizona, USA. (Picture from Figure 1 (right) in Lytle *et al.*, 2008.) (See the colour version of this figure in Plate 6.3.)

simplest and more species-poor assemblages occur in mid- to late summer as streams become trickles or a series of separate pools, or even completely dry for considerable periods (Towns, 1985; Gasith & Resh, 1999; Acuña *et al.*, 2005; Bonada *et al.*, 2006). The combination of spatial pattern of drying and its duration strongly influence the distribution and abundance of the biota (e.g. McElravy *et al.*, 1989; Pires *et al.*, 2000; Arab *et al.*, 2004; Fonnesu *et al.*, 2005; Acuña *et al.*, 2005).

In general, the fauna has a low resistance to the summer drought (Fonnesu *et al.*, 2005; Acuña *et al.*, 2005), though some species may use refuges such as burrowing into the streambed (the hyporheic zone) (Legier & Talin, 1975; Gagneur & Chaoui-Boudghane, 1991) or have drought-resistant eggs. Resilience is strong, as recovery from the summer drought with the return of flow is rapid and substantially occurs through adults flying in from persistent pools and perennial streams, with chironomids being the early dominant group (Legier & Talin, 1975; Towns, 1985; Acuña *et al.*, 2005).

During the summer drought, organic matter, both autochthonous and allochthonous, accumulates in pools and the dry stream channel (Maamri *et al.*, 1997a; Acuña *et al.*, 2005, 2007). In streams that continue to flow during summer droughts, detritus processing occurs at a higher rate than in the cool, high-flow part of the year; in streams that cease to flow, detritus processing is greatly reduced (Pinna & Basset, 2004; Pinna *et al.*, 2004; Sangiorgio *et al.*, 2006). Thus, detritus processing is slower in low-order, headwater streams that cease to flow in drought than in higher-order streams that continue to flow (Pinna & Basset, 2004).

Surprisingly, in summer drought, detritus processing may be faster in persistent pools than in flowing stream sections (Maamri *et al.*, 1997b). Detritus decomposition in pools in summer drought may rapidly cause the development of 'blackwater' conditions with accompanying deoxygenation. In an Australian stream in summer drought, two species of leaf-processing caddis larvae (*Leptorussa darlingtoni* and *Lectrides varians* (Leptoceridae)) occurred in the shallow, oxygenated areas of the pools, and in the deeper pool sections, tubificids and chironomids, adapted to low oxygen conditions, processed the detritus (Towns, 1985, 1991).

In a Spanish stream that continued to flow during summer droughts, both ecosystem respiration and gross primary production increased above the autumn-winter levels (Acuña *et al.*, 2005). However, when the summer drought caused the stream to cease flowing, both respiration and primary production went to very low levels and recovery was delayed. When flow returns after summer droughts, there can be a heavy influx from the catchment of nutrients and organic matter (both DOM and POM), and microbial activity can reach high levels (Artigas *et al.*, 2009), rendering such streams strongly heterotrophic. Detritus is the major food resource for the recolonizing invertebrates, as the development of algal biofilms may take some time (Legier & Talin, 1975; Acuña *et al.*, 2005). High photosynthetic activity occurs in early spring and winter, when the deciduous trees are bare and both nutrients and light favour primary production (Artigas *et al.*, 2009). Thus, for Mediterranean climate type streams, there is some understanding of ecosystem metabolism and detritus processing dynamics during summer drought that may partly apply to processing in supra-seasonal droughts. This information is important, as there are no studies on detritus processing during and after lengthy supra-seasonal droughts.

As drought is such a strong disturbance, one would expect adaptations in the fauna to contend with it. Adaptations to regular seasonal droughts would be indicated by traits, and such traits may also allow animals to contend with supra-seasonal droughts. Unfortunately, to date, we do not have a trait-based analysis of macroinvertebrates that successfully contend with supra-seasonal droughts.

There are two studies of traits of macroinvertebrates that contend with summer droughts in Mediterranean streams (Bêche et al., 2006; Bonada et al., 2007). Using data from long-term studies of two Californian Mediterranean streams, Bêche et al. (2006) found constancy in the representation of traits to contend with the summer drought, especially in the fauna of the intermittent stream. Traits significantly represented in the intermittent stream during the summer drought included small body size (< 2.5 mm), moderate sclerotization, protective cases or shell, spherical body shape, semi- and multivoltinism, adult stages > 10–30 days, passive aerial dispersal, parthenogenic reproduction, free single and terrestrial eggs, oviparity, deposit feeding and predation as feeding habits. Unexpected but significant traits included lifespans >1 year, gill respiration, and having macrophytes, dead animals, microinvertebrates and vertebrates as major food types.

The study of Bonada et al. (2007) differs from that of Bêche et al. (2006) in that it studied the traits of macroinvertebrates living in stream sections that were either perennial, intermittent or ephemeral. The traits positively correlated with intermittent, and the more severe ephemeral, conditions may be traits to deal with droughts. Traits positively associated with intermittent stream sections included small body size, eggs laid in vegetation, aerial active dispersal, diapause or dormancy, spiracular aerial respiration, flying and surface swimming locomotion, and microinvertebrates as food. Traits associated with ephemeral conditions –more suggestive of supra-seasonal drought traits – included large body size (4–8 cm), free isolated eggs, parthenogenesis, aquatic passive dispersal, cocoons, tegumental respiration, epibenthic and endobenthic habitat, using sediment microorganisms and fine detritus as food, and deposit feeding. The ephemeral stream traits appear to be shaped by the need to contend with extended dry periods and do not overlap with those traits for the intermittent stream sections. However, traits shared between the two studies for intermittent streams (Bêche et al., 2006; Bonada et al., 2007) do not overlap, except in the case of small body size.

Along a section of the Po in Italy, a Mediterranean climate river, some reaches became intermittent with summer droughts (Fenoglio et al., 2006), due largely to water extraction (Fenoglio et al., 2007). Species richness and total abundance declined as flow intermittency increased. Collector-gatherers were the dominant functional feeding group, with shredders and predators declining with increasing intermittency. When the stream channel was dry, Fenoglio et al. (2006) found that both larvae and adults of the dytiscid beetle *Agabus paludosus* were present in the hyporheic zone, 70–90 cm below the surface. This suggests that in Mediterranean streams with regular seasonal drying, particular faunal species may use the hyporheic zone as a refuge that may also allow persistence through a supra-seasonal drought.

At the downstream sites with long cease-to-flow periods, the dominant groups were *Group e* (small-medium size, short generation time, uni-or multivoltine, cemented eggs, crawlers, plant feeders) and *Group f* (medium size, univoltine, crawlers, attached eggs, dormant stage), which partly agree with the traits for intermittent streams found by Bêche *et al.* (2006) and Bonada *et al.* (2006). These studies outline the traits of animals that can contend with intermittent stream conditions. How much the traits indicate success in dealing with supra-seasonal droughts is still uncertain. Indeed, given the unpredictability in timing, duration and severity of supra-seasonal droughts, it may be difficult for organisms to adapt successfully to such a strong selection force.

Long-term studies of Mediterranean streams in California have produced an understanding of how seasonal and supra-seasonal droughts differ. In years of supra-seasonal drought, there is a reduction in winter precipitation and a reduction in the number of storms that produce scouring floods (McElravy *et al.*, 1989; Power, 1992; Power *et al.*, 2008). In a severe drought year (1977) in the perennial Big Sulphur Creek, species richness and abundance of Ephemeroptera, Plecoptera and Trichoptera were significantly reduced (McElravy *et al.*, 1989; Resh *et al.*, 1990). Surprisingly, in the drought, the abundance of the algal-grazing caddis larvae *Gumaga nigricula* was high and comprised 57 per cent of total invertebrate abundance (McElravy *et al.*, 1989). The cause for this dominance may be the lack of the normal scouring winter floods, which deplete *Gumaga* populations, rather than the summer conditions of low flow (McElravy *et al.*, 1989). *Gumaga* is a burrowing caddis that favours silt in depositional areas, so the lack of winter floods removing silt, combined with the encroachment of *Typha* in the channel during drought, may have favoured *Gumaga* populations (Bêche *et al.*, 2006).

However, in a small brook that dried up completely in the 1976–77 drought, the population of *G. nigricula* was eliminated (Resh, 1992). With the return of normal flow, recovery was slow. Prior to the drought, *Gumaga* had multiple cohorts. Two years after the drought, there was only a single cohort, and recovery to the pre-drought condition did not occur till 1986, ten years after the supra-seasonal drought (Resh, 1992). Thus, the severity of a single supra-seasonal drought had markedly different effects in different habitats and accordingly affected populations of the same animal in different ways.

The situation whereby lack of scouring floods favours particular species in drought years is further illustrated by a remarkable 18-year study by Power *et al.* (2008) on the South Fork Eel River in California, a perennial river in a Mediterranean climate. After normal winter floods, which scour the stream bed, *Cladophora* and diatoms flourish and are consumed by primary consumers (Power, 1992), which are part of a food web terminating in steelhead trout. In drought years in which preceding winter rains fail, stream bed scouring is reduced, allowing a 'predator-resistant armored caddis fly' *Dicosmoecus* to graze down the *Cladophora* and the diatoms (Power *et al.*, 2008). As *Dicosmoecus* are not consumed, the food chain from the alga does not go beyond the caddis larvae, forcing the consumers, during normal years, to seek other prey. Presumably, this drastic drought-created restructuring of the food web results in lower secondary production. The examples of *Gumaga* and *Dicosmoecus* being favoured in drought years illustrate the point that drought effects may be more a function of the lack of floods rather than of low flow conditions *per se*.

In a study over three separate streams, flow was negatively correlated with total abundance and taxon richness with major changes in community composition before, during and after supra-seasonal droughts (Bêche & Resh, 2007). Changes in community structure from wet to drought years were mainly due to changes in chironomid taxa and their abundance. At first order sites, drought resulted in high stability in community structure, with rapid drying producing a distinctive set of robust survivors. However, at sites on higher-order streams, community structure was less stable, due perhaps to the shift to drought conditions occurring more slowly (Bêche & Resh, 2007). This is a good illustration of the differing effects of drought in different parts of a stream system, with the differences being generated by the different rates of drying.

Further studies on these ephemeral, intermittent and perennial Mediterranean streams revealed that a supra-seasonal drought (1987–1991) did not significantly alter species richness or total abundance of invertebrates. However, pre-drought assemblage structure did differ from the drought assemblage, both of which in turn differed greatly from the post-drought assemblage. The latter assemblage was stable and, while drought did not alter aggregates such as total abundance and species richness, it did produce marked and durable changes in assemblage structure (Bêche *et al.*, 2009).

Overall, these findings indicate, not surprisingly, that in Mediterranean streams, the effects of supra-seasonal droughts are more severe and less predictable than those of seasonal droughts. Accordingly, seasonal droughts may not serve as a reliable guide to the effects of prolonged supra-seasonal droughts, though traits evolved to contend with seasonal droughts may partly help to contend with supra-seasonal droughts. Furthermore, evidence suggests that lengthy supra-seasonal droughts can leave significant and durable changes in assemblage structure in Mediterranean climate streams.

6.4.3 Dryland streams

There are many streams in areas of low rainfall that are intermittent and subject to both seasonal and supra-seasonal droughts. These include streams in the maritime and continental temperate (winter precipitation) and tropical climate (summer rainfall) zones.

In the tropical savanna regions, there is a short, wet season and a long, dry season. Consequently, streams regularly dry out – seasonal drought. However, supra-seasonal drought, with a failure of the wet season rains, does periodically occur. In Zimbabwe (then Rhodesia), Harrison (1966) studied the fauna of pools and runs in a small intermittent stream for two normal years (1962–63) and a drought (supra-seasonal) year (1964) in which rainfall was 50 per cent below normal and the stream dried out for seven months compared with the normal period of about three months.

In each year (normal and drought), the fauna was depleted with the drying. However, recolonization of the pools was rapid, with the early colonizers being nematodes, the oligochaete *Limnodrilus*, copepods (*Cyclops* spp.) and the chironomid *Chironomus satchelli*, followed by nymphs of the mayfly *Cloeon crassi* and Odonata. Within two months, the pool fauna was fully recovered. In the runs, re-colonization occurred after flow resumed and was marked by the early arrivals of abundant simuliid larvae, along with nematodes, oligochaetes, *Chironomus satchelli* and the orthoclad chironomid *Rheocricotopus capensis*. Recovery took longer in the runs than in the pools, with a distinct succession of taxa. Refuges used by the fauna included resting eggs (e.g. *Cyclops*), damp places under banks (oligochaetes, nematodes), and protected and sealed shells (pulmonate snails) ('polo-club' refuges; Robson *et al.*, 2008b). Insect re-colonization was thought to occur through adult insects flying in from elsewhere, with some species laying eggs (Harrison, 1966).

In an intermittent, continental-temperate climate, prairie stream, Fritz and Dodds (2004) found that both floods and drying greatly reduced both taxon richness and abundance and that, in contending with these disturbances, resilience was far more important than resistance. Recovery rates after drying were relatively high and were similar to those after floods. The rapid recovery after drying may have been facilitated by the rapid recovery of primary production (Dodds *et al.*, 1996).

A high rate of recovery in a prairie intermittent stream was also found by Miller & Golladay (1996). The fauna dwelling in this 'harsh' type of stream system are characterized by having 'short, asynchronous life cycles' (Fritz & Dodds, 2004). The major refuges from drying appear to have been perennial upstream reaches, with the rate of recovery of taxon richness being a linear function of distance from the nearest upstream refuges (Fritz & Dodds,

2004). Human activities, reducing perennial surface water through exploitation, may degrade and deplete refuges and thus reduce the resilience of the fauna to flow-generated disturbances.

Two intermittent streams in a rain shadow area of south-eastern temperate Australia were investigated by Boulton (1989) and Boulton & Lake (1990, 1992a, 1992b, 1992c). Sampling occurred in a severe supra-seasonal drought (1982–83) and through a normal summer low flow period. Normally, pools peaked in their species richness and abundance shortly after flow ceased and persisted through summer, with a fauna dominated by lentic taxa and predators. The riffles reached their highest species richness and abundance shortly before flow stopped. With the return of normal flows in autumn in the riffles, there was a predictable succession, beginning with simuliid and chironomid larvae and proceeding to a fauna rich in Ephemeroptera, Trichoptera, Diptera, Coleoptera (Elmidae) and Plecoptera (Boulton & Lake 1992a, 1992b).

Over summer and in the drought, there was a build-up of organic matter in the channel due to the leaf fall of eucalyptus, which peaks in summer, and to the lack of strong flows. Shredders were relatively few, and their abundance correlated with organic matter levels in the riffles, but not in the pools, where the organic matter mostly accumulated. The dominant detritivores were collector-gatherers and collector-scrapers. The 1982– 83 drought greatly depleted the abundance of detritivores, especially shredders. The availability of organic matter for the detritivores at the end of summer or after a drought is dependent on the way that flow begins. Flooding can deplete detritus availability by moving large amounts downstream.

The 1982–83 drought greatly reduced faunal densities and species richness. As the drought set in, immature individuals of some species, such as those of the stonefly *Austrocerca tasmanica* and of species of rheophilous, predatory trichopterans (Hydrobiosidae), were eliminated by the drying being earlier than normal. Species dependent on free water, such as the amphipod *Austrochiltonia australis*, were greatly depleted or, as in the case of the shrimp *Paratya australiensis*, were simply eliminated. Thus, in the next year, previously important species were either absent or only present in very low numbers. Recovery of many species in the next winter was rapid, though there were deletions, exemplifying the marked lag effects that supra-seasonal droughts, as opposed to summer droughts, may produce.

Many of the species returning after the drought used refuges, principally, the 'polo club' type of Robson *et al.* (2008). Boulton (1989) sampled a range of potential refuges and documented their role in drought survival. There were five major types of strategies to use refuges: surviving in persistent pools; in moist habitats (mats of dried algae and leaf litter, below stones and

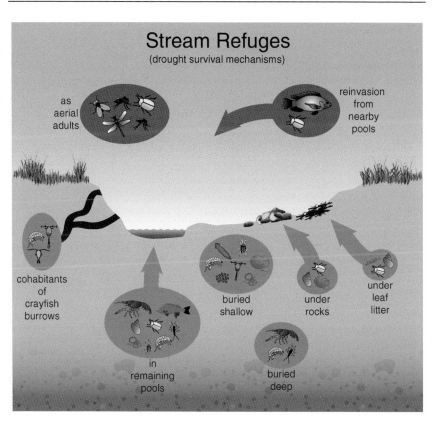

Figure 6.4 The drought refuges used by macroinvertebrates dwelling in the Lerderderg-Werribee Rivers in the sharp and severe 1982–83 drought. (Adapted from Figure 9.5 in Boulton & Brock, 1999.)

logs); moving into the hyporheic zone or crayfish burrows (pholeteros; Lake, 1977); having desiccation-resistant stages such as eggs and cysts; and migrating to permanent water bodies (Boulton, 1989; see Figure 6.4). By far the most common strategy was to survive in persistent pools, followed by having desiccation-resistant stages. The least used refuges were crayfish burrows, residing in moist habitats (under stones, logs, dry algae) and moving into the hyporheic zone (Boulton, 1989). Thus, to survive normal summer droughts and supra-seasonal droughts, the fauna of these streams used an array of refuges. However, there were still taxa that were greatly depleted.

In a comparison with the Australian situation, Boulton *et al.* (1992a) found that of about 50 invertebrate taxa in Sycamore Creek, Arizona (a desert stream – see above), only 13 used refuges in seasonal drought, and only three types of refuges were used: below dried litter, in the hyporheic zone and in dry channel substrata (desiccation-resistant eggs?).

The biota of intermittent streams, in successfully contending with significant periods of low flow or no flow have, by and large, gained adaptations that allow them to contend with supra-seasonal droughts when they arise. What is not known are the effects of extended droughts, even megadroughts, on the viability of the biota of temporary waters. As stressed by Boulton (2003) and Bêche & Resh (2007), this knowledge may only come from long-term studies.

6.5   Drying and recovery in temporary wetlands and streams

The ecosystem dynamics of what drought does to temporary lentic systems is basically unknown territory. As drying occurs in ponds, water quality deteriorates, with rises in conductivity and possible occurrence of 'blackwater events'. Finally, the system dries out, with the dry sediments containing nutrients and usually a layer of particulate organic matter, of both allochthonous and autochthonous origins, sitting on the surface. Although this dry detritus slowly decomposes, it may also become enriched in nutrients as carbon is lost (Bärlocher *et al.*, 1978).

Droughts are broken by precipitation sufficient to increase and maintain water volumes. With the drought being broken, nutrients are released from the benthic sediments and DOC enters from the sediments, the benthic detritus and from the catchment. Conceivably, bacteria levels rise and biofilms of bacteria and fungi form, to be then followed by the primary production of phytoplankton in unshaded systems. This boom fuels filter-feeding animals that consume either phytoplankton or bacteria or both. Within a short time, predators – mostly insects – arrive, mainly by migration from pools that persisted through the drought.

However, there is a danger of 'false starts', whereby enough rain falls to partly fill the system, but the filling is short-lived. Germination of seeds and hatching of eggs may occur but, with rapid drying, their lives are cut short. The damage from false starts remains to be assessed.

In temporary streams, as drought sets in, linkages are severed. First the lateral links between the stream channel and its littoral zone are broken, and then longitudinal links are disrupted and pools form, interspersed by dry channel (Boulton & Lake, 2008). Water quality in the pools deteriorates with hypoxia, high temperatures and possibly localized blackwater events with high DOC levels. The stream becomes reduced from a continuum to an assortment of patches (pools) separated by dry sections. The pools may become quite different from each other under the influence of specific local conditions, rather than being regulated by balancing force of streamflow. Following the conceptual model of Stanley *et al.* (1997), the stream as an ecosystem contracts, to become a fragmented assortment of patches. In extreme cases, such as in a long supra-seasonal drought, the stream ecosystem contracts completely and disappears, though in many cases subsurface water may remain in patches. In the subsurface water, ecosystem processes such as decomposition and nutrient transformation may continue, but the major pathways of material transfer are severed (Fisher *et al.*, 1998).

With the breaking of droughts, there is a pulse of nutrients (nitrogen and phosphorus), POM and DOC from the stream bed and from the catchment into the stream (see Chapter 5). If the drought breaks with a flood, the retention of nutrients, DOC and POM in the upstream sections may be poor, and these components may be swept downstream, possibly to fuel recovery in more retentive downstream reaches. However, if the drought breaks with steadily increasing flows, the pulse of nutrients and organic matter may be substantially retained and could thus drive recovery of both autotrophic and heterotrophic metabolism. Such metabolic stimulation may increase at first heterotrophic production, and then autotrophic production, to levels much higher than normal. Initially, with a significant lag in primary production, consumer recovery after drought is delayed in comparison to the recovery of detritivores (viz. Closs & Lake, 1994).

During the process of recovery from drought, significant trophic interactions may influence the composition of producers and their production levels. In a typical intermittent prairie stream in Kansas (Fritz & Dodds, 2004), recovery from seasonal drought was rapid when flow returned for a short period (≈ 35 days), leaving persistent pools (Murdock *et al.*, 2010). The dominant macroconsumer was the southern redbelly dace *Phoxinus erthrogaster*, which, when flow returned, rapidly migrated in numbers into the study section. With channel re-wetting, algal growth promptly commenced, but was checked by grazing macroconsumers. The initial growth of the filamentous, desiccation-resistant alga *Ulothrix* declined due to consumption, while chain-forming pinnate diatoms increased.

During the early to mid-phases of recovery, the macroconsumers, predominantly the dace, along with crayfish (*Orconectes* spp.), lowered macroinvertebrate biomass and algal biomass and productivity. This strong, top-down effect of the macroconsumers did not last, as the fish had, after about 35 days, migrated out of the stream section of the study (Murdock *et al.*, 2010). Thus, in this case, it appears that primary, rather than heterotrophic, production fuelled recovery, with the macroconsumers initially influencing biomass and production levels. This situation compares interestingly with the study of Power *et al.* (2008) (see above), where with the lack of floods, a winter drought created later conditions that favoured the population growth of a macroinvertebrate grazer, which greatly curtailed algal biomass and very significantly altered the normal food web of the river.

The trophic structure of temporary streams undergoes considerable changes while there is flow. As described by Closs & Lake (1994), detritus and algae comprise the basal resources of the food web in an intermittent Australian stream. In a short period from cease-to-flow to drying in seasonal drought, the food web goes from being in its most complex and diverse form to nothing. Rebuilding as flow returns after drought appears to be strongly driven by detritus, and algae do not appear to be important until well after flow has started. Possibly, such changes in the food web structure with summer drought occur after supra-seasonal drought, with the proviso that, during supra-seasonal drought, some key species, such as some molluscs and crustaceans, may have been eliminated.

6.6   Conclusions

Temporary waters cover both standing and flowing waters, which may be either ephemeral or episodic or intermittent. For these systems, especially intermittent systems with predictable seasonal droughts, there has been many studies documenting effects of seasonal drought on biota, but relatively few studies of the effects of supra-seasonal droughts. This imbalance stresses the need for long-term studies that track both regular seasonal droughts as well as the infrequent supra-seasonal droughts, and that in particular detect the lag and possible cumulative effects of drought.

Algae may survive drought, though this appears to be strongly dependent on the rate of drying as droughts build and on the lengths of drought. As recovery of algal-driven primary production is a key factor governing recovery of consumers after drought, the study of the impacts of drought and recovery in streams must be scaled up to the catchment level. This requirement remains an unfilled challenge, and the need to increase the spatial extent also applies to other ecosystem processes and biota.

Vascular plants have means of drought avoidance, such as the production of propagules, and means of drought tolerance that involve physiological adjustments. Supra-seasonal drought in wetlands may lead to 'terrestrialization', whereby terrestrial biota (e.g. plants and fauna) invade the domains of aquatic plants, whereas in other cases there may be wholesale replacement of particular aquatic plant species by other aquatic species. The latter illustrates the fact that in many cases with different biota, supra-seasonal drought may facilitate the expansion or invasion of some species by harming others.

The fauna of temporary waters may deal with drought either by having means of survival *in situ*, such as desiccation-resistant propagules or life stages (sedentary refuges), or by migrating away to refuges (migrational refuges) such as nearby permanent water. Different faunal groups vary in their use of these two types of refuges, with crustaceans and molluscs by and large using sedentary refuges, whereas insects and fish mainly use migrational refuges. In the case of propagules, the cumulative effects of long, supra-seasonal droughts remain uncertain.

By producing a variety of conditions in water bodies at a landscape scale, drought may actually favour the survival of some species. This particularly applies for the fauna of temporary waters, many of which are good colonizers, whilst many inhabitants of permanent waters are poor colonizers. In eliminating potential competitors and/or predators in some locations, drought may create conditions that favour good colonizers for a period. Indeed, drought and interspecific interactions, such as predation, can interact to produce novel outcomes at a landscape scale.

Antecedent conditions can greatly influence the effects of drought. Recovery from a supra-seasonal drought may take time and may not be complete before the next drought, even a normal summer drought, sets in. Thus, past droughts may mould the 'ecological memory' of a system, and their effects need to be considered in understanding the effects of individual droughts. However, there are only hints at this stage of the nature and strength of this phenomenon.

The available evidence does suggest that the biota of temporary waters, in evolving to contend with seasonal droughts, possess adaptations which do increase their resistance and resilience to the stresses of supra-seasonal drought. This capacity is, however, very dependent on the duration and the severity of supra-seasonal droughts.

Finally, while an understanding of the effects of drought on the inhabitants of temporary waters is starting to emerge, there is a great dearth of knowledge on how ecosystem processes (e.g. primary production, decomposition, denitrification) are affected by drought. Studies in this area will need to embrace catchments, rather than specific sites, and examine not only processes in the water body itself, but also processes in the terrestrial portions of the catchments that influence processes in the water body.

7

Drought, floodplain rivers and wetland complexes

Floodplains occur alongside rivers and are relatively level areas which may be regularly inundated. They are mainly constructed from the deposition of material from the river. In constrained rivers, the floodplain is narrow, but in large, unconstrained lowland rivers, the level floodplain may be very wide. The channels of lowland rivers are winding, with many meanders, and they undergo considerable lateral migration. Resulting substantially from past lateral migration of the channel, the floodplain usually contains an abundance of standing water bodies – lagoons, ox-bow lakes or billabongs, permanent and temporary wetlands, channels, distributaries and flood runners. These water bodies are linked with the mainstem channel during floods, forming a strong but temporary axis of lateral connectivity. The floods usually occur regularly, especially in regions with regular wet and dry seasons. In arid and semi-arid regions, the floods occur episodically as large irregular events (Puckridge *et al.*, 1998; Bunn *et al.*, 2006).

When flooding occurs, nutrients, sediment, organic matter (both DOM and POM) and biota are carried across the floodplain and into the mosaic of temporary and permanent water. This event stimulates a boom of primary and secondary production, fuelled by both nutrients and DOM from the river channel, and also by nutrients and DOM released from the re-wetted sediments of the floodplain. This process of high production is the 'flood pulse' of Junk *et al.* (1989) and Tockner *et al.* (2000).

The nutrients and DOM stimulate heterotrophic (bacterial) and autotrophic (phytoplankton, benthic algae) production (e.g. Ward, 1989b; Valett *et al.*, 2005; Lake *et al.*, 2006) that fuels the production of primary consumers, such as the rotifers and micro-crustaceans that hatch from eggs in rewetted sediments (Jenkins & Boulton, 2003). This is followed by

Drought and Aquatic Ecosystems: Effects and Responses, First Edition. P. Sam Lake.
© 2011 P. Sam Lake. Published 2011 by Blackwell Publishing Ltd.

booms in macroinvertebrates, fish and aquatic macrophytes, so that natural floodplains, especially those in the Tropics, can become one of the most productive environments on earth (Tockner *et al.*, 2008). In arid zones, the episodic floods generate high production, culminating in booms of fish (Bunn *et al.*, 2006; Balcombe & Arthington, 2009) and of water birds (Kingsford *et al.*, 1999). This boom in production may be accompanied by a massive increase in biodiversity (e.g. Junk *et al.*, 2006). Indeed, floodplains are stark exemplars of ecosystems that expand and contract (Stanley *et al.*, 1997) and which, in doing so, have processes that boom and subsequently decline.

Water leaves the floodplain by evapotranspiration, seepage, and by receding back into the channel. Receding floodplain waters carry nutrients, detritus and biota (from plankton to fish) back to the river channel, a subsidy that may be very important to its metabolism – a transferred 'flood pulse' which may provide a lasting stimulus to in-channel production.

The dynamics of recession and the effects of the inputs from the floodplain to the ecology of the mainstem channel appear to have been neglected. Part of the boosted production of the floodplain may be transferred to the surrounding hinterland by birds and insects in the flood and by foraging animals as the floodwaters recede (Ballinger & Lake, 2006). Needless to say, many floodplain rivers no longer have flood pulses, due to human activities greatly reducing the magnitude and frequency of flood, as well as constructing barriers and levees to isolate the river channel from its floodplain (see Chapter 11). These activities have greatly reduced the production of floodplain river systems.

The receding waters leave behind well-watered and nutrient-rich plains, which then quickly become populated by a productive expanse of plants from grasses to shrubs. In turn, this can attract a wealth of herbivores and their predators in the natural state, and livestock in the human-controlled state.

A similar pattern of pulsed production – a 'pulsed ecosystem' (Odum, 1969) – occurs in some large wetland complexes, of which the Florida Everglades is a prime example. In this large wetland complex in southern Florida, the wetlands are inundated in summer (May to October) by a moving sheet of water and become the 'rivers of grass'. In winter, this dries to plains with waterholes, lagoons and sloughs – the 'dry-down'. Rather than the plains being fed by floods from a mainstem river, the Everglades are flooded by waters overflowing from a large lake, Lake Okeechobee. In this chapter, the accumulated knowledge on the effects of drought on the biota of the Florida Everglades is outlined as a seminal example of drought and wetland complexes.

As floodplains and wetland complexes are ideal habitat for many verte-brates, such as amphibians, reptiles, mammals and birds, this chapter ends with a brief treatment of the effects of drought on wetland-dwelling vertebrates, to illustrate both the large-scale and strong lag effects of droughts.

7.1   Drought and floodplain systems

In such pulsed ecosystems as floodplains, seasonal drought may regularly come with the dry season. Supra-seasonal drought arises naturally from a failure of wet seasons that leaves the floodplains lacking in regenerating floods. Water in the channel recedes from the littoral edge and volume decreases, though the large volumes of lowland floodplain rivers can buffer them from severe drought effects. In rare cases under natural conditions, flow may cease. Little is known about the effects of drought on the biota of the mainstem river channels. The effects of flow regulation on floodplain rivers and their floodplains, and consequent interactions with drought are, how-ever, important and are dealt with in Chapter 11.

With a failed wet season, flooding may not occur or may only be minimal – a severing of lateral connectivity. This loss, without any recharging of flood-plain water bodies and an extended dry period, leads to a loss of water and a decline in water quality in floodplain water bodies. With drought, floodplain lagoons, being usually shallow, can become highly turbid (e.g. Crome & Carpenter, 1988) and may lose macrophyte species and cover (Santos & Thomaz, 2007). Water temperatures increase and may fluctuate diurnally and oxygen levels may fluctuate, but may also greatly decrease – an effect exacerbated if stratification occurs. Conductivity (salinity) usually rises as water levels drop (e.g. Briggs *et al.*, 1985) and pH may change, for example becoming more acidic in the floodplain lagoon studied by Crome & Carpenter (1988). On the Phongolo floodplain in Botswana during drought, some floodplain lagoons became saline due to the input of saline seepage, whereas others became saline due to evaporation and drying (White *et al.*, 1984).

On floodplains, leaf litter can accumulate in the basins of wetlands. As drought sets in, and with loss of water and increased temperatures, decom-position of the leaf litter occurs and 'blackwater events' may take place (Slack, 1955; Paloumpis, 1957; O'Connell *et al.*, 2000), which greatly lower oxygen levels and can cause fish kills.

As the lagoons recede, the sediments dry, along with detritus. This drying reduces microbial activity and stops anoxic processes such as denitrification. Bacteria are killed, which increases the nitrogen and phosphorus contents of the sediments. Upon re-wetting, the sediments can release a pulse of nitrogen and phosphorus that may stimulate phytoplankton and macrophyte growth (Baldwin & Mitchell, 2000).

7.2   Drought and the biota of floodplain systems

7.2.1   Vascular plants

Drought may stress floodplain trees. If it leads to a long period of no flooding and with a fall in groundwater levels, even hardy trees may die. For example, in the 1997–2010 drought in south-eastern Australia on the Murray River floodplain, there were no floods and groundwater levels dropped drastically. This resulted in the death of hitherto drought-tolerant river red gum (*Eucalyptus camaldulensis*) (Horner *et al.*, 2009; see Chapter 5).

Increases in groundwater salinity during drought can result in wetland/riparian trees being killed (e.g. Hoeppner *et al.*, 2008). Floodplain tree species differ in their tolerance to flooding and drought (Waldhoff *et al.*, 1998), and such differences may partly determine forest composition. For example, Lopez & Kursar (2007) found, in seasonally flooded forests in Panama, that the dominance of seedlings of the tree *Prioria copaifera* was partly due to the relatively low mortality of its seedlings in times of drought.

Ground cover vegetation may be greatly depleted by the combined effects of grazing and drought. Considerable damage may be done to many floodplain areas and riparian zones during drought due to the aggregation of grazing animals, both wild and domestic.

In floodplain lakes, changes in water level are known to affect the species richness and composition of aquatic macrophytes (Van Geest *et al.*, 2005). Lake Merrimajeel is a shallow billabong on the Lachlan River floodplain in western New South Wales, Australia (Crome & Carpenter, 1988). Before a drought in 1977–78, the dominant macrophyte was *Vallisneria spiralis*, along with filamentous algae and Characeae (Briggs & Maher, 1985). The drought dried up the system and, with refilling, there was a marked surge in productivity of *Vallisneria spiralis*, with *Myriophyllum verrucosum* becoming an abundant macrophyte (Briggs & Maher, 1985). Nutrients released from sediments undoubtedly stimulated the high macrophyte production.

On the floodplain of the tropical Paraná River floodplain, Santos & Thomaz (2007) investigated the macrophytes in lagoons connected with the river in high water and in disconnected lagoons. Drought decreased the macrophyte species richness in both types of lagoons from 13–18 to 7–9 species. With flooding, the species richness in the connected lagoons jumped from 7 to 20, while species richness did not change in the disconnected lagoons – a vivid example of the influence of lateral connectivity in floodplain systems. In both of these studies, the aftermath of drought was a marked increase in species richness and production.

7.2.2   Phytoplankton

There are very few reports on drought and phytoplankton in floodplain lagoons.

In two tropical floodplain lakes in India, Das and Chakrabarty (2006) compared two years: 1995, which was normal, and 1996, which had a severe drought. The drought caused an increase in salinity, total alkalinity, hardness and in the nutrients (total phosphate, total nitrogen, nitrates and nitrites). With the increase in nutrients, there was a great increase in primary production (GPP & NPP) and in phytoplankton density (Das & Chakrabarty, 2006).

A drought in 2000–2001, affecting lagoons of the Upper Paraná River floodplain, decreased connectivity between the river and the lagoons. This loss coincided with a decline in phytoplankton species richness but an increase in beta diversity between lagoons, reflecting increasing habitat differences between the isolated lagoons (Borges & Train, 2009). On the other hand, in an experiment mimicking drought effects in enclosures in a floodplain lagoon, Angeler & Rodrigo (2004) found that, with declining water depth, densities of *Synechococcus* cyanobacteria also declined rather than increased as may have been expected. It is not clear why densities dropped with declining water depth.

Drought conditions can exert strong effects on phytoplankton, altering production, species richness and composition, perhaps due to the loss of connectivity with the river and other lagoons. Furthermore, changes in phytoplankton with drought may be influenced strongly by other forces, such as interspecific interactions (grazing by primary consumers), that come into play with declining habitat space.

In the mainstem channel of floodplain rivers, drought by lowering water flow reduces suspended solids and increases light penetration (Devercelli, 2006). From a fluvial phytoplankton normally dominated by diatoms (*Aulacoseira* spp., *Skeletonema* spp.), in drought the phytoplankton may become one dominated by small flagellates (Crytophyceae, *Rhodomonas*) and chlorophytes (*Chlamydomonas*). The latter group is known to be favoured by shallow lentic conditions enriched with nutrients (Devercelli, 2006). Thus, in this case, drought exerted a dramatic effect on phytoplankton, with major changes in species composition and in functional groups.

7.2.3 Zooplankton

Floodplain rivers, with many backwaters, can harbour a zooplankton fauna, usually of small body size (e.g. rotifers, protozoans – Wetzel, 2001). A severe drought in 2003, the worst in 90 years, greatly reduced the flow of the Po River in Italy. High nutrients, combined with low flow, induced eutrophication (Ferrari *et al.*, 2006), which produced a marked increase in rotifer abundance and a decline in cyclopoids. It is interesting to note that in standing waters, when eutrophic conditions and drought coincide, the zooplankton is also dominated by rotifers rather than by the normal crustacean fauna.

In two floodplain lakes in India exposed to severe summer drought, with the loss of water and increased nutrients, there was a sharp increase in phytoplankton abundance (Das & Chakrabarty, 2006). This increase was coupled with a doubling in zooplankton abundance and there was a major shift in abundance, with an initial increase in rotifers, particularly *Brachionus* sp., which faded away during the drought. Copepod abundance persisted to become dominant toward the end of the drought, which virtually eliminated Cladocera from the zooplankton (Das & Chakrabarty, 2006). Little information, however, is available on the recovery of the zooplankton.

Drought in Brazil dried out a floodplain lake such that its basin was broken up into four small fragments (Nadai & Henry, 2009). Zooplankton species richness in the fragments dropped, with rotifers becoming dominant, largely at the expense of cladocerans. With refilling of the basin, zooplankton densities declined, presumably due to dilution, but species richness jumped by over 50 per cent with a large increase in copepods (Nadai & Henry, 2009).

In a severe drought, Lake Meerimajeel in south-eastern Australia dried out for three months (Crome & Carpenter, 1988). As the system dried, salinity rose sharply and oxygen dropped to undetectable levels. The highest densities of the diverse zooplankton occurred just before the drying – a function of habitat reduction. Just before drying was complete, many species were eliminated, but rotifers (especially *Brachionus calyciflorus* and *Asplanchna sieboldi*) dominated in very high densities ($3,769 \, l^{-1}$), surviving in water with high temperatures ($\approx 40\,^{\circ}C$) and a high turbidity. After the drought broke, the water was turbid and rich in organic detritus. Faunal recovery was slow, with low densities of adult cladocerans (*Daphnia carinata*, *Bosmina meridionalis*) and calanoid copepods (*Boeckella fluvialis*) and immature cyclopoids for many months.

Surprisingly, in the filling phase, a group of hitherto rare species (e.g. *Austrocyclops australis*, *Pleuroxus aduncus*, *Alona cf. davidi*, *Diaphanosoma excisum*, *Brachionus dichotomus*) briefly flourished. Crome & Carpenter (1988) suggested that this group consisted of early successional species using the rich organic detritus. Other studies of drought and zooplankton (e.g. Arnott & Yan, 2002; Schneider, 1997; Crome & Carpenter, 1988; Nadai & Henry, 2009) have also shown that, in the early phases of the recovery of zooplankton from drought, rare species may briefly flourish, perhaps breed and produce eggs which persist in the sediment survive until the next drought is broken.

With long dry periods, even in temporary water bodies, the viability of stored propagules declines, weakening the response to water when it comes. Boulton & Lloyd (1992) and Jenkins & Boulton (1998) showed that zooplankton hatching responses from sediments declined as the intervals between flooding events increased. In experiments, Jenkins and Boulton (2007) found that dry periods lasting longer than six years caused a very significant decline in the densities of microinvertebrates that hatched from sediments with the addition of water.

7.2.4 Benthos

A severe drought in Indian floodplain lagoons reduced density and biomass of benthic macroinvertebrates which were dominated by ostracods, oligochaetes and dipteran larvae (Chironomidae, Chaoboridae and Culicidae) (Das & Chakrabarty, 2006). However, during a severe drought in two small floodplain ponds in Iowa, USA, the benthos dominated by chironomid, ceratopogonid and chaoborid larvae reached very high densities, presumably a function of concentration from reduced habitat availability (Paloumpis, 1957). In Lake Merrimajeel, with receding of water in drought, a shifting strip of wet mud was exposed – a strip that harboured high densities of the scavenging beetles *Heterocerus* and *Berosus*, along with dipteran larvae (e.g. Dolichopodidae, Muscidae, Chironomidae and Tabanidae) (Maher, 1984). Few species survived in the dry mud. With flooding, there was a boom of the chironomid *Chironomus tepperi*, with a generation time of eight days (Maher & Carpenter, 1984). However, this species did not persist and gave way in abundance to the chironomids – *Chironomus 'alternans a'* and *Polypedilum nubifer* (Figure 7.1).

The very high initial production of chironomids with re-wetting was heavily utilized by waterfowl populations. Chironomids, as a group, appear to vary greatly in their capacity to survive drought. In a comparison with Lake Merrimajeel, in a Paraná River floodplain lagoon in drought, at least 13 taxa of chironomids (out of a total of 25 taxa) were capable of surviving in dry sediments for a month (Montalto & Paggi, 2006).

Overall, the responses of the invertebrates of floodplain lagoons to drought are poorly known. Not only are there hints of major changes in community structure, but the boom of invertebrates upon flooding may be critical to the

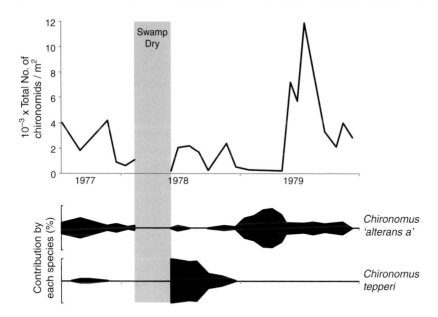

Figure 7.1 Changes with drought in the abundances of two of the dominant chironomid larvae in a floodplain lake, Lake Merrimajeel, New South Wales, Australia. (Redrawn from Figure 3 in Maher & Carpenter, 1984.)

development of a trophic web and to the maintenance of floodplain ecosystems. In temporary wetlands and floodplain systems, drought may be regarded as an essential driver for ecosystem pulsing that governs biodiversity and productivity. By drying sediments and detritus, drought primes the ecosystem for a boom of production with the release of nutrients and high quality detritus when water returns.

7.3   Floodplain rivers, fish and drought

Floodplain rivers, especially those in the Tropics, undergo regular wet and dry seasons, with a pulse of high fish production following the seasonal floods. Many fish in such rivers spend their lives between wet-season and dry-season habitats. Consequently, many fish undergo extensive migrations, longitudinal and lateral (Lowe-McConnell, 1975; Welcomme, 1979, 1985; Winemiller & Jepsen, 1998). A group of fish called 'white fish' move onto the floodplain with floods, but recede back to the main channel with floodplain drying and undergo lengthy longitudinal migrations in the mainstem channel. A second, or 'gray' group of fish comprises those that move onto the floodplain with the flood and recede back into the channel with drying, but do not undergo lengthy migrations. The 'black' group of fish is the most drought-resistant, as these fish endure the dry season and droughts in floodplain water bodies, which may become severely degraded in water and habitat quality (Welcomme, 1979, 1985; Junk *et al.*, 2006).

When droughts occur with greatly reduced flooding, recruitment of floodplain-dependent species can be severely curtailed, and growth of the fish in the river channels may be reduced due to the lack of the food subsidy from the floodplains, especially that delivered as floods recede (Welcomme, 1985, 1986, 2001; Neiland *et al.*, 1990; Laë, 1995). The main channel may serve as a drought refuge for species that migrate from the floodplain and from tributaries as the drought sets in. Water bodies on the floodplain may also serve as drought refuges for a wide array of species. Survival in these refuges may be threatened by the lowering of water quality, especially the occurrence of hypoxic conditions, and by both increased competition for food and increased predation pressure (Welcomme, 1979; White *et al.*, 1984; Merron *et al.*, 1993; Swales *et al.*, 1999) coming from piscivorous fish, birds and humans. Alternative refuges for fish in floodplain rivers can be in the channel itself (e.g. Thomé-Souza & Chao, 2004), which, in severe droughts of long duration, may become a series of pools. In such pools, stressful conditions can arise, such as algal blooms that can create hypoxic conditions.

7.3.1   Fish and the mainstem channel

Little is known about the effects of drought on fish assemblages dwelling in main river channels of floodplain rivers. The Havel River in Germany is a regulated floodplain river substantially altered by human efforts. Using commercial fish data, Wolter & Menzel (2005) found that, after drought years with low flows, yields of pike (*Esox lucius*) and pikeperch (*Sander lucioperca*) were significantly reduced. This reduction was thought to be due to low water levels decreasing habitat cover for larval and juvenile fish, notably macrophyte stands, giving rise to high mortality due to predation. In turn, lack of the young of the year may reduce the growth of their major predators, such as one-year-old pike and pikeperch (Wolter & Menzel, 2005).

Two Brazilian floodplain rivers in the Rio Negro basin were subject to a severe El Niño drought in 1997–1998 that lowered water levels by up to two metres below normal dry season levels (Thomé-Souza & Chao, 2004). In the main channels, with drought, species richness and biomass were reduced by up to 50 per cent. The benthic fish assemblages were normally dominated by species in the orders Siluriformes (catfish) and Gymnotiformes (knifefishes and electric eels). In the drought, these two orders virtually disappeared and were replaced by Characiformes (leporins and piranhas) and Perciformes (perch-like fishes). The causes for this dramatic change are uncertain, with differential migration and high predation being suggested (Thomé-Souza & Chao, 2004).

Clearly, more study needs to be done to determine the impact of drought on fish in the main channels of large rivers. The disruption of connectivity by human-imposed barriers could have damaging effects on fish that migrate along these large rivers in both normal times and in drought. Migration may be essential for recovery of fish communities after drought, and yet barriers can impede these critical migrations.

7.3.2 Drought and adaptations of floodplain fish

The fish that reside in floodplain water bodies through drought belong to the 'black' group, and many of these species have adaptations to survive harsh conditions, such as undergoing physiological torpor, the ability to aestivate or produce desiccation-resistant propagules, and the ability to breathe air from the water surface. Oxygen depletion of a floodplain water body can occur as water volume decreases and temperature and salinity increase. Fish may respond to hypoxia by (Kramer, 1983, 1987; Chapman et al., 1995; McNeil & Closs, 2007):

1 changing habitat and/or locality;
2 changing activity;
3 increasing use of aquatic surface respiration;
4 air breathing.

Faced with low oxygen levels, fish may migrate away to habitats that are oxic, though this may be a limited measure. When there are low oxygen levels, fish may increase their rate of breathing and their gill ventilation frequency and amplitude (Kramer, 1987). With hypoxia, fish may also breathe in the surface layer of water, which is replenished by atmospheric oxygen. This mode of breathing is called aquatic surface respiration (ASR) (Kramer, 1983). Some fish have morphological adaptations, such as upturned mouths and flattened heads, to carry out ASR effectively. It is worth noting that adaptations to hypoxia by fish are well developed in the Tropics, where high water temperatures, even under normal conditions, may greatly limit oxygen availability.

Fish living in floodplain lagoons exposed to oxygen stress may increase their gill ventilation rate (GVR) (Kramer, 1987) and/or resort to aquatic surface respiration (ASR) (Kramer, 1983). In experiments using fish from floodplain billabongs, McNeil & Closs (2007) found that, with increasing hypoxia, the nine species present increased their GVRs. As hypoxia strengthened, eight species then moved to aquatic surface respiration (ASR). The exception was the weatherloach (*Misgurnis anguillicaudatus*), which used air-gulping as its major form of respiration with hypoxia, and which was most tolerant of extended hypoxia. Three species – Australian smelt (*Retropinna semoni*), flat-headed galaxias (*Galaxias rostratus*) and redfin perch (*Perca fluviatilis*) – were the first to resort to ASR and were intolerant of hypoxic conditions. The exotic cyprinids (*Carassius auratus*) and common carp (*Cyprinus carpio*) were the last of the eight species to use ASR, and both of these fish tolerated severe hypoxia and even anoxia.

Thus, temperature and oxygen levels during drought may differentially allow fish species to survive and, with time change, the assemblage composition of fish in floodplain wetlands (McNeil & Closs, 2007), even allowing alien species to survive better than the native species.

Some fish that live in environments that regularly become hypoxic can breathe air. Air-breathing alone is rare (e.g. Australian and South American lungfish; Kemp, 1987; Mesquita-Saad *et al.*, 2002 respectively), and most of the fish capable of air breathing have bimodal breathing, taking up oxygen from both water and air (e.g. cichlids; Chapman *et al.*, 1995).

As systems dry out in drought, a few fish are capable of surviving by aestivating in cocoons. As the water levels drop with drought, species of African lungfish (*Protopterus annectens, P. aethiopticus, P. amphibicus*) build a burrow in the mud and line it with mucus to form a cocoon, in which they then undergo aestation with greatly reduced respiration (Johnels & Svensson, 1954; Delaney *et al.*, 1974; Greenwood, 1986; Lomholt, 1993). The other species of African lungfish (*P. dolloi*) also constructs a cocoon but, as respiration is not reduced, it has been said to undergo terrestrialization rather than aestivation (Perry *et al.*, 2008; Glass, 2008). The South American lungfish (*Lepidosiren paradoxa*), like the African lungfishes, is an obligatory air-breather and undergoes aestivation (Mesquita-Saad *et al.*, 2002; Da Silva *et al.*, 2008). The Australian lungfish (*Neoceratodus forsteri*) does not aestivate, but is capable of breathing air from the water surface (Kemp, 1987). Other fish, such as clariid catfish in Africa (Bruton, 1979) and the oriental weatherloach (*Misgurnus anguillicaudatus*) (Ip *et al.*, 2004) may aestivate in mud and wet soil.

Aestivating fish adjust their metabolism to a very low level, but they face the problem of dealing with ammonia, the normal nitrogenous waste product of fish, which is readily soluble in water. Fish such as the African lungfish can convert the ammonia to urea that is stored and subsequently excreted, whereas the oriental weatherloach, when exposed to air, can adjust the pH of their body surface to volatilize the ammonia (Ip *et al.*, 2004).

Rather than surviving drought as adults, fish may invest in desiccation-resistant propagules. This strategy is exemplified in fish in the family Cyprinodontidae, which thrive in temporary waters and have short life histories with a high level of reproduction. Faced with receding water levels with drought, many species of cyprinodontid fish mate and then lay diapausing desiccation eggs (Wourma, 1972). These eggs can hatch as soon as water is present, producing fish that mature rapidly and lay many eggs (Hrbek & Larson, 1999). Species which have this strategy occur in both South America and Africa, and the ability to produce diapausing eggs appears to have evolved independently (Hrbek & Larson, 1999).

Nevertheless, in spite of some fish having physiological adaptations to deal with drought in drying floodplain lagoons, many fish do die from physiological stress. Furthermore, as drought takes hold and reduces habitat availability and water volume, fish populations are compressed, heightening the pressure of intraspecific interactions such as competition, and interspecific interactions such as predation and competition.

One of the risks of using lagoons as refuges is that while the lagoons may persist, they may also harbour predators. Kobza et al. (2004) observed that small fish used deep pools (depth >1 m); although these pools persisted through drought, they harboured large-bodied predators, which preyed upon the small-bodied fish and depleted their numbers.

7.4   Drought, fish assemblages and floodplain rivers

As with the impacts of drought on fish in large river channels, studies on the impacts of drought on fish assemblages in floodplain water bodies are few. The Phongolo River in South Africa and Mozambique was subjected to severe drought during 1982–1984, which was accentuated by regulation from an upstream dam (Merron et al., 1993). At the peak of the drought in late 1983, only 3.7 per cent of the surface area of floodplain lakes held water (White et al., 1984) and, of 98 lakes on the floodplain, only seven retained water through the drought (White et al., 1984; Merron et al., 1985, 1993). In one lake and in some shrinking pools during the drought, salinity rose sharply due to seepage from mineralized marine sediments (White et al., 1984). Conditions in the lakes that persisted were very taxing, with high temperatures and turbidity (White et al., 1984).

During the drought, due to loss of habitat, declining food resources, increased predation and fishing pressures and declining water quality, population numbers dropped drastically (Merron et al., 1993). Before the drought in the period 1974–76, some 35 fish species were recorded in the Phongolo floodplain, and after the drought there were 30 species

(White *et al.*, 1984; Merron *et al.*, 1985, 1993). In one lake, Nhlanjane, before the drought there was a diverse fish fauna dominated by *Clarias gariepinus* and *Hydrocynus vittatus* (White *et al.*, 1984). During the drought, Nhlanjane become totally dominated by the cichlid *Oreochromis mossambicus*. The cichlid appeared to benefit during the drought by comparison with other species, in that before the drought it was estimated to comprise 22 per cent of total fish biomass, but during the drought it comprised 47 per cent of the biomass. After the drought, it dropped back to 18 per cent (Merron *et al.*, 1985, 1993). This fish is well adapted to deal with drought stresses in that it has a high salinity tolerance, it can switch its diet and eat a wide variety of food, and it can breed under drought conditions (White *et al.*, 1984).

After the drought, the dominant species in the floodplain lakes were the cyprinid *Labeo rosae* and the silver catfish *Schilbe intermedius*, with both *O. mossambicus* and *H. vittatus* declining (Merron *et al.*,1985, 1993). The dominance of *Labeo* and *Schilbe* may have been due to large numbers of these fish migrating upstream from refuges with the drought-breaking floods of the cyclone Demoina (Merron *et al.*, 1985). Drought on the Phongolo floodplain induced major changes in the fish assemblages, probably due to the harsh environmental conditions and lack of flooding to allow breeding of many species. While population numbers and biomass may have recovered, the compositions of the assemblages after the drought were quite different from those prevailing before the drought (Merron *et al.*, 1985).

Drought, by preventing flooding of the floodplain, prevents the breeding of fish that are obligatory floodplain breeders. Such a fish is the sábalo, *Prochilodus lineatus*, which supports a valuable fishery in Bolivia. The severe El Niño drought of 1990–1995 prevented flooding on the floodplains of the Pilcomayo River, stifling sábalo recruitment. This, combined with over-fishing, led to a collapse of the fishery (Smolders *et al.*, 2000; see Figure 7.2).

A three-year drought during 1994–1996 prevented flooding of floodplain lagoons of the São Francisco River, Brazil (Pompeu & Godinho, 2006). Fish were sampled from three lagoons, two of which substantially maintained their volumes and were well-oxygenated, and one which lost a large amount of its volume and became hypoxic. All three lagoons lost species, with the loss being greatest in the lagoon that lost much of its water (from 34 species to seven). Across the lagoons, the greatest loss of species occurred in the migratory fish (the 'white' and 'gray' groups) rather than in the sedentary species (the 'black' group). In the lagoon with hypoxia, four of the seven species that survived were air breathers (Pompeu & Godinho, 2006).

These two South American studies show the dramatic effects that the loss of floods due to drought can have on floodplain fish communities, with large losses in species, biomass and abundance. In relation to management, this loss may be partly or wholly generated by river regulation by dams, and it

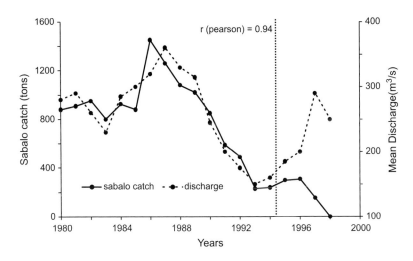

Figure 7.2 Catches of the fish sábalo (*Prochilodus lineatus*) (solid line), plotted with the mean three-year discharge (dashed line) of the Pilcomayo River of the La Plata basin, South America. A major drought coincidental with an El Niño event occurred from 1990 to 1995. (Redrawn from Figure 1 in Smolders *et al.*, 2000.)

indicates that river regulation can be a major force in the degradation of river ecosystems and their fisheries.

The Fly River is a large river in New Guinea, with a rich fish fauna totalling 128 species. It has an extensive floodplain area (45,000 km^2) with four major habitat types: blocked valley lakes, oxbow lakes, grassed floodplain and forested floodplain (Swales *et al.*, 1999). In 1993 to 1994, the Fly River region was exposed to a severe El Niño drought that resulted in low river flows and the drying of many floodplain habitats (Swales *et al.*, 1999). Fish biomass and species richness declined sharply with the drought in one oxbow lagoon and two blocked valley lakes. The lake beds were invaded by dense growths of terrestrial grasses. When the drought broke, the blocked valley lakes were joined with the main channel, but recolonization was much slower than expected. This delay was probably due to the accumulation of floating grass mats over lake surfaces, which resulted in low oxygen levels and low levels of food. The floating grass mats persisted for 18 months, producing unfavourable conditions for fish and delaying recovery from drought (Swales *et al.*, 1999). This is a clear example of how the lingering effects of terrestrialization of freshwater systems during drought can delay recovery.

The Sahelian drought that affects northern Africa commenced in the late 1950s and still continues (2010) (Dai *et al.*, 2004a; Held *et al.*, 2005); indeed, the drought may deepen through 'anthropogenic forcing' from increases in aerosol loading and from the increase in greenhouse gases (Held *et al.*, 2005). The drought has had a major effect in reducing river flows (e.g. Niger River) and lake levels (e.g. Lake Chad) in central and western Africa.

The Niger River and its tributaries support an intensive artisanal fishery and fishery data (although subject to some error) reveal a consistent drop in fishery yield with declines in river discharge (Welcomme, 1986; Laë, 1995). The fishery on the River Benue, a tributary of the Niger, starting in 1969–70 to 1985–1988, underwent a marked change in the major species present, from fish dependent on floodplain inundation for growth and breeding (e.g. *Labeo, Synodontis*) to generalist species that can grow and breed in the river channel (*Tilapia, Clarias*) (Neiland *et al.*, 1990).

With data from fish landings at Mopti in the Central Delta of the Niger River in Mali, Welcomme (1986) found that progressively, from 1966 to 1983, there was a decline in fish yield from 110,000 to 61,000 tons, and that this reduction correlated with the flood discharge (measured at Kouliokoro), that declined from 2,736 m^3/sec to 1514 in 1983. The drop in both magnitude and duration of the wet-season floods means that less floodplain area is flooded and that flooding duration is shorter. As found by Neiland *et al.* (1990), in the catch, the proportion of floodplain breeders (e.g. *Citharinus, Heterotis*) declined, while the proportion of generalist species tolerant of harsh conditions (e.g. *Tilapia, Clarias*) rose. Indeed, from 1969 to 1980, the proportion of *Tilapia* in the catch rose from 19.82 per cent to 44.09 per cent respectively (Welcomme, 1986).

With later fisheries data, Laë (1994, 1995) documented the decline in the fishery of the Central Delta due to the effects of the Sahelian drought, combined with river regulation by two hydroelectricity-generating dams. The area of delta floodplain inundated by wet season floods dropped from ≈20,000 km^2 in 1968 to 3,000 in 1984, and the total catch of the fishery dropped from 87,000 tons in 1969–70 to 37,000 tons in 1984–85 (Laë, 1994). However, fish catch per hectare of inundated floodplain rose from 40 kg to 120 kg. The reasons for this rise was suggested to be due to a switch from a fishery with a high proportion of large-bodied fish to one based on fast-growing, small-bodied fish, along with an actual increase in fishing pressure (Laë, 1995). Indeed, the number of fishermen rose from 43,000 in 1966 to 63,000 in 1989 – a function of population growth and diminishing food resources. The same switch in species composition as that found by Welcomme (1986) and Neiland *et al.* (1990) was obvious (Laë, 1995), from species reliant for growth and breeding on floodplain inundation, to species tolerant of harsh conditions such as hypoxia, and high salinities (Laë, 1995). The fishery is dominated by the tolerant species, such as clariid catfish and cichlids, and by small, fast-growing fish that breed in their first year, such as *Labeo senegalensis* (Laë, 1994, 1995).

The Sahelian drought is a megadrought. The combination of river regulation, by reducing the volume of the seasonal floods, and heavy fishing pressure, are clearly making dramatic changes to the floodplain fishery of the Niger River.

7.5   Summary

Floodplain systems and their biota have evolved with regular flooding, which generates avenues of hydrological connectivity between the river channel and the floodplain and between water bodies on the floodplain. This stimulates a boom of production – the flood pulse. Floodplain systems are a prime example of the dynamics of biodiversity and ecological processes involved in ecosystem expansion and contraction (Stanley *et al.*, 1997). The system expands with the flood pulse and then contracts, with some water lost on the floodplain (evapotranspiration, seepage) and more water receding from the floodplain into the river channel. The recession represents a major donation to the river channel of biota, detritus and nutrients, and the significance of the recession for the channel has not been studied.

Supra-seasonal droughts in floodplain rivers result from the failure of wet season floods. Not only is the flood pulse weakened or eliminated by drought, but so is the recession donation or subsidy. Also weakened is the subsidy from the floodplain to the surrounding hinterland. The strength and nature of these two subsidies, especially the recession subsidy, await elucidation.

Studies on the effects of drought on floodplain systems are relatively few and mainly come from tropical systems. No doubt this is because, in the developed temperate parts of the world, many rivers have had their floodplains developed and alienated from their rivers and have been regulated by dams designed to reduce floods, thus eliminating the flood pulse. By greatly reducing floods, humans maintain many floodplains throughout the world in a drought-like state, with abnormally long periods between diminished floods.

The elimination of floods stifles production on the floodplain and the recruitment of many biota, from macrophytes to fish. Supra-seasonal droughts have severe impacts, as they can greatly reduce the aquatic and riparian biota of the floodplain. The riparian vegetation, notably riparian trees, can be threatened as groundwater levels drop in drought. The tolerance of seedlings to drought appears to be a major factor governing the structure of riparian forests or woodlands (drought as an environmental filter). Drought changes the composition of phytoplankton, both in lagoons and in the river channel, with a shift to small-bodied cyanobacteria and flagellates during drought. Species richness and the production of macrophytes in lagoons both drop greatly in drought, but recovery after drought is marked by peaks in production and diversity, no doubt boosted by the release of nutrients from re-wetted sediments. Zooplankton species composition changes greatly in drought, with the small-bodied rotifers being favoured rather than the larger-bodied crustaceans. During recovery, as in temporary waters, rare species may be briefly favoured.

The phenomenon of brief booms of rare species in post-drought succession has been reported in temporary ponds, floodplain lagoons and streams, suggesting that there is a distinct niche for long-dormant, rapidly-growing species that thrive briefly in the opportunities created by severe drought. On the other hand, experiments have shown that the longer the drought, the more limited will be the hatching/germination responses of the floodplain biota. Recovery may be a drawn-out process as floodplain-dependent biota are not necessarily replaced by the biota that survived drought in the river channel. This particularly applies to the fish.

Supra-seasonal drought leads to a decline in fish species, abundance and biomass in floodplain water bodies, and major changes occur in assemblage composition, with 'white' and 'gray' species declining. This loss in floodplain fish gives rise to major losses of biomass in floodplain river channels and marked declines in floodplain fishery yields.

Relatively little is known about post-drought recovery in floodplain systems. From the few reports, it appears that recovery, provided that floods are restored, is quite variable. It is rapid for plankton, but drawn out for fish, and there may be a successional sequence of fish species. By alienating floodplains from their rivers and by building dams that act as barriers to migration, human activity is severely impeding recovery after supra-seasonal droughts.

7.6 Large wetland complexes with seasonal flooding

Large wetland complexes which are not necessarily connected to floodplains are subject to regular flooding, and they harbour a distinct biota adapted to the wet and dry seasons. The best documented example of such wetlands comes from the Florida Everglades. Over many years, this system has been the subject of intensive research, which has provided a good understanding of the effects of regular seasonal drying and flooding and has also provided a valuable insight into the effects of supra-seasonal drought.

7.6.1 The Florida Everglades

The Florida Everglades is large wetland complex that dominates the landscape of southern Florida. The system arises in the north, with the Kissimmee River, which drains into the large and shallow Lake Okeechobee.

Normally, the wet season occurs in summer, from May to September-October (Duever *et al.*, 1994).

In the original condition, with the wet season, the waters of Lake Okeechobee drained into the wetland plain of the Everglades and the Big Cypress Swamp (Light & Dineen, 1994). Originally, a large part of the wetland consisted of sawgrass (*Cladium jamaicense*) marsh and wet prairies (on peat or marl) (Gunderson, 1994). Being a system with a gentle gradient, water moved through the marsh system ('the rivers of grass') at a very slow speed, so that for a molecule of water to move from Lake Okeechobee to Florida Bay ($\approx$200 km) it took about eight months (Holling *et al.*, 1994). Through the marshes and prairies, there are sloughs – deeper channels, along with tree islands, alligator ponds and deep solution holes (Gunderson, 1994; Davis *et al.*, 1994).

Drainage and diversion works have now created a system of large water conservation areas that store and supply water to agriculture and urban areas (Light & Dineen, 1994; Sklar *et al.*, 2005). Consequently, the hydrology of the Everglades has been very significantly altered, and the drainage of the Everglades has been substantially diverted to either the Atlantic Ocean or the Gulf of Mexico instead of into Florida Bay (Light & Dineen, 1994; Sklar *et al.*, 2005). Normally, the water in the Everglades recedes in the winter and spring (November to May) and the wetlands may dry in parts to form a mosaic of ponds, holes and sloughs. In the wet season, the Everglades are inundated for varying lengths (hydroperiods), and water levels in the dry season are dependent on the amount of the of the previous winter precipitation.

Droughts can occur in the dry season and may extend into the wet season. It appears that droughts vary greatly in both spatial extent and locality in the Everglades complex (Duever *et al.*, 1994). In the severe drought from 1989 to 1990, 'all of the Everglades were dry for some period of time' (Trexler *et al.*, 2002). As well as greatly impacting on the aquatic life of the wetlands, droughts may also set the stage for fires, which may not just burn the vegetation and litter but may also burn the peat soils (Duever *et al.*, 1994).

7.6.2   Drought and crustaceans of the Everglades

Crayfish (*Procambarus alleni*) occur largely in wet prairies, whereas freshwater prawns (*Palaemonetes paludosus*) occur in both sloughs and wet prairies (Jordan *et al.*, 2000). When drought occurs, the crayfish survive by burrowing in the wet prairie rather than retreating to the sloughs, whilst prawn abundance in the wet prairies declines sharply (Kushlan & Kushlan, 1980). The prawns do not have adaptations to survive drought, and thus populations survive in habitats with water, such as deep sloughs, while the crayfish persist in burrows in the dry prairie habitat. When the drought breaks, prawn numbers need time to build up in order to return to normal wet season densities on the wet prairies (Kushlan & Kushlan, 1980; Jordan *et al.*, 2000). The survival of the crayfish in their burrows depends on the length of the normal hydroperiod of the wetland, and is very low in short hydroperiod (three months) wetlands, due possibly to the deeper ground-water depths in these wetlands (Acosta & Perry, 2001).

In the prairie wetlands studied by Acosta & Perry (2001), *Procambarus alleni* was the sole crayfish species, whereas in permanent wetland habitats such as sloughs, a second species, *P. fallax*, occurred (Hendrix & Loftus, 2000). In a nine-year study by Dorn & Trexler (2007) of the crayfish in two habitats – a shallow area bordering a slough and the deeper slough – *P. alleni* densities were higher in the shallow habitat and those of *P. fallax* were higher in the slough. In a period of droughts (1999–2004), densities of *P. alleni* increased at both the side of the slough and in the slough, while *P. fallax* disappeared from the side of the slough and persisted in the slough. However, *P. alleni* populations did not persist in the deep slough.

Experiments demonstrated that the burrowing *P. alleni* had a higher growth rate and a much greater tolerance of drought than *P. fallax*. In competition experiments with shelter as the limited resource, *P. alleni* outcompeted *P. fallax*. Thus, the distribution of *P. fallax* in the deep slough is explained by a lower drought tolerance, but the decline of *P. alleni* in the slough after the initial increase due to drought is more difficult to explain as it grows faster, has a larger body size and outcompetes *P. fallax*. Dorn & Trexler (2007) suggest that in the slough, with the long hydroperiods, fish popula-tions persist and that fish predation may be heavier on *P. alleni* than on *P. fallax*. This suggestion is given some support by observations by Dorn (2008) that *P. alleni* biomass was high in wetland ponds that were depleted of fish by drought.

Dorn (2008) sampled ten ponds before and after a drought in 2006 that dried out some ponds and rendered them fishless. In the fishless ponds, immediately after the drought, there was a higher biomass of invertebrates and especially of 'non-shrimp biomass' than in the ponds in which fish had survived the drought. The non-shrimp biomass was largely made up of predatory, highly mobile and invading insects (Odonata, Hemiptera and Coleoptera) and *P. alleni* crayfish. The non-shrimp biomass was negatively correlated with fish biomass.

In systems such as the Florida Everglades, drought may create wetland patches from which fish are eliminated and in which, after the drought, there is a window of opportunity for prey species to breed and recruit before fish return and the predation pressure again becomes high. The observations and experiments on invertebrates in Everglade wetlands strongly demonstrate the spatially explicit way that large-scale disturbances, such as drought, can produce major changes in community structure by selectively depleting abundant species and altering biotic interactions (notably predation).

7.6.3   Drought and fish of the Everglades

The fish of the Everglades have been divided into two groups: small-bodied and large-bodied (Loftus & Eklund, 1994; Trexler et al., 2002; Chick et al., 2004). The small-bodied fish (standard length <8 cm) are mainly poeciliids and cyprinodontids, and these dominate the fish fauna in terms of abundance and biomass, whereas the large-bodied fish group mainly consists of large piscivores such as centrarchids (e.g. *Micropterus salmoides*), bullhead catfish *(Ameiurus natalis)*,gar *(Lepidosteus platyrhincus)* and bowfin *(Amia calva)*. In normal times, the two groups of fish react differently to the fluctuations in water levels across the wetlands.

The small-bodied fish have one or two generations per year and do not migrate great distances. In the normal winter drawdown, small-bodied fish may move into refuges such as persistent pools (Kushlan, 1974a, 1980), alligator holes (Kushlan, 1974b), solution holes (Kobza et al., 2004) and sloughs (Jordan et al., 1998). Even in normal dry season drawdowns, the water quality of pools may deteriorate to such levels that fish kills occur (Kushlan, 1974a).

In a pool in the Big Cypress Swamp, Kushlan (1974a) observed a severe depletion of oxygen and an increase in carbon dioxide coincidentally with a bloom of green algae. This situation caused a severe fish kill, with only six of the 22 fish species originally present surviving and, only 0.6 per cent of the original population surviving. There was a differential gradient in mortality. All the centrarchids (e.g. *Micropterus salmoides*, *Lepomis macrochirus*, *L. punctatus*), golden shiner (*Notemigonus crysoleucas*) and bluefin killifish (*Lucania goodei*) died within four days of the onset of extreme conditions. A second group of fish, including the golden topminnow (*Fundulus chrysotus*), sailfin molly (*Poecilia latipinna*), least killifish (*Heterandria formosa*) and yellow bullhead (*Ictalurus natalis*) were drastically reduced in abundance within 7–8 days, and a final, semi-tolerant group consisting of Florida gar (*Lepisosteus platyrinchus*), mosquitofish (*Gambusia affinis*), American flagfish (*Jordanella floridae*) and the freshwater prawn (*Palaemonetes paludosus*) survived, though with heavy mortality (Kushlan, 1974a). Such a pattern of mortality illustrates the winnowing effect that both seasonal drying and the onset of drought may have on fish communities, and of how, at a larger spatial extent covering a number of pools in a wetland, high levels of community dissimilarity may be generated.

As water levels drop and the small-bodied fish become confined to pools, predation by birds may take its toll (Kushlan, 1976a, 1976b). Small fish that move into pools, hollows and shallow solution holes (depths less than 47 cm) run the risk that the holes may dry out, whereas those that move into deep, persistent pools and solution holes may be reduced by predation from large-bodied fish (Kobza *et al.*, 2004). As the small-bodied fish do not migrate great distances, and as fully inundated wetlands, with drying, fragment into a variety of water bodies of differing sizes and persistence, the variability of small fish assemblage structure operates at a local (i.e. 10 km) rather than regional scale (Trexler *et al.*, 2002). Density of small-bodied fish is correlated with the length of the hydroperiod (Loftus and Eklund, 1994), with the highest densities occurring in marshes with hydroperiods of 340–365 days per year (DeAngelis *et al.*, 1997; Trexler *et al.*, 2002).

However, a model of fish dynamics called ALFISH found that water depth was only a weak indicator of fish density, and that the availability of dry season refuges (permanent ponds) was critical to predicting fish density (Gaff *et al.*, 2004). Even if the population abundance that survives in the small refuges is a very low fraction (≈ 0.001) of the 'equilibrium fish population size', this can be sufficient to allow a rapid recovery once water re-inundates the wetland (DeAngelis *et al.*, 2010).

Large-bodied fish are mainly piscivores, and they can undertake long distance movements. As marshes dry up in drought, the large-bodied fish migrate to deep refuges such as alligator holes, solution holes and sloughs, though some fish may become trapped in the marshes and die (Trexler *et al.*, 2002; Chick *et al.*, 2004). It appears that large fish in the Everglades may be able to detect drying and migrate into deep-water refuges before escape becomes impossible (cf. Cucherousset *et al.*, 2007, Chapter 6). The variability in large fish abundance and in assemblage structure is largely at the regional level (25 to 87 km) (Chick *et al.*, 2004), rather than at the local level as for the small-bodied fish.

In increasing the severity and duration of the seasonal drying process, droughts may cause great losses of fish. The magnitude of the losses in density of small-bodied fish is correlated with the frequency of drought (or drought return time) (Trexler *et al.*, 2005). As recovery of small-bodied fish may take three to five years after a drought has ceased, frequent and relatively short droughts can limit the structure and density of the small fish assemblage (Loftus & Eklund, 1994; Trexler *et al.*, 2005). Large-bodied piscivores, on the other hand, are affected by regional factors, and thus their numbers decline with large-scale persistent drought and their recovery is slow (Kobza *et al.*, 2004; Chick *et al.*, 2004). As the piscivores recover slowly from drought, because small-bodied fish recover relatively rapidly, they may be only subject to low predation levels, at least until the large-bodied piscivores build up. Thus, drought, by reducing top consumers, may serve to shorten food-chain length and alter community structure (Williams & Trexler, 2006).

The small-bodied fish showed considerable variability in tolerance to drought and in speed of recovery (Trexler *et al.*, 2005; Ruetz *et al.*, 2005). As suggested in the response to very low water quality (Kushlan, 1974a), three species – bluefin killifish, least killifish, golden topminnow – were severely affected by drying and recovered slowly (three to four years) (Ruetz *et al.*, 2005). The very tolerant eastern mosquitofish showed no clear response to drying and recovered rapidly, whilst the American flagfish attained high densities immediately after the drought (Ruetz *et al.*, 2005).

Hydrological disturbance such as that due to drought may induce synchronous changes in populations of affected species – the Moran effect (Hudson & Cattadori, 1999). Synchrony of populations may also be created by widespread and active dispersal between populations. In assessing the Moran effect and the importance of dispersal, Ruetz *et al.* (2005) concluded that drying did synchronize the populations of four of the fish species, but not those of the American flagfish. Flagfish populations were synchronized by rapid and widespread dispersal among populations shortly after the drought ceased.

7.6.4   Summary

In large wetland complexes, fish, being dependent on the availability of free water, are particularly susceptible to the impacts of drought. In the Everglades, populations may become trapped in pools and die from water quality stress or from predation by birds. Different species have different tolerances to low water quality and different responses to predation pressure. In drought with pools vary in terms of water quality and predation pressures. Different pools may contain different species, thus generating an intricate mosaic of pools and local species composition.

Directional migration away from drying water bodies to more permanent ones is a common strategy in wetland systems, with different species appearing to use different cues to migrate. In the Florida Everglades, migration is a common strategy to avoid drought, with small-bodied fish migrating short distances and the large-bodied piscivores migrating considerable distances in search of deep water. Thus, the small-bodied fish in response to drought are subject to local (pool-specific) pressures, while the large-bodied fish are regulated at a regional level.

Recovery of fish populations from drought may be staged, in that different species with different dispersal and recruitment strategies recover at different rates. In the Everglades, the small-bodied fish recover much more rapidly than the large-bodied species; therefore, in recovery, small-bodied fish may have a window of opportunity to build high populations away from the predation pressure of large-bodied fish. With the changes in fish communities during drought and the different strategies of fish after drought, in wetland systems such as the Everglades, drought can create a dynamic mosaic of fish communities which may be spatially predictable.

7.7   Amphibious and terrestrial vertebrates

Wetland complexes, floodplain systems and riparian zones are all highly productive systems with a great variety of resources and habitat structure. It is thus not surprising to find that these areas attract and harbour high populations of a large number of vertebrates. However, reports on the effects that drought has on the amphibians, reptiles, wetland birds and mammals dwelling in aquatic ecosystems – especially wetlands – are few.

7.7.1   Amphibians

Amphibians require free water to breed, with many species using temporary water bodies rather than permanent systems, and many species being philopatric for breeding (i.e. tied to a particular site). Temporary systems may be free of predators such as fish, but they may also be unpredictable environments in terms of water availability. Drought can very dramatically reduce the availability of breeding sites at large spatial extents and for long periods, thus greatly reducing recruitment. This reduction may occur through the suppression of migration and breeding by adults due to no available water, or to the death of eggs and larvae as ponds dry up due to drought. Furthermore, adult survival during the drought may be threatened by the desiccation of the habitats in which they are dwelling. This hazard may operate in long supra-seasonal droughts.

In two five-year studies in Florida, Dodd (1993, 1995) investigated the populations of two amphibians, the striped newt, *Notophthalmus perstriatus* (1993) and the eastern narrow-mouthed toad, *Gastrophyrne caroliniensis* (1995). When the studies began in 1986–1988, the study pond held water for short periods; however, between 1988 and 1990, drought increased in intensity and the pond dried up. If water was present, newts migrated to the pond to breed, but this was only successful in 1987, with a few metamorphosed juveniles being produced (Dodd, 1993). Numbers of newts declined, but striped newts are long-lived (15 years for males, 12 years for females), and thus individuals in refuges outlasted the drought and successfully bred afterwards, replenishing the local population (Dodd, 1993).

Toad populations were studied in the same pond. Reproduction was only successful at the beginning of the study (1985–86), and the adult population steadily declined (Dodd, 1995). In contrast to newts, the toads are relatively mobile and may maintain populations in a variety of habitats; so, when the drought broke, mobile individuals that survived in persistent refuge habitats migrated and bred to replenish the population (Dodd, 1995). A variety of refuge habitats, such as forests, can thus aid survival during drought (Piha *et al.*, 2007).

These two studies at the same site show that adult survival is critical to weathering a drought, and that this may be accomplished in two different ways (long lifespans or migration from persistent refuges).

Drought may either prevent breeding pools from filling or greatly reduce the period in which pools hold water. In either case, for amphibian populations, especially in philopatric species, recruitment may be severely curtailed. In ponds in Zimbabwe, a short severe drought reduced the species breeding in them and the abundance of tadpoles by 50 per cent (Muteveri & Marshall, 2007). In studying a population of the mole salamander, *Ambystoma talpoideum*, in ponds in South Carolina, Semlitsch (1987) found that when severe drought in 1980–81 dried out ponds before metamorphosis, out of 33,019 eggs laid in a pond, only three metamorphosed juveniles emerged. However, in the following normal year, recruitment was successful. In 12 temporary wetlands in southern Florida, Babbitt and Tanner (2000) found that with a severe drought in 1993, the wetland system was completely dry and none of the 11 resident amphibian species bred. In the following years, the wetlands filled and breeding returned.

Drought can reduce the densities of predators such as odonatan nymphs, dytiscid and hydrophilid beetles and fish. Blair (1957) observed populations of cricket frogs (*Acris crepitans*) and bullfrogs (*Rana catesbiana*) in a pond in Texas before, during and one year after a drought that dried out the pond. The bullfrog population was 'apparently extirpated' by the drought, whilst the cricket frog population dropped from a pre-drought level of ≈ 310 to only 15–36 immediately after the drought. However, these survivors bred in the pond free of predators after the drought and, with a good food supply, they produced a breeding population of ≈ 600 by summer. The extirpation of the bullfrogs was thought to be due to a lack of suitable refuges, whereas a few cricket frogs could survive in cracks in the pond bottom. Survival of young after the drought was high and these animals could mature in 2–4 months, thus allowing a rapid recovery.

In a range of ponds, Werner *et al.* (2009) censused populations of two species of frog, the chorus frog (*Pseudacris triseriata*) and the spring peeper (*P. crucifer*) for a period of 11 years, during which a severe drought occurred. Drought in drying out ponds eliminated many predators, which gave rise to an increase in regional population size and the number of ponds colonized by the chorus frog, while the relatively high regional population size and number of ponds colonized by spring peepers remained fairly constant (Werner *et al.*, 2009). Drought did not change the colonization or extinction probabilities for the spring peepers, but it did increase the colonization probability and decrease the extinction probability of the chorus frog.

Long supra-seasonal droughts may have lasting effects on amphibian populations, at least in normally well-watered environments. Palis *et al.* (2006) studied a population of the flatwoods salamander, *Ambystoma cingulatum*, associated with a pond in Florida. The study period covered a drought (1999–2001) and the year after the drought (2002). Over the entire period, the numbers of immigrating adults declined, and no larvae or juveniles were produced. Palis *et al.* (2006) suggest that the reason for the decline is the attrition of the adult population due to no recruitment, and continuing mortality of the adult population, with the adult lifespan being no longer than four years. Thus, drought duration in comparison with lifespan may be a key factor governing population persistence. This applies not only to amphibial populations, but also to many other species populations where recovery is dependent on breeding by surviving adults.

In some systems, amphibians may recover from drought with great success. In an isolated wetland in South Carolina and after a prolonged drought (2000–2003), Gibbons *et al.* (2006) found, in the year following the drought, a remarkable recovery of a diverse amphibian assemblage with 24 species (17 anurans, seven salamanders), with one species, the southern leopard frog, *Rana sphenocephalus*, being particularly successful. The strong recovery was attributed to four factors: a relatively long lifespan in many species; a great reduction in predators; a high productivity of larval food; and the maintenance of forest cover (refuges) around the wetland (Gibbons *et al.*, 2005).

Management measures and drought may interact. In western North Carolina, ponds were monitored for both wood frog *(Rana sylvatica)* and spotted salamander (*Ambystoma maculatum*) (Petranka *et al.*, 2003). A drought, combined with an outbreak of the pathogen *Ranavirus*, greatly reduced larval survival and reduced the adult population of the wood frog. However, the salamander breeding population was unaffected – a function, possibly, of differences in lifespans (2–3 years for the frog, up to 32 years for the salamander) (Petranka *et al.*, 2003).

Land use changes may augment the effects of drought. Precision land-levelling creates a uniformly flat topography, thus eliminating undulations that may house breeding pools for amphibians. In Arkansas, such habitat change, combined with drought, has greatly reduced the abundance and distribution of the Illinois chorus frog (*Pseudacris steckeri*) (Trauth *et al.*, 2006). In Finland, the reduction in landscape habitats, both aquatic and terrestrial, by agricultural development, has served to heighten the effects of a severe drought on populations of the common frog, *Rana sylvatica* (Piha *et al.*, 2007).

Frogs dwell successfully in arid environments, with many species breeding shortly after rainfall events, having rapid larval growth rates and aestivating in the extended dry periods (Bentley, 1966). In adapting to contend with arid conditions, such frogs may be both resistant and resilient to drought, provided that they have long enough lifespans to survive the long periods between rainfall events. Burrowing and aestivating between rainfall events is widespread among desert anurans. Such burrowing frogs in Western Australia may build cocoons in their burrows (8 species) or burrow without cocoon-building (16 species) (Tracey *et al.*, 2007). With rain, large number of frogs may emerge, so much so that 'in the desert of Western Australia the number of frogs emerging after rain has been so vast as to interfere with the passage of trains, which are unable to maintain traction on rails made slippery by crushed frogs' (Bentley, 1966).

In many parts of the world, amphibian populations have declined, with climate change being regarded as being a contributing cause, along with disease (e.g. Chytridiomycosis) and increased ultraviolet-B radiation, habitat loss and pollution (Beebee & Griffiths, 2005).

A striking example of the effects of climate change expressed mainly as increases in drought comes from a study of amphibian populations in Yellowstone National Park in Wyoming (McMenamin *et al.*, 2008). This park rates as the one of the best and longest protected parks in the world. Over the past 60 years, temperature has risen, precipitation has decreased and 'drought has become more common and more severe than at any time in the past century' (McMenamin *et al.*, 2008). A wide survey carried out in 1992–93 was repeated in 2006–2008, and the later study has revealed that there has been a sharp increase in the number of permanently dry ponds. In the 'active ponds' that remain, both amphibian species richness and populations have declined very significantly (McMenamin *et al.*, 2008). Not only may drought alone diminish amphibian populations, but its effects may be synergistically heightened by human interventions such as habitat loss through urbanization and agriculture (e.g. Piha *et al.*, 2007).

7.7.2   *Reptiles and mammals*

Many reptiles dwell substantially in lakes and wetlands. Such animals include lizards, snakes, tortoises, turtles, crocodiles and alligators. Many mammals, such as otters, pigs, buffalo and hippopotamuses, also live in and around standing water bodies. Little has been published on the effects of drought on these animals.

In terms of dealing with drought, reptiles can either migrate to water bodies with water, or stay and become dormant (aestivate) in the drying water body or in terrestrial surrounds. The challenge in aestivation is the need to have sufficient body reserves to survive for the duration of the droughts.

In a wetland complex in south-eastern Australia, Roe and Georges (2007) observed that, as wetlands dried in drought, eastern long-necked tortoises (*Chelodina longicollis*) moved considerable distances to permanent wetlands. Inversely with flooding, there was migration back to the temporary wetlands. In southern Western Australia, the endangered western swamp tortoise (*Pseudemydura umbrina*), that dwelt in temporary winter-filled wetlands, is now reduced to only a few wetlands (Burbidge & Kuchling, 1994). In summer, with the wetlands dry, the tortoise aestivates, and in low rainfall years the females do not reproduce. As drying and droughts are increasing in this region, survival of this species is acutely threatened (Burbidge & Kuchling, 1994).

In many wetlands, several turtles may exist. In an Iowa pond, with drying due to drought, two of the common painted turtles (*Chrysemys picta*, *Chelydra serpentina*) moved to permanent water, while the yellow mud turtle (*Kinosternon flavescens*) stayed and aestivated (Christiansen & Bickham, 1989). Ellenton Bay is a shallow wetland complex in coastal South Carolina and it has been the focus of an extended research effort on amphibian (e.g. Gibbons *et al.*, 2006) and reptile populations. Gibbons *et al.* (1983) found that with drought drying out the wetland in 1980–1981, turtles showed three different strategies. Two species (*Pseudemys scripta* and *P. floridana*) emigrated and their reproduction in the following year was limited. Two other species (*Sternotherus odoratus*, *Deirochelys reticularia*) did not emigrate and had very limited reproduction after the drought, while one species, the mud turtle *Kinosternon subrubum*, stayed and its post-drought reproduction was unaffected.

Wetlands can have high levels of abundance and species richness of snakes that may be differentially affected by drought. At Ellenton Bay, the snake populations have been monitored in a long-term project. With a drought drying the wetland from 1988 to 1990, common species of aquatic snakes varied in response (Siegel *et al.*, 1995). *Nerodia fasciata* left mainly with the drying, while *Seminatrix pygaea* left later when drying had eliminated its prey, the fish *Gambusia affinis* and the salamander *Ambystoma talpoideum*. An uncommon third species, *Nerodia floridana*, left in small numbers and was not seen in the wetland for five years after the drought. Two years after the drought, both *N. fasciata* and *S. pygaea* had returned in low numbers to the wetland and, by 1995, *N. fasciata* in high numbers was the dominant species (Siegel *et al.*, 1995). The drought had greatly reduced the aquatic snake populations, altered the pattern of relative abundance, and caused a lag in recovery.

Ellenton Bay suffered another severe drought from 2000 to 2003 (Willson et al., 2006), with the snakes monitored shortly after the drought. Again, different drought-survival strategies were revealed. Cottonmouths (*Agkistrodon piscivorus*), which normally migrated to and from the wetland, were relatively unaffected by the drought and reproduced normally. *Nerodia fasciata* abundance declined greatly and this species did not reproduce during the drought, while, as before in the previous drought, *N. floridana* was eliminated (cf. Siegel et al., 1995).

Seminatrix pygaea survived the drought largely by aestivating in the wetland (Willson et al., 2006; Winne et al., 2006), even though it is a small snake requiring aquatic prey and with high rate of evaporative water loss. This snake is unusual in that it feeds during pregnancy that benefits the reproductive output – a strategy called 'income breeding', as opposed to the more normal 'capital breeding', where breeding needs a threshold condition (Winne et al., 2006). Hence, when the drought breaks and aquatic prey such as frogs and salamanders are available, the snake can breed and rapidly build up numbers (Winne et al., 2006).

The work at Ellenton Bay has shown that amphibians and reptiles have different strategies for dealing with drought. Key to these strategies is the presence in the wetland system of permanent water bodies, of intact terrestrial vegetation and the provision after the drought of suitable prey.

7.7.3   Waterbirds

Large numbers of bird species dwell in the habitats of wetlands, both permanent and temporary. Many bird species are restricted to standing waters, while other may only be temporary inhabitants. By partly or completely drying lentic systems, droughts can have major effects on bird populations. These can operate through loss of habitat, the loss of food resources and adverse physical-chemical conditions (e.g. high temperatures, low water quality). In addition, certain species may be threatened by the intensification of adverse intra- and interspecific interactions, such as competition and predation.

As water bodies dry and water levels drop, prey (especially fish) can become concentrated and attract large numbers of predatory birds, notably waders (e.g. herons) and divers (e.g. cormorants) (e.g. Kushlan, 1976b; David, 1994). However, in the Florida pond studied by Kushlan (1974a), drought triggered adverse water quality conditions, producing a fish kill. Fish concentrations that were normally available to waders with the coming of the dry season were absent, and wader numbers were low.

By lowering water levels, drought may deprive water birds of important foraging areas, especially littoral vegetated areas. Many waterbirds, especially ducks and coots, depend on aquatic vegetation for foraging and thus, with the drying of a wetland, they are forced to migrate. Dropping water levels also affect prey, especially sedentary species, which are consumed by water birds. Drought in Lake Balaton caused the mass mortality of littoral mussels and forced mussel-feeding waterbirds to feed elsewhere (Balogh et al., 2008) and a winter drought in Lake Constance depleted bivalve beds depriving over-wintering birds of a food resource (Werner & Rothhaupt, 2008).

In drought, water birds may move to refuges. The decline of a particular prey species due to drought can have damaging effects on their predators. In the Florida Everglades, the snail kite (*Rostrhamus sociabilis*) feeds almost exclusively on the apple snail (*Pomacea paludosa*). Drought either kills the snails or induces their aestivation, and thus reduces the food supply of the kite. The snail kites may not breed, may die or disperse to other wetlands (Bennetts et al., 1994). Indeed, the species is highly mobile, so dispersal to refuges is its major way of dealing with drought (Beissinger & Takegawa, 1983). However, with changes in water management in the Everglades, the drought-free refuges have been greatly reduced, threatening the persistence of the kite. Modelling suggests that the viability of snail kite populations is threatened if the interval between droughts becomes less than 4.3 years (Beissinger, 1995).

In a short drought (1996–97), breeding wood storks (*Mycteria americana*) moved their foraging for food from freshwater wetlands to estuarine wetlands. Breeding was less successful, perhaps because the storks could only forage in the estuarine wetlands at low tides (Gaines et al., 2000).

With drought, parental survival may not be greatly affected but breeding may fail. In an Alabama swamp, a short drought severely limited the breeding success of white ibis and cattle egrets (Dusi & Dusi, 1968). In an agricultural area with many farm dams in south-eastern Australia, breeding of maned geese was severely limited by a short but severe drought (Kingsford, 1989).

In drought-prone Australia, it is perhaps not surprising to find that many species of waterbirds are highly mobile (Roshier et al., 2001; Kingsford & Norman, 2002). Hence, when wetlands – even desert wetlands – are flooded and rapidly become productive, waterbirds are quick to arrive, feed and breed (Kingsford et al., 2010).

When drought strikes, waterbirds may migrate to distant productive wetlands and human-maintained wetlands such as reservoirs. However, human-regulated impoundments, with their low productivity, are far from ideal drought refuges (Kingsford et al., 2004). Recently, south-eastern

Australia was locked in a megadrought, but northern Australia, both inland and coastal, had heavy rains, thus providing many flooded wetlands that offered refuges for waterbirds at the continental scale. Coastal habitats, such as lagoons and estuaries also serve as drought refuges for inland waterbirds (Kingsford & Norman, 2002).

7.7.4 Summary

Amphibians, newts, salamanders, frogs and toads need water to breed, and thus a major effect of drought is to check or eliminate successful breeding. To survive drought, adults require refuges. In human-inhabited areas, refuges may be depleted. Key to surviving supra-seasonal droughts is lifespan, as long droughts may be longer than the lifespans of resident amphibians. Areas where amphibians do survive severe droughts may have long-lived species with adequate refuges, and productive water bodies free of predators when droughts break.

Reptiles associated with or dependent upon water bodies have two strategies to contend with drought. They may migrate away from the drying water body to persistent water bodies, or they may aestivate in or around the drying water body. Both strategies appear to impair reproductive effort after the drought has broken.

In the case of birds associated with water bodies, drought gives rise to loss of habitat, foraging areas and of food resources. Some species remain around drought-affected water bodies and breeding is eliminated or greatly reduced. Other species migrate to refuges, and these migrations may be very lengthy – across continents.

8

Drought and perennial waters: plants and invertebrates

Chapter 6 dealt with drought and temporary waters; in contrast, this chapter is concerned with the effects of drought on perennial water bodies. There are links between the two chapters because, in extreme droughts of long durations, the effects of drought may be very similar between perennial and temporary systems.

This chapter explores the effects of drought on biota with a focus on algae, vascular plants, and invertebrates. Unfortunately, due to a lack of research, the effects of drought on ecological processes in perennial systems are only fragmentarily known, though one can speculate on some of the outcomes. As there are numerous studies of drought and the fish of perennial waters (Lake *et al.*, 2008), this will be the subject of the next chapter.

Perennial waters are, by and large, those water bodies that persist through drought. They are buffered from the drying of drought either by having a large volume of water or by having persistent inflows from their catchments. In the case of lentic systems, persistence during drought is related to having a low surface area to volume ratio or having maintained inflows, whereas in lotic systems, persistence is maintained by having inflows which may be coming from groundwater sources. Large lakes are buffered from drying by their large volumes, and large rivers by the summing of their tributaries. In extreme droughts, perennial or permanent water bodies may dry out – the semi-permanent category of Chase & Knight (2003).

As described in Chapter 5, the morphology of water bodies exerts a strong influence on how and where the effects associated with the loss of water during drought are exerted. In shallow systems, the effects of the drawdown due to drought are more severe than those that occur in steep-sided systems – most likely a function of habitat loss. Loss of the surface area covered by water usually produces a loss of habitat for benthic taxa, as well

Drought and Aquatic Ecosystems: Effects and Responses, First Edition. P. Sam Lake.
© 2011 P. Sam Lake. Published 2011 by Blackwell Publishing Ltd.

as those animals that forage in these areas. Consistent abiotic effects in both lentic and lotic waters include increases in conductivity and temperature, changes in ion, nutrient, pH and DOC concentrations, changes in turbidity and decreases in oxygen concentrations (see Chapter 5).

Most droughts become evident in summer. In the case of supra-seasonal droughts, they continue because of a failure in winter or wet-season precipitation. Most studies have focused on the effects of drought in the summer or dry season time of the year rather than the effects of drought in winter, but one exception to this comes from the study of a winter drought in central Canada by McGowan *et al.* (2005). This study is also exceptional in being one of the very few experimental studies of drought on aquatic ecosystems. In winter droughts, with the water level dropping, the littoral zone may be subject to freezing and ice scouring, adversely affecting the shoreline biota.

There are marked differences between both lentic and lotic systems in terms of their biota, for example in the algae (phytoplankton vs. benthic algae) and in the invertebrates (zooplankton vs. mobile benthos). There are also similarities in the composition of the biota (e.g. macrophytes, benthos, fish). Similarities also arise in how biota contend with the stresses of drought.

Most studies of drought and aquatic biota have focused on particular biota; there have been very few studies in which the effects of drought have been investigated across the physico-chemical to the biotic realms in any one ecosystem. In terms of lentic systems, two long-term studies stand out, both in Africa, and they are focused on Lake Chilwa (Kalk *et al.*, 1979) and Lake Chad (Carmouze *et al.*, 1983; Dumont, 1992). In the case of Lake Chilwa, the study was complete in that the three phases of drought (pre-drought, during and post-drought) were covered, whereas in the case of Lake Chad the coverage was limited to pre-drought and early periods of an ongoing drought.

Similarly, there have been few studies of drought that covered two or more stream components. Studies looking at multiple components in lotic systems include those of Paloumpis (1957), Larimore *et al.* (1959), Cowx *et al.* (1984), Canton *et al.* (1984), Griswold B.L. *et al.* (1982), Adams & Warren (2005) and Power *et al.* (2008). In only two of these studies were links between the components described or commented upon. Larimore *et al.* (1959) commented that the post-drought recovery of the macroinvertebrates was sufficient to provide food for fish. In their complex long-term study, Power *et al.* (2008) described how drought affected algal production, grazer abundance and predators. Clearly, drought may alter trophic links between species and trophic levels, but this highly interesting area remains unexplored, with the exception of the illuminating study by Power *et al.*

To illustrate the ecosystem-wide effects of supra-seasonal drought on a lake ecosystem, the example of Lake Chilwa is more informative than that of Lake Chad, as in this lake, the drought continued without recovery being studied.

8.1   Drought and lentic systems

8.1.1   Drought in Lake Chilwa

Lake Chilwa in Malawi is a shallow, mildly saline, endorheic lake, which in normal years occupies a surface area of about 2,000 km^2 when full at the end of the wet season. There is evidence that the lake reached very low levels in 1879–1880, 1900–1901, 1913–1916, 1920–1922, 1960–1961, 1966–68 and 1973–74 (Kalk, 1979a; Lancaster, 1979). The extreme drought of 1966–68, when the lake dried out completely is comparable with the drought of 1913–1916 (Lancaster, 1979; see Figure 8.1.)

Even in a 'normal' year, conductivity of the lake varies considerably from $\approx$1,000 to 2,500 μScm^{-1} (McLachlan, 1979a). When drying occurred in the 1967–68 drought, the lake became increasingly turbid, very salty (conductivity $\approx$14,000 μScm^{-1}) and very alkaline (pH $\approx$ 10.8), with the cations dominated by carbonate and bicarbonate rather than chloride. Calcium and magnesium concentrations fell due to precipitation, and the receding lake was surrounded by 'a belt of soft deep mud' with salt crystals on the surface, which was 'virtually impenetrable' (Figure 8.2) and which, as it dried out, produced 'a noxious odour' (McLachlan, 1979a).

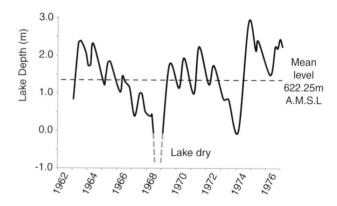

Figure 8.1 The levels of Lake Chilwa from 1962 to 1976 on a gently sloping shore at Kachulu. (Redrawn from Figure 3.1 in Kalk *et al.*, 1979.)

Figure 8.2 The shore of Lake Chilwa as the lake dried, showing the cracked, salt-encrusted surface of the mud and a retracting waterway used to gain access to the body of the lake. (From a slide kindly provided by A.J. McLachlan.) (See the colour version of this figure in Plate 8.2.)

When full, the margins of Lake Chilwa were lined with *Typha domingensis*, but with the severe 1966–68 drought the *Typha* disappeared, leaving a large area of exposed mud that was so 'unfavourable' that 'only three (plant) species survived' (Howard-Williams, 1979) – a grass (*Diplachne fusca*), a sedge (*Cyperus laevigatus*) and a shrub (*Aeschynomene pfundii*).

With the drying, the lake became dominated by planktonic, filamentous cyanobacteria (*Arthrospira*, *Spirulina* and *Anabaenopsis*), with *Arthrospira* being most abundant (Moss, 1979). Under these conditions with high water temperatures, oxygen levels fluctuated greatly from being supersaturated by day, to very low at night (McLachlan, 1979a). drought in eutrophic and hypertrophic lakes, especially in the Tropics (such as Lake Chilwa and Lake Chad), readily appears to induce dense blooms of cyanobacteria.

In normal times, Lake Chilwa supports a dense population of zooplankton, dominated in the hotter months by the cladoceran *Diaphanosoma excisum* and the copepod *Tropodiaptomus kraepelini*. In the cooler months, *Daphnia barbata* dominates at the expense of *Diaphanosoma* (Kalk, 1979b). In 1966, as drought set in, no nauplii or young cladocerans or rotifers were to be found, and only declining adult populations existed. It appears that signals from the drying – such as decline in depth and rises in conductivity (up to 6,000 µS cm^{-1}), chlorinity, alkalinity and pH – all served to induce the production of resting eggs. This step is a clear example of seeking refuges as drying sets in. Shortly before drying at the peak of the drought in late 1968, conductivity was at lethal levels and anoxia occurred, reducing the zooplankton to a few individuals.

With the loss of the littoral vegetation, the density of the swamp-dwelling and hitherto very abundant chironomid *Nilodorum brevipalpis* collapsed (McLachlan, 1979b). The decline in species richness of the benthos was closely linked with the rate at which the water levels dropped over three years (Figure 8.3). The offshore substrate became immense areas of inhospitable mud with the remaining water reaching very high conductivities. As the water levels dropped further, larvae of the midge *Nilodorum brevibucca* disappeared, to be followed by the corixid *Micronecta scutellaris*. Strandlines of dead corixids up to 30 cm high formed along the windward shores (McLachlan, 1979b). This was followed by the mass death of the snail *Lanistes ovum*, the white shells of which littered the dry shores of the lake (McLachlan, 1979b). Near the end of the drying, in 1968, only one species, the larvae of the hydrophilid beetle *Berosus vitticollis*, persisted. As the lake retreated, a vast area of mud was exposed, until finally the lake became completely dry (McLachlan, 1979b).

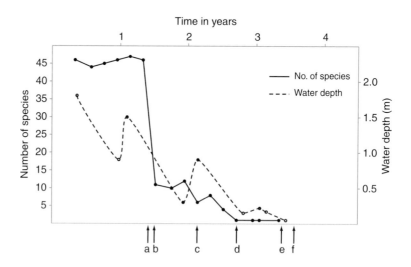

Figure 8.3 The declines in the depth of the lake and in species richness of the macroinvertebrates. **a** and **b** mark the losses of the swamps and of the chironomid *Nilodorum brevipalpis*; **c** the loss of live snails (*Lanistes ovum*); **d** loss of the chironomid *Nilodorum brevibucca*; **e** the mass death of the corixid *Micronecta scutellaris*; **f** the lake is dry. (Redrawn from Figure 9.4 in Kalk *et al.*, 1979.)

The lake normally harbours 30 species of fish, including 12 species of the cyprinid genus *Barbus*. Until the drought, the lake supported a productive fishery based on clariid catfish, *Barbus* spp. and *Sarotheradon shiranu* (Furse *et al.*, 1979). With the drought developing and lake volumes declining, the first fish deaths involving the cichlid *Sarotheradon* were generated by strong winds stirring up the fine sediments in bottom mud, such that concentrations of $12\,g\,l^{-1}$ were reached – a level deemed lethal to *Sarotheradon*. In addition, the fine organic-rich sediments stirred by high winds appeared to have caused oxygen levels to drop (Furse *et al.*, 1979). With the fine suspended sediments, high water temperatures and oxygen stress, large numbers of *Barbus* subsequently died. As lake volume declined and nutrients increasing, blooms of cyanobacteria occurred, with consequent deoxygenation, causing further stress for fish and particularly for the remaining *Sarotheradon* and *Barbus*. As the lake dried up and fish populations crashed, vertebrates such as birds, reptiles and mammals (e.g. otters) left or attempted to leave the lake (Figure 8.4).

Perennial streams flowing into the lake served as refuges for fish migrating out of the lake (Furse *et al.*, 1979). This escape was especially used by fish that normally dwelt in shoreline swamps rather than in open water. As the lake became dry, the only fish surviving were clariid catfish, which sought refuge by burrowing into the mud of the lake and undergoing aestivation.

Figure 8.4 An exhausted clawless otter, *Aonyx capensis*, attempting to move across the drying lake, is intercepted and is about to be killed and subsequently eaten. (Slide kindly provided by A.J. McLachlan and which appears as Figure 4.4(a) in Kalk *et al.*, 1979.) (See the colour version of this figure in Plate 8.4.)

When water returned in 1968 and 1969, the lake became very turbid, with low phytoplankton densities. On the wet mud there was a surface film of green algae, to be followed in 1971 and 1972 by high densities of the cyanobacteria *Anabaena* and *Anabaenopsis*. This stage in the post-filling phase then gave way to a diverse community of diatoms, chlorophytes, euglenophytes and cyanobacteria (Moss, 1979), indicating that lake levels and water quality exerted a very strong influence on phytoplankton community structure, and that its recovery from drought was lengthy and did not follow the pathway that drought produced in its impact.

With the filling of the lake, initially there was a scanty zooplankton. However, as time progressed (late 1968), there was a great increase in the density and diversity of rotifers – presumably a function of the rapid hatching of their resting eggs, their short generation time and the fact that adults of some rotifer species can survive drying (Kalk, 1979b). Again, as in other cases (e.g. Crome & Carpenter, 1988; Chapter 7), it appears that when harsh conditions deplete microcrustacean populations, rotifers briefly flourish, free from competition and predation (by cyclopoids). However, by the end of the year after the drought had broken, the rotifers were great depleted, presumably by predation from the cyclopoid *Mesocyclops*. At this time, *Moina* – a small cladoceran – became dominant, to be itself replaced by early 1970 by *Diaphanosoma* and *Daphnia*.

In the first year after the drought, zooplankton abundance was high but, in subsequent years, abundance declined to be well below 'normal' levels. This checking of zooplankton population growth, in a way similar to what happened post-drought in Lake Okeechobee (Havens *et al.*, 2007), may have been due to fish predation. With the breaking of the drought, the catfish *Clarias* bred after surviving the drought, in refuges in isolated lagoons and inflowing streams, and by aestivating in mud. *Clarias* juveniles are planktivores, and a high density of juvenile fish was quickly reached. In the second year after the drought, a second fish species, *Barbus palidinosus*, bred and also produced large numbers of planktivorous juveniles (Kalk, 1979b). This successful breeding by resilient fish survivors which produced profuse offspring, themselves released from predation pressure, could have checked the return of the zooplankton to 'normal' levels of abundance and diversity.

With the lake bottom and the marginal swamps being flooded in 1969, snails – particularly *Lanistes ovum* – appeared in abundance, along with dense populations of the midge *Chironomus transvaalensis*. These larval midges were confined to a narrow belt of water on the outer edge of the rising water, and McLachlan (1979b) suggests that while this animal can tolerate relatively high salinities, it appears to be intolerant of the combination of high salinity and fine silt such as occurred with the re-filling. After the re-filling reached the marginal swamps with an abundance of dead vegetation, *C. transvaalensis* was replaced by *Nilodorum brevibucca*. As the detritus decomposed and was consumed by the chironomids, the benthic fauna biomass was also reduced. The recovery of the marginal vegetation (e.g. *Typha*) took years (Howard-Williams, 1979).

It appears that the snails recovered as mature animals which had survived the drought by aestivating in the mud of the lake bottom. On the other hand, the recovery of the insect fauna appears to be due to recolonization from small number of survivors in other water bodies, such as springs and nearby rivers within 'a hundred kilometre radius of Chilwa and some as close as 20 kilometres from the south shore' (McLachlan, 1979b). Overall, water levels returned to pre-drought levels in one year; recovery to pre-drought levels of species richness of the benthos was closely linked to the rise in water level, and thus was relatively rapid. (Figure 8.5). As McLachlan (1979b) summarizes: 'much of the littoral benthos is . . . obliterated at intervals of a few years and would have to recolonize the swamps afresh from other bodies of water each time this happens'.

With the return of water to the lake, the clariid catfish were among the first species to repopulate the lake, to be followed two years later by *Barbus*. This was followed by *Sarotheradon*, which took three years to build their

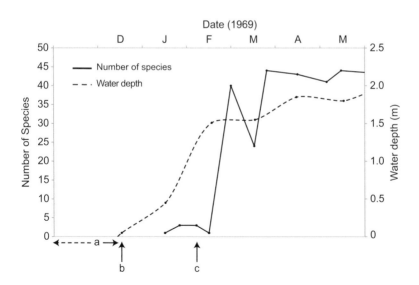

Figure 8.5 Depth of Lake Chilwa as re-filling of the lake occurs and recovery of the macroinvertebrate fauna indicated by increases in species richness. **a**: the dry lake period; **b**: when refilling starts; **c**: when marginal swamps were flooded. (Redrawn from Figure 9.5 in Kalk *et al.*, 1979.)

populations up to pre-drought levels. Thus, the succession of dominant fish was the inverse of the sequence of their demise in the drought.

In summary, drought created very taxing conditions (high salinities, high temperatures, low oxygen) for the aquatic biota. The shoreline became bereft of littoral vegetation, high densities of fine sediment occurred in the water column and vast areas of inhospitable mud were exposed. The phytoplankton became dominated by cyanobacteria, which further lowered the water quality and no doubt contributed to the loss of the crustacean zooplankton. The benthos of the lake (chironomids, corixids, snails), along with the fish, were drastically reduced by the drought, though some taxa (e.g. snails, clariid catfish) persisted in refuges in the lake.

Recovery of the phytoplankton was marked by blooms of cyanobacteria. This was followed some years later by a more 'normal' phytoplankton with diatoms, chlorophytes and euglenophytes. The zooplankton was initially dominated by rotifers, which gave way to a microcrustacean zooplankton; both of these groups persisted as resistant propagules in the lake sediments. This fauna may subsequently have been depleted by heavy predation from larval fish.

Snails which aestivated through the drought and a hitherto uncommon chironomid dominated the early recovery of the benthos. In turn, with the decline in available detritus, the chironomid declined, to be replaced by the original dominant species. Recovery of the benthos was rapid, with many species migrating in from surrounding waterways that had persisted. The fish recovered relatively slowly, with the aestivating clariid catfish breeding shortly after the water returned. It took 2–3 years for the populations of cichlids to return.

The comparability of the effects of drought in Lake Chilwa with those created by drought in other lakes is limited, due to the lake being moderately saline when full and to the fact that it dried up completely. However, the Lake Chilwa case study does provide a clear account of the major changes in water quality and habitat availability that induce changes in the biota. The drought produced a loss of biota and some steps in this loss, especially the decline in water quality, briefly favoured particular biota (e.g. cyanobacteria, rotifers). Similarly, after the drought, different groups recovered at different rates, with some taxa (e.g. cyanobacteria, rotifers) also being briefly favoured. For the littoral macrophytes and the fish, but not the macrobenthos, the recovery from drought had a distinct lag.

8.1.2   Drought in Lake Chad

Lake Chad is a shallow endorheic lake in central Africa, just south of the Sahel zone and surrounded by Nigeria, Niger, Chad and Cameroon. It is fed by rivers from the south and south-west, with the Chari River providing the major input. Being shallow, it can fluctuate greatly in water surface area. Its water surface area was $25,000\,km^2$ in 1964, but dropped to $10,315\,km^2$ in 1974 and declined further to occupy only $1,350\,km^2$ in 2000 due to both an ongoing megadrought and excessive water extraction (Carmouze & Lemoalle, 1983; Compère & Iltis, 1983; Coe & Foley, 2001; Pearce, 2007).

The lake consists of two basins; northern and southern. Both basins, when full, have extensive archipelagos of flooded dunes, with their crests forming islands, reed islands and open water. The lake was studied in a major research effort from 1964 to 1978 (Carmouze *et al.*, 1983).

In 1973, drought set in and the lake started to contract (Carmouze & Lemoalle, 1983), with the basins separating from each other (see Figure 5.3). The period before 1973 became known as the 'Normal Chad' and the drought period as the 'Lesser Chad'. During this latter period, the northern basin dried, while the southern basin retained water, albeit with a great drop in area (Carmouze *et al.*, 1983). In the north basin, as drying set in, turbidity rose sharply and oxygen concentrations varied greatly, 'with frequent periods of anoxia' (Carmouze *et al.*, 1983). Salinity rose greatly from $1,000\,mg\,l^{-1}$ to $3,000\,mg\,l^{-1}$. The drought converted both basins into eutrophic systems (Compère & Iltis, 1983).

Unfortunately, in the case of Lake Chad, the drought continued after the study finished in 1978, and continues to this day (e.g. UN News Service, 15 October 2009). Thus, we have initial results for the drought-induced decline of the lake ecosystem, but no data on the full effects of the drought. There has been no recovery, and the lake could be set for an 'ecological catastrophe' while its people could suffer a 'humanitarian disaster' (UN News Service, 15 October 2009). Indeed, to offset this disaster, there are plans to divert water from the Ubangi River a tributary of the Congo River in the Central African Republic, to Lake Chad (Pearce, 2007).

Different biota of an aquatic ecosystem react in various ways to the imposition of drought. In the next sections, we examine the various differences and similarities in the way that biota react to and recover from supra-seasonal drought.

8.2 Phytoplankton in lakes

As water levels and volumes of water bodies drop, and physico-chemical conditions in the water column change, phytoplankton and levels of primary production change. As detailed above, during drought, conductivity (salinity), pH and nutrient levels may all change, with nutrient levels and ratios, in particular, strongly affecting the phytoplankton.

In both Lake Chad and Lake Chilwa, drought promoted eutrophication and produced blooms of cyanobacteria. In Lake Chad, with the loss of water from the two basins, algal biomass increased greatly, with the northern basin acquiring the 'characteristics of a eutrophic pond' before drying, and the southern basin moving from being mesotrophic to eutrophic (Compère & Iltis, 1983). It was estimated that in 1971, the lake occupied 18,135 km^2 and that the algal biomass was 40,800 metric tons, but by 1975, in the drought, the area was 11,315 km^2 and the algal biomass was a staggering 244,135 metric tons (Compère & Iltis, 1983).

Similarly, in eutrophic reservoirs in the 'Drought Polygon' of north-east tropical Brazil, drought increased conductivity, alkalinity and pH and induced a sharp stratification in oxygen concentrations (Bouvy *et al.*, 1999, 2003). In the severe 1998 drought, chlorophyll-*a* concentrations steadily rose, with the phytoplankton dominated by the cyanobacterium *Cylindrospermopsis* (Bouvy *et al.*, 1999, 2000, 2003), which produces potent toxins that can affect other biota and greatly alter food web structure. In the many reservoirs with *Cylindrospermopsis* blooms, the cyanobacteria were not affected by grazing pressure from microcrustaceans and rotifers (Bouvy *et al.*, 2000). However a two-year study in one drought-affected reservoir did show that, during and after a *Cylindrospermopsis* bloom, zooplankton (rotifers, copepods) densities rose and rotifers and copepods were both consuming and removing the filaments of the cyanobacteria, such that they could be consumed by other zooplankton species (Bouvy *et al.*, 2001).

In a subtropical hypertrophic lake, Hartbeespoort Dam in South Africa, a severe drought from 1982 to 1987 greatly reduced the lake volume from 100 per cent to 21 per cent (Zohary *et al.*, 1996), but only slightly changed the high nutrient levels. These conditions gave rise to massive blooms of the cyanobacterium *Microcystis aeruginosa*, which dominated the phytoplankton for most of each of the five drought years, except for a spell each spring when a mixture of algal species occurred.

With the return of normal seasonal rains, the lake refilled in 1987 and phosphorus and nitrate concentrations declined. However, with phosphorus concentrations declining more rapidly than nitrate concentrations, there was an increase in the TN/TP ratio (by weight). The decline in phosphorus and the increase in the TN/TP ratio were held to be a reason for the decline in *Microcystis* densities (Zohary *et al.*, 1996), such that the blooms had disappeared by 1988, two years after the drought broke. As *Microcystis* declined, the chlorophyte *Oocystis* and the diatom *Cyclotella* became important species in the phytoplankton. Furthermore, in this time, new genera of algae (e.g. the chlorophytes *Crucigenia*, *Ankyra*, *Kirchneriella* and *Golenkinia*) were detected (Zohary *et al.*, 1996).

As in Brazilian and African eutrophic lakes, drought in the South African system gave rise to cyanobacteria blooms, which totally changed phytoplankton production and drastically altered the trophic structure of the lakes. Furthermore, with the breaking of the drought and changes in both competition and predation levels, new and rare algal species had a chance to flourish briefly.

In eutrophic lakes in temperate and Mediterranean climate zones, with a decrease in volume due to drought, nutrient concentrations, notably phosphorus, may rise and phytoplankton production dominated by cyanobacteria may reach high levels that are potentially harmful (e.g. Nõges & Nõges, 1999; Nõges et al., 2003; Beklioglu & Tan, 2008). Lake Võrtsjärv, in Estonia, a large, shallow and eutrophic lake- was exposed to a severe drought in 1995–96 that lowered the lake level by 50 per cent and greatly increased turbidity (Nõges & Nõges, 1999). Phosphorus concentrations increased and phytoplankton increased dramatically, being largely made up of cyanobacteria, with the nitrogen-fixing *Aphanizomenon skujae* and *Planktolyngbya limnetica* being key species (Nõges & Nõges, 1999). In this lake, the filamentous cyanobacteria *Limnothrix redeki* and *L. planktonica* normally dominate but, with low water levels and consequential higher light intensities and low N/P ratios, *Aphanizomenon* and *Planktolyngbya* became dominant (Nõges et al., 2003). In Lake Emir, Turkey, drought in 2001–2002 increased concentrations of both nitrogen and phosphorus, which led to cyanobacteria blooms that depleted oxygen levels, especially in the hypolimnion (Beklioglu & Tan, 2008), and produced fish kills.

As opposed to eutrophic systems, the ecological responses to drought in oligotrophic systems seem to be quite different. In the 1980s, a period of droughts occurred on the Boreal Shield of Ontario, Canada, specifically in the region of the Experimental Lakes Area (Schindler et al., 1990; Findlay et al., 2001). The lakes are dimictic (stratifying twice a year) and oligotrophic. Their responses to drought, both abiotic and biotic, were temporally coherent (viz. Magnuson et al., 2004). With drought, the length of the ice-free season, the depth of the thermocline, Secchi depth and water residence time increased as precipitation and direct runoff declined (Schindler et al., 1990). With the decline in direct runoff, the DOC concentrations in the lakes declined. As DOC declined, so did light attenuation, resulting in an increase in the depth and volume of the euphotic zone (Findlay et al., 2001). Concentrations of nitrogen and phosphorus were low and did not change significantly with drought. However, both biomass and species richness of the phytoplankton increased during drought.

There were major changes in the phytoplankton assemblage structure, with dinoflagellates (*Peridinium* and *Gymnodinium*) and some chrysophytes (*Dinobryon* and *Chrysochromulina*) (golden algae) increasing greatly in abundance. These groups contain flagellated species that can be autotrophic as well as heterotrophic (i.e. mixotrophic) – capable of consuming bacteria. With the deepening of the euphotic zone, there was an increase in mixotrophs compared with autotrophs, with mixotrophs comprising 20–40 per cent of algal biomass before drought and mixotrophs increasing to 50–60 per cent during drought. This increase was maintained after the drought (Findlay *et al.*, 2001).

It appears that the mixotrophs may migrate into deeper nutrient-richer levels of the lakes, consume bacteria and return to the euphotic zone, enriching this zone with phosphorus by recycling. With access to nutrients through mixotrophy and migration, the mixotrophs may out-compete autotrophic algae in the drought-affected lakes, and this superiority is maintained for some time after the drought (Findlay *et al.*, 2001).

Similar changes in algae were found in an oligotrophic mountain lake in southern Spain (Villar-Argaiz *et al.*, 2002) with a severe drought in 1995. This drought reduced the depth of the lake from a maximum depth of ≈ 14 m to a minimum of 1.5 m and greatly decreased the volume, which facilitated the remobilization of nutrients, especially phosphorus, from the bottom sediments. Nutrient remobilization, along with the greatly reduced volume and inputs of nutrient-rich dust from Africa, served to increase phosphorus concentrations in the lake (Villar-Argaiz *et al.*, 2001). In the drought, phytoplankton diversity and species richness increased, with both ciliates and mixotrophic flagellates becoming common. The zooplankton community was marked by an increase in rotifer abundance and a decrease in calanoid copepods (Villar-Argaiz *et al.*, 2002). Drought thus induced major changes in the grazing food chain of the lake. The drought abruptly ceased with heavy rains in 1996, followed by a big increase in phytoplankton and ciliate biomass and a marked decline in zooplankton biomass. In the next year, the ciliates virtually disappeared and the crustacean-dominated zooplankton was restored to pre-drought levels (Villar-Argaiz *et al.*, 2002).

Further evidence of major changes in phytoplankton communities in oligotrophic lakes comes from an alpine lake, Green Lake in Colorado (Flanagan *et al.*, 2009). The region around the lake suffered a supra-seasonal drought from 1998 to 2002 and sampling of the lake occurred from 2000 to 2005, during and after the drought. The drought produced moderate increases in summer water temperature, conductivity, acid-neutralizing capacity and in certain ions (e.g. calcium, potassium) (Flanagan *et al.*, 2009). Phytoplankton densities (as biovolume) were low, and dominating the phytoplankton were two hitherto rare species, a chlorophyte (*Ankyra* sp.) and a diatom (*Synedra* sp.). The diatom appeared to reduce silica concentrations significantly. These two species almost disappeared after the drought.

Equally remarkable was the marked increases in phytoplankton densities after the drought (2003) – densities which were maintained for the next two years. In 2003, Chrysophyta (*Chrysococcus* sp.) dominated, giving way in 2005 to dominance by the chlorophyte *Chlamydomonas* spp. (Flanagan *et al.*, 2009). Interestingly, the high densities and dominant species in the post-drought algal community were ascribed to nitrate concentrations augmented by deposition of atmospheric nitrate deposition in the catchment (Flanagan *et al.*, 2009). As noted in Chapter 5, high concentrations of nitrate may enter streams and lakes after lengthy periods of drought. Further, as the atmospheric nitrogen deposition is created largely by human activities, drought as a disturbance may exacerbate the effects of atmospheric contamination.

In some cases, such as in dystrophic lakes with high DOC and low nutrient concentrations, phytoplankton growth may not change. James (1991), for example, found that in dystrophic lakes in Florida, phytoplankton levels remained low during a drought and were perhaps inhibited by the high levels of dissolved organic carbon.

Drought, besides greatly changing phytoplankton community structure, may also greatly alter the spatial distribution of phytoplankton communities and their functions. When drought lowered the water levels of Lake Okeechobee, there was an increase in the spatial variation of phosphorus and chlorophyll-*a* (Maceina, 1993; Phlips *et al.*, 1997). When the lake was at its normal high level, parameters of photosynthesis such as α^B (light-limited rate of photosynthesis), $P_m^{~B}$ (maximum photosynthetic rate at optimal irradiance) and E_k (the light intensity at the onset of light-saturated photosynthesis) were uniformly spread across the lake. However, during a drought, these parameters differed significantly within the lake, with α^B and $P_m^{~B}$ being significantly influenced by the variation in depths (Maki *et al.*, 2004). Thus, as water levels fall, drought in standing waters may increase spatial heterogeneity in both biota and ecological processes.

Overall, there is clear evidence that the changes in phytoplankton assemblages in lakes and wetlands depends very strongly on the pre-drought trophic state – ecological memory. If the lake is even mildly eutrophic, drought can readily push the phytoplankton assemblage from one of mixed diversity into one dominated by a small number of cyanobacteria species. The extent to which the lake volume is decreased appears to regulate the chemical composition of the lake, especially in terms of nutrient availability. The changes in nutrients may be a function of concentration due to loss of volume and to the decrease in depth, allowing mixing to remobilize nutrients from bottom sediments. Dominance of phytoplankton by cyanobacteria, as expected, can cause major changes in the limnetic consumers and the trophic structure. Selected species of zooplankton can consume the cyanobacteria, and their densities may increase as a result (Bouvy *et al.*, 2001; Wilson *et al.*, 2006).

In oligotrophic systems, drought is unlikely to push nutrients to levels that induce eutrophication. Changes in nutrient availability can occur that induce changes in algal trophic status (autotrophy to mixotrophy), which in turn change the trophic structure of the system. In oligotrophic systems, relatively small changes in water quality may favour some species – even hitherto rare species – which can briefly flourish during the drought. In shallow systems, with the decline in volume with drought, uniform whole-lake conditions may be changed to a state of high spatial heterogeneity. In general, recovery from drought by phytoplankton is fairly rapid, with successional changes in species composition.

8.3   Zooplankton

Studies of the responses of zooplankton to drought are few compared to those focusing on phytoplankton (their food) and fish (their predators). Nevertheless, the responses of zooplankton to drought vary considerably, perhaps because they are subject to control by changes in water quality, phytoplankton and the level of predation. As they are a key primary consumer in standing waters, declines in zooplankton may benefit phytoplankton populations and stress secondary consumers, such as fish. In some cases, even though there may be substantial changes in environmental variables in lakes with drought, the zooplankton may not be affected. For example, in the experimental drought study of McGowan *et al.* (2005), even though there were significant changes in water chemistry and macrophytes, no changes were detected in the phytoplankton and zooplankton.

In Lake Okeechobee, after a human-induced drawdown followed by a drought in 2000–2001, as depth decreased, so did chlorophyll-*a* concentrations and suspended non-organic solids (Havens *et al.*, 2004). Prior to the drought, the zooplankton was dominated by calanoid copepods, with cladocerans and rotifers being significant. With the drought, the biomass of cladocerans and rotifers declined dramatically, while the copepod biomass was unchanged. The increase in fish abundance due to habitat compression and their selective predation on the cladocerans may have caused the crash in cladoceran biomass.

Havens *et al.* (2007) suggest that predation by cyclopoid copepods contributed to the sharp decline in the rotifer biomass. The copepod *Arctodiaptomus dorsalis*, which dominated the post-drought zooplankton, appears to be a species that can escape fish predation. There was a massive post-drought increase in submerged vegetation, which in turn favoured fish recruitment, and the increased fish predation pressure on zooplankton induced a major change in the structure of the zooplankton assemblage, which persisted for least four years after the drought (Havens *et al.*, 2007).

In Lake Chilwa, drought created major changes in the zooplankton assemblage structure, with rotifers becoming dominant after the drought (Kalk, 1979b). Subsequently, a microcrustacean-dominated assemblage arose and increased to high population densities. Predation by larval fish from the breeding of drought survivors (e.g. clariid catfish) then reduced the zooplankton populations. Thus, as also in Lake Okeechobee (Havens *et al.*, 2007), successful breeding by resilient fish which produced profuse off-spring, themselves possibly released from predation pressure, could have checked the return of the zooplankton to 'normal' levels of abundance and diversity.

In Lake Chad, during the pre-drought period (1964–1971), in the open water of the southern basin, zooplankton density was relatively low and dominated by copepods – calanoids and cyclopoids – and toward the end of each year, there was a peak in zooplankton biomass (Saint-Jean, 1983). Meanwhile, the northern basin had an even lower density of zooplankton than the southern, with less temporal variation. Cyclopoids dominated the abundance, while cladocerans dominated the biomass (Saint-Jean, 1983). With the drought, the water level dropped to divide the lake into two basins – north and south – and conductivity and turbidity both rose to high levels. Data on zooplankton are only available for the south basin (Saint-Jean, 1983). In the early stages of the drought (1973), Cladocera declined, the dominant calanoid copepod *Tropodiaptomus* was replaced by *Thermodiaptomus* and cyclopoid copepods increased. As the drought strengthened in 1974, rotifers increased greatly. Overall, zooplankton biomass declined sharply, probably due to deteriorating water quality, marked by episodes of hypoxia (Saint-Jean, 1983; Bénech *et al.*, 1983). Thus, drought caused considerable changes in species composition and dominance and a decline in biomass.

From all of the accounts, notably from shallow lakes, it appears that zooplankton undergo considerable changes in species composition and abundance with drought. These changes, such as the transient shifts to rotifer dominance, may be due to changes in food supply (viz. grazing-resistant cyanobacteria), or deteriorating water quality (e.g. increases in salinity and hypoxia), or changes in predation pressures (e.g. fish predation). Recovery from drought may be marked by major changes in predation pressures and competition. In post-drought conditions, zooplankton assemblages may differ considerably from pre-drought assemblages, and rare species may briefly flourish as recovery proceeds.

8.3.1 *Drought, lake acidification and plankton*

Many lakes in North America and Europe have been damaged by acid rain due to sulphur dioxide emissions from fossil-fuel burning and from smelters. Many lakes, such as in eastern Canada, have extensive peatlands and wetlands in their catchments, with acid rain caused to accumulate sulphur in the form of sulphides. As explained earlier (Chapter 5), when the wetlands are exposed to El Niño droughts and sediments dry and become oxic, sulphides can be converted to sulphates. With the rains breaking the droughts, sulphate-enriched water can flow into lakes, re-acidifying them. Furthermore, when lake levels drop with drought, exposed sediments containing sulphides may then be oxidized and can enter the lake as sulphates when the drought is broken (Yan *et al.*, 1996; Arnott *et al.*, 2001).

Lakes in eastern Canada affected by acid deposition were recovering after sulphur emissions were greatly reduced (e.g. Keller & Yan, 1991, 1998). In the period from 1977 to 1987, Swan Lake, near the Sudbury smelter in Ontario, Canada, showed distinct indications of ecological recovery (Arnott *et al.*, 2001). In the phytoplankton, the acid-tolerant dinoflagellates were replaced by chrysophytes, but both richness and diversity (H′) were well below the levels of unaffected reference lakes. In the rotifers, the dominant acid-tolerant species *Keratella taurocephala* was replaced by acid-sensitive species and both richness and diversity (H′) were near those of reference lakes (Arnott *et al.*, 2001). The crustacean component of the zooplankton had changed considerably during recovery, to be dominated by calanoid copepods in 1987.

In 1986–87, eastern Ontario was affected by a severe drought. With the drought breaking in 1988 and the return of inflows from surrounding wetlands, fresh influxes of acid into lakes reversed the recovery from acidification. Thus, Swan Lake was re-acidified, with the pH falling from 5.8 to 4.5, metal concentrations rising and DOC levels dropping dramatically. With the drop in DOC, transparency increased and the Secchi level increased from 4.3 m to 7.6 m (Yan *et al.*, 1996; Arnott *et al.*, 2001). The phytoplankton community was dramatically changed by the re-acidification, with the chrysophytes dropping in abundance and both phytoplankton richness and diversity declining. In the rotifers, the acid-sensitive taxa declined sharply and the acid-tolerant *Keratella taurocephala* returned to dominance.

Unexpectedly, with the drought breaking and re-acidification of the lake, the zooplankton richness briefly increased in 1988 from 10 to 18 species. The drought in lowering the lake created dry sediments, increased water temperatures and light availability with depth. An experimental study by Arnott & Yan (2002) indicated that four cladoceran species hatched from dried sediments; the emergence of six species (4 cladocerans, 1 cyclopoid, 1 calanoid) was influenced by temperature; and the hatching of three species (2 cladocerans and 1 calanoid copepod) was affected by light (Arnott & Yan, 2002). However, the acid-sensitive taxa that emerged did not persist in the lake, and in subsequent years the richness declined to levels prior to the re-acidification event (Arnott *et al.*, 2001).

In this case by changing lake conditions directly and indirectly to produce re-acidification, drought induced major changes, reducing the diversity and productivity of planktonic communities. By inducing the premature hatching of acid-sensitive crustaceans, drought may have set back future recovery by depleting the egg bank in the sediments.

Diatoms are well preserved in lake sediments, and the presence of particular species in sediments can reflect the pattern of pH fluctuations in lakes. Using diatoms in sediments, Faulkenham *et al.* (2003) found that a lake in which the diatoms reflected acid conditions had extensive wetlands, whereas a lake with no wetlands did not appear to have been subject to drought-induced re-acidification. Thus, the presence or absence of catchment wetlands may determine the likelihood of re-acidification in those areas subject to sulphur deposition from acid rain.

8.4   Macrophytes of lentic systems

In large, shallow water bodies, littoral plant communities are greatly altered by water recession. In the drought in Lake Chad ('Lesser Chad', 1972–75), in the north basin, the rapid loss of water prevented vegetation development, and plant formations of the 'Normal Chad' disappeared with the shoreline recession (Iltis & Lemoalle, 1983). Few plant formations remained, with some *Typha australis* and sedges. In the south basin, *Phragmites australis* populations receded and forests dominated by the semi-terrestrial shrub Ambatch *Aeschynomene elaphroxylon* developed, with meadows of *Vossia cuspidata* on the dry sediments (Iltis & Lemoalle, 1983). In Lake Chilwa, normally the margins were lined with *Typha domingensis*, but *Typha* disappeared with the severe drought, leaving a large area of exposed mud colonized by semi-terrestrial plants – a grass (*Diplachne fusca*), a sedge (*Cyperus laevigatus*) and a leguminous shrub (*Aeschynomene pfundii*) (Howard-Williams, 1979).

With changes in water levels and water quality due to drought, changes in emergent and submerged aquatic vegetation are to be expected. Changes in aquatic vegetation during and after drought may be very marked. Lake Okeechobee, was deliberately drawn down by 2000, and this was followed by a drought that reduced the lake level further (Havens *et al.*, 2005). Before and during the drought, the shoreline emergent vegetation was dominated by bulrush, *Scirpus californicus*. However, conditions during recovery stimulated the growth of two hitherto uncommon species, the spikerush (*Eleocharis* spp.) and knot grass (*Paspalidium germinatum*), which, along with *Scirpus*, dominated the shoreline vegetation (Havens *et al.*, 2005).

Thus, major changes occur in the littoral vegetation of lakes in drought, with the normal littoral plants dying back to be replaced by invading semi-aquatic and terrestrial species, especially if the supra-seasonal drought is prolonged. Recovery after re-filling is a slow process.

As for littoral semi-aquatic plant communities, drought can induce major changes in aquatic macrophytes. In Lake Okeechobee, the drawdown and severe drought saw marked changes in the submerged vegetation (Havens *et al.*, 2004, 2005). Before the drought, the submerged vegetation was dominated by a low biomass of *Potamogeton*, *Vallisneria* and *Hydrilla*. With the drought, plant biomass increased and was totally dominated by *Chara*. Following the drought, *Chara* declined, possibly due to a short period of high winds and high wave energy that churned up this rootless plant. With the disturbance to *Chara*, a window of opportunity arose for the development of a productive assemblage of *Vallisneria*, *Hydrilla* and *Potamogeton*, with a biomass much greater than that existing before the drought.

A similar situation occurred in Pukepuke Lagoon, a shallow, permanent dune lake in the North Island of New Zealand (Gibbs, 1973). A severe drought (1969–1970) exposed a large amount of the lagoon bed, which was colonized by two semi-aquatic macrophyte species, *Veronica anagallis-aquatica* and *Ranunculus fluitans*. In 1971, with the return to normal water levels, the charophyte *Chara globularis* dominated the lagoon, but by 1972 the *Chara* beds had disappeared. This drastic decline coincided with a massive phytoplankton bloom, which declined in the winter of 1972. In the following spring, the macrophytes *Potamogeton pectinatus* and *P. crispus* recruited and rapidly dominated the lagoon, occupying about 90 per cent of its area (Gibbs, 1973).

As occurred in Lake Okeechobee after drought, *Chara* became the first dominating plant, only to drastically decline and be replaced by flowering macrophytes (Havens *et al.*, 2004, 2005). In the experimental winter drought in Canada reported by McGowan *et al.* (2005), submerged vegetation was dominated by *Ceratophyllum demersum* and, after the drought, macrophyte biomass increased greatly and was dominated by *Potamogeton pectinatus*. These three examples illustrate that with drought, aquatic macrophyte diversity and biomass can be greatly reduced, while after drought, the new conditions may create a change in dominance and an increase in biomass. The increased post-drought biomass may be a response to elevated nutrient levels, notably phosphates, with sediment re-wetting.

In shallow lakes, changes in water level and in light availability with drought can greatly alter macrophyte and phytoplankton communities. This links in with the concept of alternative states of shallow lake ecosystems – either turbid and micro-algae dominated, or clear and macrophyte dominated (Scheffer, 1998). For example, in a small oligotrophic reservoir in tropical Africa, primary production was normally dominated (55 per cent) by macrophytes, principally *Potamogeton octandrus* (Thomas *et al.*, 2000). In a severe drought (1997–98), the macrophytes were stranded and died as the water levels dropped. Primary production then became dominated by microphytobenthos and phytoplankton, a situation that persisted throughout and after the drought (Thomas *et al.*, 2000; Arfi *et al.*, 2003).

Changes in the supply of nutrients to lakes due to drought can change macrophyte communities. In a chain of hard water kettle lakes in Michigan, USA, under normal conditions, the lake at the top of the chain was eutrophic, with high phytoplankton, non-rooted macrophyte abundances and high turbidity. Lakes lower in the chain had lower nutrient levels, with lower levels of phytoplankton and non-rooted macrophytes, but a higher level of rooted macrophytes (Hough *et al.*, 1991). In a severe drought (1987–88), stream inflows and nutrient inputs to the eutrophic Shoe Lake were reduced, with the loading rates of nitrogen and phosphorus being reduced by 80 per cent (Hough *et al.*, 1991). Consequently, phytoplankton levels were reduced by ≈ 50 per cent and light transparency improved. This increase in light availability favoured the growth of rooted macrophytes (e.g. *Nymphaea tuberosa*, *Myriophyllum exalbescens*), whereas the non-rooted macrophytes (e.g. *Ceratophyllum demersum*, *Utricularia vulgaris*, *Najas flexilis*) declined greatly (Hough *et al.*, 1991).

Where shorelines recede with drought, aquatic macrophytes may be replaced by terrestrial species, which are slowly replaced when the drought is over. In lakes that retain water during drought, there may be major changes in aquatic macrophytes, with *Chara* being favoured in some cases during drought at the expense of rooted macrophytes. Drought appears to be a force inducing alternative stable states, changing lakes from being clear and macrophyte-rich to lakes with high turbidity and dominated by phytoplankton and possibly *vice versa*. Increased nutrient availability can occur post-drought, so that aquatic macrophyte production may be high, though the persistence of this production is uncertain due to the lack of post-drought data. Through the stresses of drought, macrophytes in lentic systems may undergo major changes in composition and production that undoubtedly exert strong effects on the pathways of ecosystem recovery after drought.

8.5   Benthic littoral fauna

As drought sets in, water levels drop, exposing the littoral fauna to drying. The littoral fauna of lakes comprises two main groups: those which are sedentary or have only limited mobility; and those that are mobile and can move with the water level as it recedes. Compared with those on the biota of the water column, accounts of the effects of drought on the benthos of permanent standing waters are few.

Molluscs, especially bivalves, may suffer with declining water levels. Small bodies of water can dry up in drought and cause molluscs, such as gastropods, to disappear or have greatly reduced distributions (e.g. McLachlan, 1979b; Koch, 2004). In the littoral zone of Lake Constance, the invading Asian clam *Corbicula fluminea* occurs in very high densities, making up to 90 per cent of the littoral biomass. In 2005–2006 there was a severe winter drought, marked by a drop in lake level and freezing temperatures. The effects of exposure and very low temperatures combined not only to kill clams exposed on the dry shore, but also clams in the sub-littoral, where only one per cent of the original population survived (Werner & Rothhaupt, 2008).

In Lake Balaton in Hungary, Sebestyén *et al.* (1951) found that on steep stony shores, a severe drought in 1949, with a 60 cm drop in water level, appeared to have had little effect on the fauna. However, on shores which were relatively flat, there was marked mortality, with up to 100 per cent mortality of the exotic zebra mussel *Dreissena polymorpha*. With the decline in water levels, there was little evidence for the mobile fauna 'moving along with the retreating water' (Sebestyén *et al.*, 1951). In a later drought in 2000–2003 in Lake Balaton, littoral populations of zebra mussel were again greatly depleted (Balogh *et al.*, 2008).

Lake Balaton has been invaded by three Ponto-Caspian amphipods, two species of *Dikerogammarus* and *Chelicorophium curvispinum* (Muskó *et al.*, 2007). Prior to the 2000–2003 drought, *Chelicorophium* occurred in high densities in stony littoral zones but, during and after the drought, *Dikerogammarus* became dominant (Muskó *et al.*, 2007).

As described before, drought in Lake Chilwa greatly depleted the littoral benthos – principally chironomids and the snail *Lanistes ovum* (McLachlan, 1979b). Littoral benthos, especially molluscs, may be depleted by drought, no doubt with deleterious effects on their normal consumers, such as waterbirds.

In one of the few long-term studies of drought, a ten-year study by Gérard (2001) and Gérard *et al.* (2008) on the gastropods and their trematode parasites in Combourg Lake, a shallow lake in France, revealed how an interspecific interaction, namely parasitism, is altered by drought.

The lake was subjected to two severe droughts in 1996 and 2003. Prior to the first drought, the lake contained 15 gastropod species, with planorbid snails dominating species richness and abundance (*Gyraulus albus* and *Planorbis planorbis* being the two most common species), followed by lymnaeid snails (*Radix peregra*) (Gérard, 2001). The level of trematode parasitism was 5.13 per cent, with 11 snail species infected. The 1996 drought reduced the snail fauna to two species with no parasite infection. After this drought, snails recolonized the lake with initially hygrophilic and amphibious species, followed within nine months by the aquatic species, a boom in planorbid and lymnaeid snail abundance and a low level of parasite infection – only 0.36 per cent in five host species (Gérard, 2001). While the snail assemblage made a post-drought recovery, it is clear that the parasite assemblage had been greatly diminished by the drought, illustrating that drought, by acting as a winnowing force, can substantially change populations and assemblages by drastically altering interspecific interactions.

Over the ten-year study period, snail species richness and abundance peaked in the spring of 1997 (Gérard *et al.*, 2008). Between the snail peak of 1997 and the summer of 2001, the snail assemblage collapsed, probably due to the toxic effects of cyanobacteria outbreaks. Gérard *et al.* (2008) suggest that members of the snail assemblage were adapted to the stress of drought and were capable of recovering from drought, but the toxic effects of cyanobacteria presented a stress to which the snails were poorly adapted. The results from the 1996 drought indicate that drought can be a powerful force influencing the strength of inter-specific interactions. Indeed, in this light, as Everard (1996) has suggested, drought may be differentially beneficial to particular taxa and assemblages.

The only complete account of recovery from drought in lacustrine benthos comes from the Lake Chilwa study by McLachlan (1979b). As the previous account in this chapter described, the recovery of the benthos was relatively rapid. With the re-flooding, the snails which survived the drought by aestivating returned, to be followed some time afterwards by the insects (chironomids, corixids), which mostly flew in from refugia of persistent water considerable distances from the lake itself.

Thus, in lentic waters, especially shallow lakes, the dropping water levels due to drought may strand important species (e.g. molluscs, oligochaetes, chironomids) and may cause changes in water quality, habitat structure and resources that greatly reduce species richness and abundances. Some species (e.g. snails, ostracods) may have desiccation-resistant mechanisms to survive the drought in the water body (in-site refuges), whereas many species, notably insects, may seek refuges in other water bodies that persist through the drought. If the drought is severe and does produce significant drying of the lake bed, recovery may be staggered in terms of marked differences in the rates of species returning. Drought may alter interspecific interactions such as parasitism.

8.6   Drought in perennial lotic systems

8.6.1   Benthic algae and macrophytes

Reflecting the importance of stream algae as a food resource and key habitat, there have been more studies of attached stream algae than of algae in standing waters. With the low flows in drought, increasing nutrient concentrations (see Chapter 5) may serve to promote algal growth (Freeman *et al.*, 1994; Dahm *et al.*, 2003). In a severe drought in Portuguese streams, Caramujo *et al.* (2008) found that biofilm biomass (as chlorophyll-*a*) increased fivefold, and in chalk streams in drought, epiphytic algae may proliferate to such an extent that growth of the host macrophytes is greatly reduced (Wright *et al.*, 2002a) by up to as much as 80 per cent (Wade *et al.*, 2002).

Conversely, drought may also limit algal growth. In a perennial Californian stream in a Mediterranean climate, drought arises through the failure of winter rains and a subsequent lack of scouring floods (Power *et al.*, 2008). In the following summer, this lack of floods (and more importantly, of scouring flows) allows armoured caddis larvae (*Dicosmoecus gilvipes*) to build up and graze down the attached algae, mainly *Cladophora* (Power *et al.*, 2008). Drought by producing conditions that favour a consumer leads to a lack of *Cladophora*, which, as outlined in Chapter 6, has important repercussions on the food web structure.

Drought may change the composition of algal assemblages. In Portuguese streams, before a drought, Chlorophyceae (green algae) and Bacillariophyceae (diatoms) dominated the biofilm (Caramujo *et al.*, 2008). During drought, the diatoms became dominant at the expense of the green algae. In an experiment (not a drought) involving the dewatering of stream mesocosms, Ledger *et al.* (2008) found that, in the normal undisturbed state, the green encrusting alga *Gongrosira incrustans* dominated the epilithon. With dewatering, the green alga declined and the vacant space on the stones was replaced by mat-forming diatoms. In an English acid stream, Ledger & Hildrew (2001) found that the epilithic biofilm dominated by coccoid green algae and diatoms remained intact when the channel dried up in a severe drought, and past-drought recovery was rapid – within three days.

Macrophytes occur in stream channels, along the littoral zone of streams and in wetlands periodically connected with river channels. Aquatic plants have a variety of means, involving reproductive propagules, vegetative structure and fragments, to survive drought and recover (see Chapter 6). The detritus from the dieback of aquatic and semi-aquatic plants can be a valuable resource for consumers when a drought breaks. Recovery from drought is usually marked by the succession of different plant assemblages, and recovery can be marked by the very high production of early successional plants.

The literature on lotic macrophytes and drought is dominated by work on the macrophytes of English chalk streams. These streams are groundwater-fed and are impacted in drought when the water table is lowered by both natural and human means, such as the extraction of groundwater (Wright & Berrie, 1987; Bickerton et al., 1993; Agnew et al., 2000; Westwood et al., 2006). Chalk streams are usually clear, nutrient-rich and very productive ecosystems, supporting rich plant, invertebrate and fish assemblages (Berrie, 1992).

In a small chalk stream – Waterston Stream in Dorset – with normal flow, the dominant macrophyte was *Ranunculus* (water crowfoot) (Ladle & Bass, 1981). In the drought of 1973–74, streamflow dropped dramatically and *Ranunculus* cover disappeared. Prior to the channel drying, two species – watercress (*Rorippa nasturtium-aquaticum*, now *Nasturtium officinale*) and Fool's watercress (*Apium nodiflorum*) – briefly flourished and persisted through the four-month dry spell. With the return of flow, *Apium* became the dominant for the next seven months of normal flow, to be steadily replaced by *Ranunculus*. With the subsequent 1975–76 drought, *Apium* once more became dominant (Ladle & Bass, 1981). Drought served to deplete a normally dominant plant and favour a transient dominant.

In the same drought, Wright & Berrie (1987) found, in intermittent sections of a chalk stream, that macrophytes and the invertebrates, while greatly depleted, rapidly recovered after the drought. However, in a perennial section, siltation due to low flows during drought greatly reduced the cover of macrophytes (*Ranunculus* and *Callitriche*). With the breaking of the drought, recovery was delayed (Wright & Berrie, 1987) – an example of drought interacting with another disturbance (siltation) to influence recovery.

Surveys of macrophyte communities in chalk streams helped to produce a scheme of 13 different community types (Holmes, 1999; Westwood et al., 2006). The community types are aggregated into four major groups along a gradient of tolerance to dry conditions: 'Perennial', 'Winterbourne', 'Ditch' and 'Intermittent'. With drought, community types may move from the 'Perennial' and 'Winterbourne' groups to the 'Ditch' and 'Intermittent', with the latter containing Community Type 13, that consists of terrestrial grasses and herbs. Across a range of 118 sites on 24 rivers, immediately after the

1989–1992 drought, Holmes (1999) found that the drought pushed communities to the terrestrial end of the gradient, but with the return of 'normal' flow, recovery took no longer than two years. In the headwaters of chalk streams, Westwood *et al.* (2006) observed similar macrophyte responses to drought in 'normal' streams (e.g. River Lambourn), and similar recovery lags (about two years) to those found by Holmes (1999). However, in streams with heavy groundwater extraction (e.g. River Misbourne), recovery from drought was incomplete (Westwood *et al.*, 2006).

During drought, stream channels, and many temporary waters, may be invaded by terrestrial plants (terrestrialization) (Holmes, 1999; Westwood *et al.*, 2006). In creeks in south-east Australia, aquatic macrophytes such as *Triglochin procera*, disappeared with drought and the channels were invaded by terrestrial plants such as *Alternanthera*. (P. Reich, L. Williams, personal communication) The process of terrestrialization undoubtedly has important consequences on recovery from drought.

Macrophytes are an important habitat and food resource for stream invertebrates. During drought, macrophyte cover as a habitat declines and, correspondingly, invertebrate densities (especially chironomids) on the remaining macrophytes may rise to high levels (e.g. Wright & Berrie, 1987; Wright & Symes, 1999). The effects of drought on the trophic links between stream algae and macrophytes and invertebrate consumers have been scarcely studied and these links remain an important research area.

8.7 Stream invertebrates and drought

Macroinvertebrates occupy the vital links between their processing of primary production and detritus and their consumption by their predators – the secondary consumers. They are a major consumer of periphytic algae, and of macrophytes to a much lesser extent, as well as being the major force in the breakdown and transformation of detritus, both autochthonous and allochthonous.

As drought progresses in streams, there is a loss in flow volume that causes habitat reduction and a loss in connectivity with, as Boulton (2003) stressed, the crossing of a number of thresholds (see Chapter 5). As volume declines, depth decreases, producing changes in flow paths and in the hydrodynamics of habitats such as riffles and runs. The decrease in stream velocity can consequently deplete populations of animals, such as those living in cascades, riffles and torrents, which rely on high velocities and well-oxygenated water. Usually associated with this decline in flow, the stream withdraws from the littoral edge, affecting animals that inhabit streamside vegetation.

Following the reduction in lateral connectivity, there is the loss of longitudinal connectivity

With the cessation of surface flow, riffles, runs and glides disappear, and pools with varying persistence are formed. However, subsurface water in the hyporheic zone may persist through the drought. In pools, physico-chemical conditions steadily deteriorate (Chapter 5), biotic interactions intensify and lentic species, especially predators, are favoured. Finally, with progressive drying, the remnant pools disappear, leaving a dry stream bed, though there may be free water below the surface. As drying occurs, the stream channel may be invaded by terrestrial scavengers that consume dead and dying stream macrobiota.

By far, the greatest number of drought studies on streams focuses on the loss of surface water, which as drought builds leads to a lowering of the water table and of flow below base flow levels. Surface water drought may progress to groundwater drought, affecting those streams that rely for the major part of their flow volume on groundwater aquifers (Wood, 1998; Wood & Petts, 1999; Wood et al., 2000). Groundwater-dominated springs and streams have stable flow regimes, and groundwater droughts affecting streamflow occur with the depletion of the aquifer. This depletion lags behind surface water drought and is due to a failure of recharge. Conversely, the breaking of groundwater drought which requires aquifer recharge usually lags well behind the breaking of surface water drought.

8.7.1 Drought and the benthos of groundwater-dominated streams

When aquifers are drawn down in drought, springs may cease to flow and disappear. Erman & Erman (1995) sampled 21 cold springs for caddis flies in the Sierra Nevada, California, before and during a supra-seasonal drought (1987–1992). Springs fed from deep aquifers were rich in ions, especially calcium, and species richness was positively correlated with calcium concentrations. During drought in those springs that underwent great fluctuations in flow, species richness declined, and springs that dried lost their species. Thus, in causing springs to shrink or dry, drought served as a major force to regulate trichopteran species richness and, no doubt, invertebrate diversity. The pattern of recovery is not known.

English chalk streams are groundwater-dominated, highly diverse and productive ecosystems. In drought, these systems may continue to flow, though small headwater streams (bournes) may dry in severe droughts. As these streams derive most of their flow from aquifers, short-term surface water droughts may be buffered by the continued provision of groundwater (Berrie, 1992). However, groundwater extraction by humans may exacerbate drought and cause small chalk streams to stop flowing (e.g. Wright & Berrie, 1987; Agnew *et al.*, 2000).

Two chalk stream systems have been the focus of long-term investigations. One of the chalk streams is the Winterbourne, a tributary of the River Lambourn, which in turn is a tributary of the River Kennet in Berkshire; the other is the Little Stour in Kent.

In the upper sections of the River Lambourn, during the 1975–76 drought, macrophyte cover and macroinvertebrates were greatly depleted, whereas at the lower perennial sections, the effects were less severe (Wright & Berrie, 1987). At perennial sites during both the 1975–6 and the 1996–97 droughts, there were few changes in the macroinvertebrate families on a range of substrates (e.g. gravel, silt), or in the macrophytes, though in some cases, some families were lost, as in the case of the macrophyte *Berula* in the 1975–76 drought (Wright & Symes, 1999; Wright *et al.*, 2002b). However, there were changes in abundance and, hence assemblage structure. This took the form of increased densities of some families, in particular the Chironomidae and Ceratopogonidae, and greatly reduced densities of some rheophilic families – Baetidae, Caenidae and Simuliidae (Wright & Symes, 1999; Wright *et al.*, 2002a, 2002b). After the droughts ended and normal flows returned, the recovery of the macroinvertebrate assemblages at the perennial sites was rapid – within one year (Wright & Symes, 1999; Wright *et al.*, 2002a, 2004).

A similar picture of considerable resilience of chalk stream biota to drought comes from the Little Stour (Wood, 1998; Wood & Petts, 1994, 1999; Wood *et al.*, 2000; Wood & Armitage, 2004; Stubbington *et al.*, 2009a). A severe drought in 1989–1992 caused 'upstream' perennial sections to become silted at the channel margins and in the stream bed, and resulted in two sites drying up. In the downstream regulated sections, reduced flow persisted (Wood & Petts, 1994, 1999). Samples from the upstream sites in 1992, compared with post-drought samples (1993–1995), indicated that drought reduced species richness and total abundance (Wood & Petts, 1994, 1999). Taxa at upstream sites reduced by drought included *Lymnaea peregra*, *Gammarus pulex*, *Baetis rhodani*, Tanytarsini spp. and the exotic *Potamopyrgus jenkinsi* (*antipodarum*), and only a few species appear to have been eliminated. In the downstream regulated sections, the fauna containing some lentic taxa (e.g. Corixidae, Haliplidae, Dytiscidae) were not affected by drought.

Recovery at the upstream sites commenced with the return of normal flows and continued, notably in terms of abundance, for two years (Wood & Petts, 1994, 1999). Analysis of data from the final year of the 1988–92 drought, and from a second one (1995–1996), indicated that the two droughts had severely reduced abundance of the total community, particularly the amphipod *Gammarus pulex* and two molluscs, *Potamopyrgus jenkinsi* (*antipodarum*) and *Bithynia tentaculata* (Wood, 1998; Wood *et al.*, 2000; Wood & Armitage, 2004). As before, at the upstream sites, recovery took at least two years, which was dependent on the rate at which the aquifer was recharged (Wood & Armitage, 2004).

On the other hand, at the sites on the Lambourn and Kennett Rivers and the Little Stour, where flow may drop but not cease, recovery from drought was relatively rapid, with a good supply of colonists from in-channel refuges. In the Little Stour River, at both sites where drought led to very low flows and those where flow became intermittent, the hyporheic zone clearly acted as a refuge for benthic invertebrates (Stubbington *et al.*, 2009a; Wood *et al.*, 2010), with a marked increase in benthic taxa, especially in *Gammarus pulex*, in the zone at the time when flow was very low (Wood *et al.*, 2010). By contrast, in another groundwater-fed stream, the River Lathkill in Derbyshire, as the stream dried, some taxa (trichopteran larvae (*Stenophylax* sp.), stonefly nymphs (*Nemoura* sp.) and adult hydrophilid beetles (*Anacaena globulus*)) were able to survive in fine-grained sediments of the hyporheic zone. However, the lack of more suitable coarse sediment may have excluded other animals (e.g. *Gammarus*) (Stubbington *et al.*, 2009b). Thus, in many flowing waters, the hyporheic zone may be a reliable drought refuge, whereas in other systems, such as those with clogged hyporheic zones, or none, escape into the stream bed may be limited.

The susceptibility of small chalk streams to the effects of drought is further confirmed by the study of Ladle & Bass (1981) on the Waterston Stream in Dorset. The drought in the summer of 1973 was short, severe, and it dried up the stream bed for four months. Overall, the fauna, in the summer after the drought, was significantly reduced, with the amphipod *Gammarus pulex* being eliminated. Examining Table 1 of Ladle & Bass (1981) reveals that some groups were greatly depleted by the drought, including Tricladida (4 spp.), molluscs (except for *Lymnaea* sp.), leeches (4 spp.), water mites (1 sp.), *Gammarus pulex*, Plecoptera, the mayfly (*Baetis rhodani*), the trichopterans (*Agapetus fuscipes*, *Silo pallipes*, *Sericostoma personatum*) and dipteran species (*Dicranota* sp, *Macropelopia* sp., *Paramerina* sp., *Brillia modesta*, *Polypedilum* sp.). Densities of several taxa were largely unchanged, including the Nematoda, oligochaetes (e.g. *Nais elinguis*, *Stylaria lacustris*, Enchytraeidae), the trichopteran *Limnephilus* sp., Coleoptera, and dipterans such as *Bezzia* sp., *Eukiefferiella* sp., and *Thienemanniella* sp.

Several taxa with resistant eggs or resting stages, such as the cladoceran *Alona affinis*, ostracods and the mayfly *Ephemerella ignita*, recovered rapidly.

Species capable of surviving in the stream bed, such as *Lymnaea* sp., *Eiseniella tetraedra*, *Asellus aquaticus* and *A.meridianus*, also became a major part of the post-drought assemblage. Similarly, taxa with very mobile colonizing adults, such as simuliids, flourished after the drought. One year after the drought, the community was still depleted (Ladle & Bass, 1981), and recovery was very incomplete. Indeed, the post-drought macroinvertebrate community was very different from the pre-drought one, and in losing some key predators (e.g. Tricladida, Hirudinea), shredders (e.g. *G. pulex*) and grazers (e.g. *Agapetus fuscipes*), the trophic structure was much altered.

In chalk streams, if drought causes habitat and resource loss but flow persists, species richness and abundances are reduced but recovery after the drought is relatively rapid. However, if flow ceases and the channel dries, the effects of drought in these normally perennial and stable systems can be marked, with significant losses in population densities and richness, and the post-drought recovery relatively slow and incomplete. If droughts are close together, the effects are severe and recovery is impaired. In one study, the species lost in the drought were mainly predators, which could reduce the regulation of prey species in recovery and alter the trophic structure.

8.7.2 Drought, invertebrates and precipitation-dependent perennial streams

Most streams are mostly dependent on precipitation (rain, snow) falling on their catchments. By far the greatest number of studies on drought and lotic invertebrates has focused on small streams, with very few studies on large rivers and their floodplains. Most of the studies have been short-term and have focused on drought effects on community/assemblage composition and structure.

Small headwater streams may be particularly susceptible to the effects of drought, as they lack the buffering capacity of large rivers. Montane streams, with their steep gradients, may also be more susceptible to drought than are lowland streams with gentle gradients and groundwater inputs from their water tables.

Before dealing with studies of macroinvertebrate assemblages, a review of certain groups provides valuable insights into the effects of drought at the species population level rather at the aggregate levels of assemblages and communities. Both lotic crustaceans and molluscs are relatively sensitive to supra-seasonal drought and are relatively easy to census. In terms of assessing the effects of drought on invertebrate populations in flowing waters, these groups provide interesting examples.

Many streams in non-seasonal tropical rainforest regions, such as Puerto Rico, are perennial and may be occasionally disturbed by hurricanes/cyclones and supra-seasonal droughts (Covich *et al.*, 1996, 2003, 2006). Both atyid and palaemonid shrimps can be abundant in tropical low-order streams. In Puerto Rican streams, atyid shrimps occur at higher altitudes than their major predators, the palaemonid shrimps *Macrobrachium carcinus* and *M. crenulatum* (Covich *et al.*, 1996).

A drought in 1994–1995 in a small stream caused flow to stop, with dry riffles and shrinking isolated pools (Covich *et al.*, 1996, 2000, 2003). In the drought, the two atyid shrimps, *Atya lanipes* and *Xiphocaris elongata*, reacted differently to the physical properties of the pools. *Atya* densities were negatively correlated with pool depth and width, whereas *Xiphocaris* densities were positively associated with pool depth. *Atya* densities were positively associated with the coefficient of variation in pool depth, whereas *Xiphocaris* densities had a negative association with this coefficient. In this way, low flow through drought acted as an agent of habitat partitioning for these shrimp species.

With the shrinking of the pools, densities of both shrimp species rose sharply, and reproduction in *Xiphocaris* was reduced (Covich *et al.*, 2000, 2003). Following the drought, the densities of both shrimps reverted to pre-drought levels, (Covich *et al.*, 2003). In the drought, the densities of *Macrobrachium* spp. reached their lowest level in 14 years of sampling (1988–2002) (Covich *et al.*, 2006). It is not clear why *Macrobrachium* densities declined with drought, but it is possible that, when they were confined in shrinking pools, intra-specific competition, cannibalism and/or predation took their toll. Recovery from the drought was rapid, with their densities being higher after the drought (1995–2000) than even before it. It is surprising to see that the big floods generated by two hurricanes/cyclones (1989 and 1998) had no significant effect on *Macrobrachium* densities (Covich *et al.*, 2006), compared to the effects of drought.

Crayfish are a major component of streams, especially slow-flowing lowland streams. As omnivores, they may be a critical link between primary consumers and tertiary consumers.

Georgia, USA, suffered two droughts in the 1980s: 1981 and 1985–86. Crayfish population responses to these droughts were recorded by Taylor (1983, 1988) in a study (1979–1986) at second-order and fifth-order sites in a stream with a shifting sand bottom. Initially, with the 1981 drought, two crayfish species (*Cambarus latimanus* and *Procambarus spiculifer*) were studied. The drought 'produced no detectable adverse effects' on the *Cambarus* populations (Taylor, 1983), no doubt due to their prodigious burrowing ability. On the other hand, the *Procambarus* populations, with their poor burrowing ability, were reduced by the drought.

The relative density and body size of *Procambarus* are positively associated with water depth (Taylor, 1983). With a decline in depth during drought, there was a reduction in the amount of refuges for *Procambarus*, which presumably increased losses to predation and cannibalism. Drought selectively reduced densities of adults, and thus changed body size distribution, increasing the juvenile-to-adult ratio. The drought in 1981 gave rise to extinction at one second-order site and, in the later 1985–86 drought, extinction occurred at another second-order site. Recovery in abundance after the 1981 drought took at least two years, but at the second-order site, 'the population did not return to its pre-drought abundance levels' (Taylor, 1988). Mean body size did not recover to pre-drought size after the 1981 drought but remained lower until the 1985–86 drought, which further reduced mean body size.

As summarized by Taylor (1988) in relation to drought, *Procambarus* populations are strongly influenced by the 'habitat characteristics' of water depth and the availability of refuges, and by past history – notably, how long has elapsed since the last drought.

The above study illustrates the differential susceptibility of rather similar species in the same environment. A further example of differing effects of stream drying on crayfish species is given by Larson *et al.* (2009). A native crayfish, *Orconectes eupunctus*, is far less tolerant of desiccation than an invading species, *Orconectes neglectus*, and hence, in drought, its tolerance of drying may allow *O. neglectus* to oust or greatly reduce *O. eupunctus* populations (Larson *et al.*, 2009).

In 1999–2000, a supra-seasonal drought in Mississippi, USA, reduced streamflow and caused many streams to dry. Of the 12 streams studied by Adams & Warren (2005), five flowed during the drought and the other seven stopped flowing. Seven species of crayfish, dominated by *Orconectes* sp. cf. *chickasawae*, lived in the streams. Drought did not alter the populations in the flowing streams, but it greatly reduced the populations in the dry streams. With the breaking of the drought, crayfish populations took about ten months to show signs of recovery. However, by the summer after the drought, crayfish populations at the dry sites had increased to be on average double those of pre-drought levels (Adams & Warren, 2005) – a somewhat familiar pattern whereby recovering populations, such as zooplankton, may overshoot pre-drought levels. As in Taylor's (1983, 1988) work, post-drought mean body size was significantly lower than pre-drought levels, and recovery was facilitated by both reproduction and migration.

This same drought threatened freshwater mussel populations and assemblages (Haag & Warren, 2008). Both small headwater streams and reaches of large streams were sampled before and after the drought.

Of the five small streams, only one dried out completely. Of the large stream sites, none stopped flowing, though water levels dropped and shallow areas were exposed, with consequential high mortality of mussels (Haag & Warren, 2008). In spite of this, at the large stream sites there were no detectable changes in mussel population density or assemblage composition but, at the small stream sites, native mussel populations declined by 65–83 per cent, with some rare species (e.g. species in the genus *Villosa*, *Lampsilis straminea*) becoming locally extinct. In the small streams, the population declines for species were generally similar across the assemblages, with the consequence that rare species were greatly reduced or eliminated. In contrast, the hitherto common invader *Corbicula fluminea* suffered very heavy mortality (Haag & Warren, 2008). While stranding of mussels was a cause of death, the effects of low oxygen levels, high temperatures and high biological oxygen demand caused by very low flows were the major causes (Haag & Warren, 2008). Recovery was not fully studied, but the point was made that in some creeks, recovery may be impaired by the presence of dams and reservoirs that limit migration of fish carrying glochidia.

The severe drought in Mississippi studied by Adams & Warren (2005) also affected Georgia, with a prolonged drought from 1999–2002 (Golladay *et al.*, 2004; Gagnon *et al.*, 2004). The Flint River basin in southern Georgia harbours 22 species of native unionid mussel, with three species listed as endangered, one listed as threatened, six species regarded as being 'of special concern' and one exotic species *Corbicula fluminea* (Gagnon *et al.*, 2004).

In the summer of 2000, nine selected sites in the Flint River basin were surveyed weekly for mussels (Gagnon *et al.*, 2004). Cumulative mortality varied from 13 to 93 per cent, with some mortality due to predation by terrestrial animals. As depth was reduced, flow velocity also declined, along with oxygen levels. Mortality of mussels increased below a flow velocity threshold of 0.01 m/s. Above a level of 5 mg/l of oxygen, mortality was low (0–13 per cent), but it increased (0–24 per cent) in the range of 3–5 mg/l, and mortality was high below 3 mg/l (0–76 per cent). Low oxygen levels due to low flows and high temperatures appeared to be the major cause of mortality, with riffle-dwelling species being very susceptible (53 per cent mortality due to low oxygen levels). Mortality was highest at the medium-sized stream sites, especially in riffle-dwelling species. In general, mortality was lower at the downstream mainstem sites, where flows were maintained, and at the low-order small stream sites, where the resident species were resistant, having adapted to frequent low flow conditions.

Over the same basin in the summer of 2001, 21 sites with pre-drought records were resurveyed (Golladay *et al.*, 2004). Mussels were defined in terms of conservation status into three groups: stable; of special concern; or endangered. With the drought, sites were defined as flowing or non-flowing. At non-flowing sites (ten of the 21), abundances of stable species declined significantly, as did taxon richness, whereas there were no significant differences in abundances of endangered or special concern species. At the flowing sites, the median abundance of stable species actually increased, with no significant changes for species of special concern or endangered species. The greatest declines in abundance occurred at middle-order sites on the major tributaries – reaches under pressure from irrigation water extraction (Golladay *et al.*, 2004). Low mortality was positively associated with the presence of logs, and it appears that scouring around logs created depressions that acted as refuges for mussels during low flows. The exotic invader *Corbicula* cannot tolerate low oxygen levels and suffered mass mortality. Their ensuing decomposition may have generated ammonia concentrations that were high enough to seriously threaten native unionid mussels (Golladay *et al.*, 2004; Cherry *et al.*, 2005).

Both large crustaceans and bivalve molluscs vary greatly between species in their susceptibility to drought. In the crayfish, species that survived drought were those that were good burrowers or moved to persistent water. Recovery was achieved through migration and high rates of reproduction. Species were differentially affected by drought, and drought may facilitate invasion by hardy exotic species. Bivalve molluscs, on the other hand, being sedentary, died in drought from being stranded or from being exposed to low quality water, principally that with low oxygen levels. Susceptibility to drought not only varied between species, but also between sites within catchments. Recovery tends to be slow due to migration, relying on the glochidial stage in fish and the movements of fish being restricted by natural and human-imposed barriers. As a group, bivalves are particularly harmed by drought.

Perennial, small, low-order headwater streams can cease to flow in supraseasonal drought, with consequential reductions in abundance and species richness. In a Polish mountain stream, drought reduced populations and species richness of Ephemeroptera, Plecoptera and Trichoptera (Kamler & Riedel, 1960), with the most resistant species being the caddis fly *Chaetopteryx villosa* (Limnephilidae). No information is available on post-drought recovery.

In a Swiss montane stream, a severe drought in 2003 increased water temperature by about 2 °C and greatly reduced flow (Ruegg & Robinson, 2004). Prior to the drought, the fauna was dominated by chironomids, while afterwards the dominant groups were predatory Turbellaria (notably *Crenobia alpina*) and chironomids. Recovery in this system, in terms of species richness, appeared to be rapid, albeit with significant changes in assemblage structure (Ruegg & Robinson, 2004).

Hynes (1958, 1961) carried out a four-year study of a small montane stream, the Afon Hirnant in Wales, during which a 'severe' drought (by Welsh standards) occurred in 1955. At one site, the drought caused the stream to dry out. In the dry period of about ten weeks, all the trout at the site died, and most of the insects in their immature instars were eliminated. Survivors included turbellarians, naidid and enchytraeid oligochaetes, ostracods, copepods (*Cyclops, Canthocamptus*), some coleopterans (*Helmis maugei, Hydraena gracilis*), some chironomids and Hydracarina – all of which presumably survived in the stream gravel. Eggs of some stoneflies (*Leuctra* spp.) and mayflies survived and hatched after the drought ceased. Nymphs and larvae of other species of mayflies, stoneflies and caddis flies did not reappear until the next year and hatched from eggs laid by adults flying into the site.

The drought appeared to have locally eliminated one species, the mayfly *Rhithrogena semicolorata*. However, it appeared to have favoured two species of stonefly and one mayfly species, possibly because of a reduction in interspecific competition and predation (Hynes, 1961). Two *in situ* refuge-seeking strategies appear to be moving into the gravel streambed or having desiccation-resistant eggs. Recovery was greatly aided by flying adults migrating from persistent stream sections. A key finding is that the timing of a drought in relation to the life history of the inhabitants can greatly affect the strength of the impact.

Similar results to those of Hynes (1958, 1961) were obtained by Morrison (1990) after a drought in small Scottish streams which dried them out and reduced them to series of pools. At the end of the drought, survivors included nematodes, oligochaetes and chironomids. In all the burns (streams), shortly after the drought, stonefly nymphs (*Leuctra* spp.), simuliids and *Plectrocnemia* caddis larvae were early colonizers. Indeed, the Plecoptera had the highest species richness and appeared to make a good recovery after the drought, especially in comparison with the Ephemeroptera, in which recovery was quite patchy (Morrison, 1990). Similarly to Hynes (1961), two refuge-seeking strategies were identified – moving into the moist gravel of the streambed and having desiccation-resistant eggs (e.g. *Leuctra* spp., chironomids). Flying adults from streams that persisted were also important.

Cowx *et al.* (1984) assessed the impact of a severe drought (1976) on the fauna of Afon Dulas, a small upland, Welsh stream. Invertebrate densities were reduced by about 40 per cent. As the reduction in habitat and drying of the streambed coincided with the hatching period of some mayfly and stonefly species, many small nymphs died – a further example of the timing of drought being critical. Molluscs (e.g. *Lymnaea peregra*) were greatly reduced. As in the other studies, with the breaking of the drought, the early successful colonists were oligochaetes, chironomids and simuliids (Cowx *et al.*, 1984).

Recovery was deemed to be complete by 1978 (two years after the drought), with aerial migration being the major form of recolonization.

Colorado suffered a severe drought from 1974 to 1978 with a 'record-breaking winter drought' in 1976–1977, resulting in very low flows in streams draining the Colorado Rockies (McKee *et al.*, 2000). Near the end of the drought in 1978, Canton *et al.* (1984) began sampling a montane stream, Trout Creek. The drought reduced streamflows to 'negligible levels' and reduced invertebrate abundance by around 50 per cent compared with a normal year.

With the breaking of the drought in 1979, there was a marked increase in total abundance. The drought triggered differing responses in the invertebrates. Some taxa (*Hydroptila* sp., *Ophiogomphus severus*, *Cricotopus* sp, *Eukiefferiella* sp., *Palpomyia* sp. and *Tipula* sp.) were all at higher densities in the drought than after the drought broke (1979). In particular, the odonatan predator *Ophiogomphus* may have been favoured by the increase in prey density due to habitat shrinkage. Many species, including *Baetis* spp. and *Glossosoma* spp., were greatly reduced in abundance by the drought, but recovered rapidly with its breaking. One previously common species, the mayfly *Rhithrogena hageni*, was very rare in the drought and was not encountered after it (cf. *Rhithrogena semicolorata* – Hynes, 1958, 1961).

Recovery was marked by a large increase in simuliids and baetid mayflies (Canton *et al.*, 1984). Normally the dominant functional feeding groups in the creek were collector-gatherers and filterers but, in drought, the dominance shifted to shredders and predators. Again as in the previous accounts, drought reduced abundance and changed assemblage structure and, even though overall recovery appeared to be rapid ($\approx$1–2 years), there were species that disappeared and even some that appeared with drought.

Georgia, USA, underwent a severe drought from 1998 to 2002 – a drought that was broken by Tropical Storm Hanna, which returned streams to 100 per cent flow in just three days (Churchel & Batzer, 2006). The high flows of the storm no doubt exacerbated the impacts of drought. Sampling started immediately after the storm in upland streams. Recovery was assessed by the shape of taxon richness time curves rather than by reference to pre-drought conditions. The curves levelled off by around 165 days, and prior conditions in the drought – either wet or dry – did not affect the shape or endpoints of the curves. The structure of the assemblages did differ with substrate, with those in sandy sections differing from those on gravel substrate (Churchel & Batzer, 2006). Drought (and the storm) affected all sites similarly but, in recovery, substrate type played a major role in determining assemblage composition. The recovery is perplexing in that the drought was broken sharply by the high flow event, making it difficult to partition the effects of the two disturbances. However, floods breaking droughts are not uncommon.

After the same drought studied by Churchel & Batzer (2006), Griswold *et al.* (2008), in a 'wetland-fed' stream and a 'seep-fed' one, sampled invertebrates in the last year of the drought (2001) and afterwards (2002–2006). In both streams, flow was greatly reduced by the drought. Following the drought, species richness steadily increased from 2002 and levelled off in 2004, while abundance rose greatly after the drought but then dropped significantly in the next two years (2005–2006). This decline is interesting, as the Standardized Precipitation Index (SPI) indicated in 2005–2006 that there was to be a subsequent return to drought conditions. Persisting through drought, there was a 'core set' of pool-dwelling taxa, comprising Ceratopogonidae, Chironomidae (*Parametriocnemus, Polypedilum, Tanytarsus, Tribelos, Zavrelimyia*), cambarid crayfish, Tabanidae (*Chrysops*) and Tipulidae (*Pilaria*). *Zavrelimyia* seems to be a drought specialist, as it also boomed immediately after drought in the stream studied by Ledger & Hildrew (2001). Faunal succession occurred during recovery, with additions in the second year including tanypod and orthoclad chironomid larvae, and phantom crane fly larvae (*Bittacomorpha*, Ptychopteridae). In 2004–2005, new taxa included Trichoptera (*Lepidostoma*), Hemiptera (*Microvelia*) and Odonata (*Boyeria*). In the last two years, changes in the community structure suggested anticipation of the impending return to drought.

During recovery from the drought, Griswold *et al.* (2008) examined changes in nine traits: 'size, body shape, body armouring, voltinism, resistance to desiccation, mobility, rheophily, habits and feeding preferences'. Rather than the traits reflecting large-scale disturbance, such as drought, Griswold *et al.* (2008) suggest that the traits reflected local habitat conditions, which are in turn strongly influenced by flow conditions and water quality generated by drought. Initially, after the drought, important traits in early colonizers included small body size, body sclerotization and armouring, tubular bodies and being common in the stream drift. With recovery, these traits gave way to other traits such as soft bodies, poor resistance to desiccation, being rare in stream drift and animals that crawl or cling to the substratum.

Community persistence was determined by measuring similarity between two consecutive years. With the breaking of the drought, persistence steadily increased (i.e. similarity between years increased), until 2005 and 2006, when persistence declined – possibly a response to the oncoming drought. Analysis of similarities in trait representation indicated that, in recovery after drought, trait persistence increased and then declined in the last two years, as the SPI score indicated impending drought.

In lowland streams and rivers, with their gentle gradients and pool sections, the effects of drought may be ameliorated in severity. Major changes in species assemblage structure occurred in a perennial lowland Danish stream after it dried up in the 1976–77 European drought (Iversen *et al.*, 1978). Oligochaetes and chironomids were the dominant fauna and, although reduced in abundance, they did survive the drought. Favoured by the drought, in terms of increased abundance, were the mayfly *Nemoura cinerea* and the elmid beetle *Elmis aenea*. As found in other studies (e.g. Ladle & Bass, 1981), *Gammarus pulex* disappeared, to be replaced by the isopod *Asellus aquaticus*. Three species of leeches were eliminated and, after the drought, there was a brief increase in coleopteran species (Iversen *et al.*, 1978) – presumably a function of their high mobility. Thus, in this case, the effects of drought were marked and, as it was only a short study, no conclusion can be reached on recovery.

Drought in 1985–86 dried up a small sandy stream in South Carolina, USA, for two spells of six months and led to a build-up of particulate organic matter in the channel (Smock *et al.*, 1994). It also led to a drastic depletion of the aquatic invertebrates – mainly the amphipod *Gammarus tigrinus*, chironomids and ostracods. Recovery was rapid, with chironomids and ceratopogonids recolonizing as adults flying in from other streams, and by the amphipods migrating upstream from the top part of an estuary (Smock *et al.*, 1994). Seeking refuge in the hyporheic zone in the stream was not possible, as this zone was anoxic. Recovery in terms of abundance from the flood generated by Hurricane Hugo in 1989 was at a similar rate to that of recovery from the drought in late 1985 (Smock *et al.*, 1994).

Lone Oak is a small acid stream in southern England. In a drought in 1995, the stream dried completely for nine weeks (Ledger & Hildrew, 2001). Fortunately, there were data on the benthic algae and macroinvertebrates before the drought. The epilithon was dominated by diatoms and coccoid green algae and remained viable during the dry period. Upon re-wetting, the algae, in terms of chlorophyll-*a*, cell density and biomass, recovered quickly to pre-drought levels within three days. The macroinvertebrate fauna recovered rapidly to pre-drought numbers of taxa and densities, reaching a peak in 26–38 days. A major but short-lived change to the post-drought macroinvertebrate assemblage was the massive dominance of the hitherto-rare tanypod chironomid *Zavrelimyia* sp. (Figure 8.6), which made up 56 per cent of the faunal density by day 38 (see Griswold *et al.*, 2008), only to decline to zero when the stream was flooded. The post-drought community was significantly different from the pre-drought one but, interestingly, subsequent spates served to re-assemble the community and put it on a trajectory converging with the pre-drought configuration (Ledger & Hildrew, 2001). In this case, the community recovered

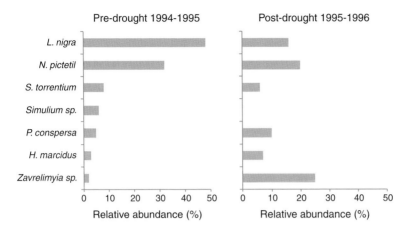

Figure 8.6 Relative abundances of the seven most common macroinvertebrate species in Lone Oak stream before and after the 1995 summer drought. Note the dramatic change in the relative abundance of the chironomid *Zavrelimyia* sp. before and after the drought. (Redrawn from Figure 5 in Ledger & Hildrew, 2001.)

quite rapidly, although, with re-wetting, there was the eruption of a rare species – possibly a poor competitor and a predator, especially of small chironomids and stoneflies favoured by summer temperatures (Hildrew *et al.*, 1985).

In Illinois, USA, a severe drought in 1953–54 caused many small streams to stop flowing and some to dry completely. One creek, Smiths Branch, was the subject of an intense study by Larimore *et al.* (1959), focusing on the fish and macroinvertebrates. There were pre-drought data for the fish but not for the invertebrates, which were studied 'primarily because of their importance as food for fish' (Larimore *et al.*, 1959). As the stream dried, pools formed and steadily declined in size and in water quality, as indicated by low oxygen, high carbon dioxide levels and extreme temperatures. In the pools, conditions so deteriorated that crayfish 'were seen leaving these foul-smelling pools' and the only insects left were rat-tailed maggots (*Eristalis* sp.), which thrived in the putrescence along with oligochaetes. Deoxygenation was associated with the development of dark brown water or 'blackwater' in the pools. The invertebrates surviving on the stream bed included gerrids, along with crayfish that survived under flat rocks, and snails, isopods, stonefly and mayfly nymphs, coleopteran, trichopteran and dipteran larvae, all of which burrowed into the damp sand and gravel of the streambed. In addition, Larimore *et al.* (1959) noted the exodus of animals flying and crawling out of the stream, presumably in pursuit of more persistent water bodies.

When the streambed was dry, Larimore *et al.* (1959) noted the presence of terrestrial predatory and scavenging insects feeding on dead and dying organisms. Similarly, Boulton & Lake (1992b) and Lake (2003) have observed scavenging and predatory insects (e.g. ants, carabid beetles) and vertebrates (e.g. birds, foxes) moving into stream beds as they dried. Williams and Hynes (1976) referred to this group of predators and scavengers as the 'clean-up crew'.

With the return of flow in late winter 1954, early inhabitants were gerrids and water beetles (Hydrophilidae, Dytiscidae), followed in spring by a boom of chironomid larvae, accompanied by stoneflies (*Allocapnia* sp.). By mid-summer, the chironomids had declined and there was an increase in crayfish (*Orconectes propinquus*, *O.virilis*, *Procambarus blandingii*) and mayfly nymphs (*Ephoron leukon* and *Caenis* sp.). However, this partial recovery was set back by a return to intermittent flow in autumn 1954. With the return of flow in the winter of 1954–55, the above pattern was repeated, but the succession progressed further with the addition of abundant snails (*Fossaria obrussa*, *F. parva*, *Ferrissia* sp.), beetles (notably elmids) and caddis larvae (Hydroptilidae, Hydropsychidae).

Two main avenues of recolonization were suggested by Larimore *et al.* (1959): flying insects coming from persistent water bodies with some laying eggs, and invertebrates coming in on the stream drift with the resumption of flow. Judgement on the scale of recovery cannot be readily made, as the makeup of the pre-drought community was unknown. The judgement was made, however, that 'the invertebrate population became re-established soon enough to serve adequately as food for ingressing fish' (Larimore *et al.*, 1959).

8.8   Stream macroinvertebrates, droughts and human activities

Human activities may serve to exacerbate the effects of drought in streams that continue to flow through drought. Groundwater extraction in chalk stream catchments is one example, and extraction of water for irrigation and attendant construction of weirs can exacerbate the effects of drought (see Chapter 11).

The inputs of wastewater and sewage can alter the impacts of drought. In the severe 1982–83 drought in Australia, water quality in the La Trobe River, Victoria, declined markedly, with low oxygen levels and increased conductivity (Chessman & Robinson, 1987). The decline was due to low drought flows not adequately diluting sewage and industrial wastes. However, the fauna was not affected, possibly due to the long-term effects of waste water disposal that had winnowed the fauna to become tolerant to the adverse conditions.

The invertebrate fauna of the River Roding in Essex was affected by the 1975–76 severe drought (Extence, 1981). This river received considerable volumes of treated sewage, generating organic pollution that was exacerbated by the low drought flows. The drought caused the stream to shrink, which combined with the low stable flow and the pollution to produce an accumulation of silt and detritus and a bloom of *Cladophora* on the stream bed. Accordingly, taxa such as Ephemeroptera and cased Trichoptera were absent, and molluscs (e.g. *Potamopyrgus* (*antipodarum*) *jenkinsi*, *Bithynia* spp., *Ancylus fluviatilis*) and Chironominae larvae were greatly reduced. The new conditions boosted the abundance of such taxa as *Asellus*, Orthocladiinae larvae, Tubificidae and *Hydropsyche angustipennis* – taxa tolerant of mild organic pollution. Thus, in this situation, drought combined with organic pollution to produce a major change in community structure, along with a boost in total abundance.

Unwise land management on catchments may generate salinization of streams and wetlands, particularly in southern Australia. When drought reduces flow, water quality may decline. In the Wimmera River in Victoria, in the severe 1997–2000 drought, the increase in salinity in downstream reaches exacerbated the effects of drought (Lind *et al.*, 2006), producing a fauna dominated by chironomid larvae, copepods and ostracods.

8.9   Drought, invertebrates and streams at a large spatial extent

As to be expected, most drought studies are serendipitous, in that droughts occur in studies with other aims. Furthermore, most studies are carried out at particular sites or small groups of sites; studies that increase the spatial extent of investigations to entire catchments are few. The studies of Gagnon *et al.* (2004) and Golladay *et al.* (2004) on the effects of severe drought on mussels at nine and 21 reaches (respectively) in the Flint River catchment in Georgia, USA, are fine examples of drought impact research at a catchment level. However, as stressed earlier, droughts are not local events, but steadily develop to occupy very large spatial extents across many catchments and bioregions. There are few studies that describe the effects of droughts on aquatic ecosystems at the spatial extent of the droughts. As in many parts of the world, there are and have been large-scale monitoring programs of flowing waters, their condition and their invertebrate fauna, targeted analysis of the collected data from these studies could determine the effects of drought at the large scale.

In the region of Otago, New Zealand, Caruso (2002) used monitoring data to ascertain the effects of drought on flow, water quality and macroinvertebrates. The macroinvertebrate data were in the form of the macroinvertebrate community index (MCI), developed for New Zealand stony streams, and in which the macroinvertebrates were ranked on the basis of their sensitivity to pollution and nutrient enrichment (Stark, 1993). While this index may indicate the effects of drought exerted through changes in nutrient levels, it is a rather indirect indicator of the effects of drought. These effects include such stressors as high temperatures, reduced flow velocities and low oxygen levels. The MCI values at a regional level for the streams in Otago did not change significantly from values gained from extensive pre-drought sampling (Caruso, 2002), but there was some evidence that the MCI did decrease slightly as flows fell below the mean annual low flow values. Accepting the value of the MCI as indicative of macroinvertebrate community structure, there are indications that severe drought only had minor effects on the invertebrates of the Otago stony streams.

In Victoria, Australia, since 1990, the Environment Protection Authority has been monitoring water quality and macroinvertebrates at 1,400 sites, including 250 intact reference sites, to determine river health (Rose *et al.*, 2008). Severe drought started in 1998 and continued until mid-2010. In the study of Rose *et al.* (2008), results from 1998 to 2004 were compared with pre-drought samples. Sampling was carried out in riffles and in the edge 'toiche' zone. The data were analyzed to produce four indices (SIGNAL, AUSRIVAS, EPT and family richness).

At a state-wide level, changes due to drought were not detected in either EPT or family richness, but changes were detected in SIGNAL scores of both riffle and edge samples and in AUSRIVAS scores for the edge samples. The results for the riffle samples were affected by the lack of sampling at sites where riffles had dried up in the drought. In terms of both edge and riffle habitats, the bioregion most affected by drought was B4 – the streams of the cleared hills and coastal areas of central and eastern Victoria (Wells *et al.*, 2002). Edge sample results from streams in B5, the very dry western and inland plains, did not show any effect – presumably these streams are inhabited by an entire invertebrate fauna well adapted to harsh dry conditions.

Streams in forested bioregions were also affected by drought, with the effects being more marked in the forested foothills region than in the upland forests. In the edge samples at drought-affected sites, pollution-tolerant fauna (e.g. Veliidae, Gerridae, Culicidae, Stratiomyidae) increased at the expense of sensitive taxa (e.g. Coloburiscidae, Psephenidae, Hydropsychidae, Helicophidae). In the riffles, sensitive taxa, such as Helicophidae and Empididae, were reduced, while pool-dwelling taxa such as Calamoceratidae and Podonominae benefited (Rose *et al.*, 2008). Overall, the study clearly shows the value of long-term sustained monitoring at the large scale, not only for detecting changes in water quality and biota due to anthropogenic stresses, but also for detecting the effects of prolonged disturbances such as drought. The study also reveals the differential effects that drought may exert on streams in different bioregions within a large region – the state of Victoria.

Using data on the large-scale effects of drought on Australian aquatic insect families, Boulton & Lake (2008) devised a scheme for the families reduced by the various sequential stages during drought in streams. In their table, six stages are listed, beginning with 'Decreases in flow/volume' and ending with 'Pools dry: taxa with desiccation-resistant stages or able to survive in moist stream bed'. The listing of families is provisional, but it does illustrate the point that taxa adapted to tolerate low water quality and degraded stream conditions are well adapted to tolerate the severe stresses exerted by drought. Drought-tolerant families include families from three insect orders – Coleoptera, Hemiptera and Diptera – all of which include many taxa tolerant of pollution and habitat degradation. The list is basically a compilation of the relative tolerances of insect families and, as such, it is a list of the ranking of families in terms of resistance to drought. It is not necessarily indicative of resilience – the capacity to recover after drought.

8.10 Summary: drought and stream benthos

Drought can have severe effects on the abundances and species compositions of benthos in perennial streams. These effects are a function of the duration and severity of the drought that governs the loss of water and of the geomorphology of the stream which governs the habitats and stream sections heavily affected by drought, such as sections where flow ceases. Clearly, some invertebrate groups are more intolerant of drought than others. The EPT (Ephemeroptera, Plecoptera, Trichoptera) along with Hirudinea, amphipods and many molluscs, are particularly susceptible, while hardy survivors include species in the Turbellaria, Oligochaeta, Hemiptera, Coleoptera and Diptera. The timing of the drought is critical, as drought can deplete populations of developing aquatic larvae and nymphs. The effects of drought can be heightened by 'blackwater' events and organic pollution and nutrients from human activities.

Recovery from drought is strongly influenced by the characteristics of the drought, in particular its duration. Recovery is partly dependent on recruitment from refuges both *in situ* (e.g. desiccation-resistant eggs, hyporheic zone) or from elsewhere (e.g. stream drift, flying insects). Rates of recovery vary greatly, with several cases of lost species and reconfigured assemblages. The reconfiguration may alter post-drought functional feeding groups, species traits and trophic structure, but this fascinating area remains unstudied. For example, drought may eliminate top predators such as fish, and thus release predatory insects from both competition and predatory pressure, which in turn may produce a novel but transitory trophic structure. During recovery, as for other groups such as zooplankton, hitherto rare species may briefly boom.

Invertebrates are very suitable for the assessment of large-scale effects of drought – the scale at which drought occurs. The few such studies with molluscs and general benthos have indicated geographical areas with severe effects and types of streams, habitats, species and groups that are particularly affected by drought.

8.11   General conclusions

By and large, it appears that, depending on the strength of droughts, their biotic effects are more severe in perennial systems than in temporary systems. This difference reflects the fact that successful biota in temporary systems have adapted to periods of low or no water.

The initial conditions of a water body prior to drought occurring can exert a strong influence on its ecological state after drought. For example, the effects of drought on oligotrophic lakes are much more muted than those that occur in eutrophic lakes. Furthermore, drought may occur when an ecosystem is recovering from the effects of another disturbance (often anthropogenic), and such timing may be very damaging. As will be seen in Chapter 11, human impacts on aquatic systems can lead to initial conditions which exacerbate the impacts of drought.

Droughts occur differentially at a large spatial extent. In running water, drought affects different stream orders, with low-order headwater streams being usually more heavily affected than the high-order, floodplain river channels. Small to medium perennial streams exposed to drought can cease to flow, with pools forming and the development of water quality problems. Thus, stream continua are transformed into linear series of heterogeneous patches, which stresses the need for drought studies to work at the appropriate streamscape level rather than at the level of individual sites.

In both lentic and lotic systems, as drought sets in and volumes decline, there is an increased risk of deteriorating water quality, with such changes as lowered dissolved oxygen and increased salinity and nutrient concentrations. Nutrients may be remobilized by increased mixing due to declines in depth and by the re-wetting of dry sediments in re-filling. Thus, especially in lentic systems, lowered water quality with drought may cause major changes in the biota.

Conversely, with the breaking of a drought and refilling and recovery, increased nutrients and inorganic ions from the catchments and the dry water body itself can alter recovery. This is well shown in lakes where, as volume declines, the normal phytoplankton can be replaced by cyanobacteria and the normal crustacean zooplankton declines while there can be a transient boom of rotifers. This pattern may emerge again when the drought breaks, with cyanobacteria and rotifers appearing with re-filling, to be in turn replaced by the pre-drought phyto- and zooplankton.

Different biota have different tolerances of drought conditions. Thus, in Lake Chilwa, there was a distinct sequence of species loss in the fish, which was mirrored in the recovery. In streams, invertebrate groups which are diagnostic of good water quality and habitat, such as Ephemeroptera, Plecoptera, Trichoptera (EPT), shrimps and molluscs (mussels), can be greatly depleted in drought, with some species becoming locally extinct. On the other hand, taxa such as oligochaetes, coleopterans, hemipterans and many dipterans can tolerate the harsh conditions of drought.

In recovery from drought (as described in Chapter 5), with the wetting of stream catchments and the return of flow, large amounts of ions, nutrients and dissolved and particulate organic matter may be swept into and along streams. The full effects of this influx, which may be sudden, remain to be described. However, it does appear that, in recovery, detritus and biofilms are the major food resources for the fauna (detritivores, primary consumers) before attached algae and their grazers return.

In both lentic and lotic systems during drought, hitherto rare species may become common. Such outbreaks suggest that in plant and animal communities, there can be species which can capitalize on environmental conditions that are hostile to the normally occurring biota.

Drought may cause unexpected changes in communities due to changes in predators and competitors, which become evident in drought recovery. For example, top predators (e.g. fish) may rapidly repopulate and deplete recovering zooplankton populations to low levels. Alternatively, the temporary loss of predators may allow a transient boom in prey populations. In other instances, drought produces conditions that favour invading species.

In perennial lentic systems, both algae (phytoplankton) and macrophytes undergo major changes with drought, whereas the changes may be minor in many perennial lotic systems. In some cases, the macrophyte assemblage after drought may be quite different from that existing before drought. Indeed, drought as a disturbance may cause a transition in lentic systems to an alternative state (stable?) mediated largely through the macrophytes.

Key to the survival of biota during drought in both lentic and lotic systems is the provision of refuges utilized by biota with the appropriate life history adaptations. Many refuges for particular stages, especially eggs, seeds and cysts, occur at sites and habitats *in situ*. Examples include desiccation-resistant eggs in the dry mud of lake bottoms or on dry floodplains. Other refuges involve migration to sites such as deep pools that persist through the drought. For all of these refuges to allow successful recovery, connectivity is a critical requirement. Connectivity has three forms: longitudinal connectivity along flowing waters and between streams and lakes; lateral connectivity, such as between floodplains and their river channels (both of these are major and critical pathways for successful recovery from drought); and vertical connectivity between surface water, the hyporheic zone and groundwater (which is increasingly being seen as important in drought recovery in some systems).

There are very few comprehensive studies of drought, from pre- to post-drought conditions. The study of Lake Chilwa, incorporating investigations of lake morphology, water quality, littoral vegetation, phytoplankton, zooplankton, benthos and fish, stands out as the only example of an integrated holistic account of the impacts of supra-seasonal drought on a lake.

Long-term studies of drought at the large spatial extent, the extent at which droughts operate, remain unexplored. It would be a great bonus if such studies were to be initiated, and if they were to incorporate an integrated approach, from hydrology through primary producers to top predators. Another obvious gap in research on drought and aquatic ecosystems is an understanding of how drought affects the dynamics of ecosystem processes, such as how it alters energy pathways in food webs.

9

Drought and fish of standing and flowing waters

Freshwater fish live in a wide variety of water bodies, from small springs to large floodplain rivers, and from small ponds to large lakes. In most of these systems, fish play an important part in biomass production, in regulating biodiversity and in strongly influencing trophic structure. In some stable systems fish, as top predators, may exert strong top-down control on trophic structure, which may result in trophic cascades (Carpenter *et al.*, 1985; Power, 1990).

Fish display a wide variety of feeding mechanisms, and different species harvest different food resources. They may feed on a wide variety of food resources, including detritus, algae, plankton, benthos, other fish and terrestrial prey subsidies (e.g. fruit, grasshoppers). Fish may range from having very specialized food requirements to being opportunistic omnivores, and most fish change their diets with age. Of all the biotic components of freshwater systems, fish are the most exploited by humans, as well as being key prey for other terrestrial predators (e.g. birds and mammals). Fish are thus a vital component of many, if not most, aquatic ecosystems.

Freshwater fish, with very few exceptions, require free water to live. Thus, they are particularly vulnerable to the loss of free water due to drought, and they share this vulnerability with a few other taxa, such as decapod shrimps. Not only is free water required to survive a drought, but water quality needs to be tolerable. In most cases, supra-seasonal drought, as water volumes decrease, will produce significant changes in water quality such as high temperatures, increased salinity and low oxygen levels, which may result in fish kills.

As mentioned earlier (see Chapter 6), a few fish live in temporary waters. Species that successfully do so display one of two attributes – the capacity to resist desiccation (e.g. aestivation of adults, diapausing eggs) or being able to migrate away from the temporary waters as drought sets in. Fish of

Drought and Aquatic Ecosystems: Effects and Responses, First Edition. P. Sam Lake.
© 2011 P. Sam Lake. Published 2011 by Blackwell Publishing Ltd.

floodplain rivers, especially the 'black group', can survive the harsh conditions of drought in floodplain lagoons (Welcomme, 1979; see Chapter 7). In this chapter, we are dealing with the effects of drought upon fish that live in permanent lentic and lotic systems.

In drought, freshwater fish use a wide range of localities as refuges. These may be specialized, such as small springs, to simply large, deep pools. In most cases, migration to the refuges is required and, with the breaking of the drought, migration and dispersal are required for effective recovery. Thus, a vital necessity besides the availability of refuges is the requirement for effective avenues of connectivity.

Understandably, given their cultural and economic importance, there have been many studies of the impacts of drought on fish – more than on any other freshwater group (Lake *et al.*, 2008). It is surprising to note, then, that while the ecology of fish is a major part of the literature on the ecology of lentic systems, studies on the effects of drought on lentic fish are few compared with those addressing drought and fish in flowing waters (viz. Matthews & Marsh-Matthews, 2003). In studying the effects of drought on flowing water systems, the greatest emphasis has been on fish dwelling in low-order streams rather than large rivers (viz. Matthews & Marsh-Matthews, 2003). In this chapter, the treatment of drought and freshwater fish will start with droughts and lentic systems, progress to lotic systems and finish with a small section on fish genetics and drought.

While many drought studies have been made at the local site level, the key to understanding the full effects of drought lies in adopting a landscape approach to addressing fish populations in streams and rivers within catchments (Schlosser & Angermeier, 1995; Fausch *et al.*, 2002). Needless to say, this approach has rarely been adopted in drought studies.

9.1   Drought and fish of permanent lentic systems

The number of studies on fish responses to droughts in standing waters is limited, and long-term studies are very few. Most studies of the effects of drought on lentic fish have been carried in tropical or subtropical areas.

For fish, as a lentic system is exposed to drought, there are major changes in habitat availability, water quality, food resources and biotic interactions. As a water body loses water, the water level may pull away from the normal shoreline, stranding littoral plants and sedentary fauna (e.g. mussels) and diminishing habitat and food availability for fish, such as access to rocky shorelines and undercut banks. In lakes, the loss of the littoral zone can alter fish population and community structure. For example, in Lake Constance, Fischer & Ohl (2005) found that intra-specific competition in turbot

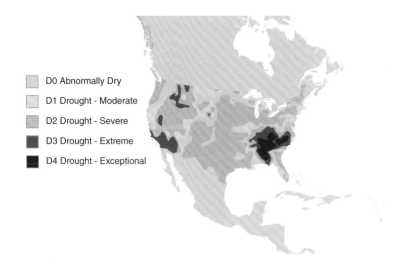

	D0 Abnormally Dry
	D1 Drought - Moderate
	D2 Drought - Severe
	D3 Drought - Extreme
	D4 Drought - Exceptional

Plate 1.2 An example of the output from the Drought Monitor, showing the extent and severity of drought in central and southern USA on October 9, 2007.

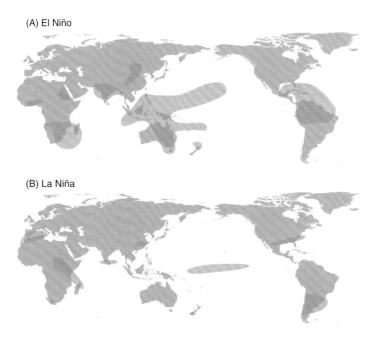

Plate 1.3 (a) Map of the world, indicating regions liable to incur drought conditions with an El Niño event. (b) Map of the world, indicating regions liable to incur drought conditions with a La Niña event. (Adapted from Allan *et al.*, 1996.)

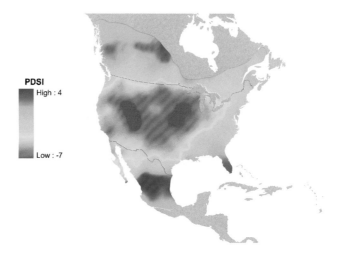

Plate 1.4 The spatial extent and severity of the Dustbowl drought in 1934, with regions in drought depicted by the Palmer Drought Severity Index PDSI (in red with negative values) and wet regions (positive PDSI and in blue). (Drawn using data from Cook, E.R., 2000.)

Plate 6.3 Bands of adult dryopid beetles (*Postelichus immsi*) plodding upstream to persistent water as drought develops in the Santa Maria River, Arizona, USA. (Picture from Figure 1 (right) in Lytle *et al.*, 2008.)

Plate 8.2 The shore of Lake Chilwa as the lake dried, showing the cracked, salt-encrusted surface of the mud and a retracting waterway used to gain access to the body of the lake. (From a slide kindly provided by A.J. McLachlan.)

Plate 8.4 An exhausted clawless otter, *Aonyx capensis*, attempting to move across the drying lake, is intercepted and is about to be killed and subsequently eaten. (Slide kindly provided by A.J. McLachlan and which appears as Figure 517(a) in Kalk *et al.*, 1979.)

Plate 9.1 A persistent pool on Middle Creek, Victoria during the 1997–2010 drought. This pool was the only refuge for kilometres, both upstream and downstream, and it harboured an abundant fish fauna consisting of at least 1,153 flathead gudgeons (*Philypnodon grandiceps*), 66 Australian smelt (*Retropinna semoni*) and 11 mountain galaxiids (*Galaxias olidus*), along with 687 yabbie crayfish (*Cherax destructor*). (Photo courtesy of Paul Reich.)

(a)

(b)

Plate 9.2 Fish kills on the Manaquiri River, northwest Brazil, in the severe drought in 2009. (Photos sourced from Reuters.)

increased with decreasing habitat, whereas stone loach simply became more catholic in their habitat requirements.

Related to this lowering of water level is the likelihood that loosely compacted sub-littoral sediments exposed to wave action may be stirred up, increasing the concentration of suspended fine sediments that may serve to lower water quality. In Lake Chilwa, as drought set in, wave action stirred up fine silt that proved to be a direct stress to fish by damaging gill epithelia. The silt also caused oxygen concentrations to decrease appreciably; indeed, 'it was capable of deoxygenating sixteen times its own volume of aerated water' (Furse *et al.*, 1979).

With drought and water loss, the salinity (conductivity) of the water body can increase, along with changes in alkalinity and pH. In Lake Chilwa, for instance, conductivity normally fluctuated between $1,000–2,000\,\mu S\,cm^{-1}$, but with the onset of drought (1966–67), conductivity reached a maximum value of $12,000\,\mu S\,cm^{-1}$, a value accompanied by an increase in alkalinity (pH 9.3–9.6). These conditions were highly stressful to the fish (McLachlan, 1979a; Furse *et al.*, 1979). With water volumes dropping in lakes and increased perturbation of benthic sediments, nutrient concentrations, (nitrogen and, especially, phosphorus) may rise and have the undesirable effect of triggering phytoplankton blooms, which may greatly lower oxygen levels at night. Furthermore, the blooms may consist of cyanobacteria, and microcystins may be produced and bioaccumulated by fish (e.g. Magalhães *et al.*, 2001), becoming a serious health hazard to humans (Malbrouck & Kestemont, 2006).

When hypoxia occurs in standing waters due to drought, it is most likely caused by a number of concurrent drivers. For example, as Lake Chilwa receded in drought, there were oxygen deficiencies due to the increase in silt, along with high water temperatures. Furthermore, cyanobacteria blooms caused oxygen stress at night (Furse *et al.*, 1979). In shallow lakes (e.g. Lake Chad in the drought beginning in 1973), with the reduction in volume and the development of dense aquatic macrophyte stands, stress-generating thermal and oxygen stratification can occur (Carmouze *et al.*, 1983). As waters recede, detritus from dead plants can accumulate in the bottom of the water body, augmenting microbial decomposition and increasing both carbon dioxide concentrations and alkalinity (e.g. Kushlan, 1974a; McLachlan, 1979a).

Suriname, in north-eastern South America, is subject to El Niño-induced droughts. Mol *et al.* (2000), using data on fish harvested from a system of brackish lagoons, assessed the effects of droughts and found that they greatly reduced fish abundance in the lagoons. During the severe 1997–98 drought, there was a sequence in fish death, with ariid catfish (*Arius* spp.) dying first, to be followed by snook (*Centropomus* spp.), tilapia (*Oreochromis mossambicus*), mullet (*Mugil* spp.) and tarpon (*Megalops atlanticus*). Dead and dying fish provided a feast for wood storks (*Mycteria americanus*). High temperatures linked with low oxygen and high salinity ($40\,g\,l^{-1}$) appeared to be the causes of the fish deaths. As evidenced by the lack of breeding nests, the long-lived callichthyid armoured catfish did not breed in drought years. Even though there were fish kills and the lagoons dried up completely, recovery occurred in the next wet season. The rapid recovery was largely due to the migration of fish (e.g. mullet, tarpon) which reproduce at sea and migrate into the lagoons.

In comparison with the rapid recovery of fish populations observed by Mol *et al.* (2000), through maintenance of connectivity and marine refuges, Piet (1998) found that drought produced major changes in the fish assemblage of a tropical reservoir in Sri Lanka. Piet's four-year study gathered data before, during and after a drought. The hydrological drought was exacerbated by water extraction for irrigation, and it resulted in the reservoir becoming completely dry for two months in 1992. The reservoir harboured nine fish species. As water levels dropped, the rich littoral habitat was lost and turbidity rose due to increased suspended silt and seston. Food availability for the fish declined, their condition deteriorated and mortality increased. Several species (e.g. *Barbus* spp., *Oreochromis* spp.) underwent major changes in diet, reflecting the loss of the littoral macrophyte zone. As the drought developed, the abundant small pelagic cyprinids (*Amblypharyngodon melettinus* and *Rasbora daniconius*) declined to very low numbers, but the abundances of benthic species (e.g. *Glossogobius giuris*, *Mystus* sp.) were unaffected.

After the drought, turbidity was low, with little seston, and the pelagic cyprinids stayed at very low levels. The assemblage structure of fish after the drought was very significantly different from those existing before the drought, with marked changes in dominance in feeding guilds and species composition (Piet, 1998). Thus, the drought altered the assemblage structure by principally altering food resources, and this change persisted for at least two years after the drought.

Unlike benthic fish, pelagic, short-lived and small-bodied fish appear to be particularly susceptible to the impacts of drought. Lake Kariba is a large impoundment on the Zambezi River on the border between Zambia and Zimbabwe. The lake has a relatively low water residence time, and nutrient retention from the nutrient-poor Zambezi and Sanyati Rivers is low (Marshall, 1988). The lake is warm, and stratification occurs for 7–8 months of the year. When mixing occurs in June and July, nutrients from the hypolimnion move to the surface, triggering sequential blooms of phytoplankton, zooplankton and planktivorous sardine (*Limnothrissa miodon*), which is the major component of the lake's fishery (Marshall, 1988). The sardine fishery is thus dependent on nutrients from destratification and from inflowing rivers; high river inflows produce high sardine catches, while catches are low in years of low inflows. In drought years (1982, 1983, 1984), catches were halved compared with normal years. The decline of the fishery in drought is mainly due to the low nutrient levels, which inhibits phytoplankton and zooplankton production (Marshall, 1988). In this case, drought did not produce stressful abiotic conditions but stifled limnetic primary production, which greatly reduced food availability for the fish.

Drought may create conditions that favour particular species. The severe 2000–2002 drought in Lake Okeechobee, Florida, lowered water levels and depleted the submerged aquatic vegetation, which was replaced by beds of *Chara*. With this change, largemouth bass (*Micropterus salmoides*) failed to recruit (Havens *et al.*, 2005). When the lake refilled, dense stands of submerged and emergent vegetation flourished and led to a greatly enhanced recruitment of the bass. As mentioned in Chapter 8, the boom of young bass produced a major change in the zooplankton. In this case, by creating new conditions afterwards, such as a boom in aquatic vegetation, drought may indirectly alter fish and plankton populations, species composition and trophic structure.

Dramatic and well-documented changes in fish populations and assemblages come from the long-term studies of two African lakes, Lake Chad (Bénech *et al.*, 1983) and Lake Chilwa (Kalk *et al.*, 1979) (see Chapter 8). Both of these lakes suffered from severe droughts. In Lake Chad, the drought was preceded by a 'drying up' phase from 1972 to 1974 and was followed by a severe period of drought called the 'Lesser Chad' that started in 1974 and continued until 1978, when the research ceased. The drought is still ongoing (Coe & Foley, 2001). Lake Chilwa was exposed to a severe drought that started in 1965 and continued until 1969, and recovery was assessed. For Lake Chad, there are only data before and during the drought.

In Lake Chad, with the coming of the drought, the lake became divided into two basins – north and south – separated by a shallow barrier with dense vegetation (see Chapters 5 and 8). The south basin comprised two major parts: the open water and the shallow southeastern archipelago, with abundant aquatic macrophyte growth.

Lake Chad had a rich and highly productive fish fauna consisting of ≈120 species, with only one endemic species, *Alestes dageti* (Alestiidae). In the north basin, water levels from 1973 to 1975 fell dramatically, accompanied by a sharp rise in conductivity and extreme fluctuations in oxygen concentrations. Even though flood waters did flow into the south basin, the barrier of dense vegetation named the 'Great Barrier' prevented flood waters from entering the north basin from 1974 onwards. As the drought set in, a major cause for fish mortality was due to an increased fishing effort, along with mortalities due to 'tornadoes' (Bénech *et al.*, 1983). The latter occurred when the water levels dropped to below 2 m and high winds whipped up fine sediments into the water column, lowering oxygen levels and damaging fish gills.

Thus, early in the drought, the abundance of hitherto important species which were sensitive to anoxia, such as *Heterotis niloticus*, *Hydrocynus brevis*, *Citharinus citharus*, *Tetraodon fahaka*, *Pollimyrus isidori* and *Mormyrus rume*, crashed. Along the windward shores of the lake, after the 'tornadoes', there were strandlines of dead fish. Nevertheless, populations of other fish, such as the catfish *Synodontis schall* and *Brachysynodontis batensoda* (both Mochocidae), three species of the cichlid *Sarotherodon* spp. and the air-breathing *Polypterus senegalensis* were not greatly affected.

As the drying of the north basin continued, fish retreated to the deeper centre of the lake, including Nile perch (*Lates niloticus*), *Sarotherodon* spp., *Synodontis schall*, *Brachysynodontis batensoda*, *Alestes baremoze* and *Alestes dentex*. By 1975, as the lake dried up, the fish assemblages changed to a 'marshy fish community' comprising *Polypterus*, *Brienomyrus niger*, *Sarotherodon* and large numbers of clariid catfish, which are all adapted to low water quality. Finally, with the northern basin drying completely, clariid catfish burrowed into the bottom mud.

In the southeastern archipelago with the decline in water level, macrophyte decomposition set in, lowering water quality. Storms stirred up fine sediments that with the decomposition created extensive areas of anoxic water. Not surprisingly, abundant fish sensitive to anoxia, such as *Hydrocynus forskallii*, *Citharinus citharus*, *Hemisynodontis membranacea*, *Lates niloticus*, *Alestes dentex*, *Synodontis frontosus* and *Labeo senegalensis* disappeared, whilst species tolerant of anoxia such as *Polypterus senegalensis*, *Clarias* spp. and *Brienomyrus niger* remained and dominated the assemblage (Bénech *et al.*, 1983).

In the open water of the southern basin, floodwaters, though reduced, came into the lake. This part of the lake did not suffer from the suspension of fine sediments but, with the volume declining and nutrients increasing, there were phytoplankton blooms that in turn depleted oxygen levels. Initial changes as the lake volume declined were severe drops in the abundance of hitherto important species such as *Citharinus* spp., *Mormyrus rume*, *Bagrus bayad*, *Alestes dentex*, *Hydrocynus brevis* and *Labeo senegalensis*. Species that had previously been rare (*Synodontis clarias*, *Schilbe uranoscopus*, *Distichodus brevipinnis*) started to increase with the increase in anoxic patches in the lake (1972–73). There was also a group of resilient fish (e.g. *Synodontis schall*, *Hydrocynus forskalli*, *Polypterus bichir*) whose abundance did not change with drought. Finally, there was a group of small-bodied fish (e.g. *Pollimyrus isidori*, *Siluraodon auritus*, *Icthyborus besse*) that were rare before the drought but became abundant with the drought.

Thus, the changes to the fish assemblages were complex in the south basin, with transition from the 'lacustrine' species to the 'marshy' species being strongly influenced by migration into and from the inflowing rivers, principally the Shari (Bénech *et al.*, 1983). Exacerbating the effects of drought in both basins was the sustained fishing effort (Bénech *et al.*, 1983; Carmouze *et al.*, 1983). The drought continued after 1978 and no further data are available.

As drought strengthened its grip in Lake Chad, there was thus a loss of fish species and a drastic reconfiguring of the fish assemblages. Much of these changes appear to have been driven by abiotic drivers generated by the drought, rather by changes in biotic interactions. Though not documented, the marked changes and reductions in invertebrate prey may have had a significant influence on the changes in the fish assemblages.

Lake Chilwa suffered an extreme drought in 1966–68, when the lake dried out completely (Lancaster, 1979). Normally, the lake has 30 species of fish, including 12 species of the cyprinid genus *Barbus*. Until the drought, the lake supported a productive fishery based on clariid catfish, *Barbus* spp., and *Sarotherodon shiranus* (Furse *et al.*, 1979). With drought developing, lake volumes declining and strong winds stirring up fine benthic sediments, fish kills involving the cichlid *Sarotherodon* occurred. The fine, organic-rich sediments appeared to have also caused oxygen levels to drop (Furse *et al.*, 1979). Note that the 'tornadoes' in Lake Chad had similar effects. This combination of fine suspended sediments, high water temperatures and oxygen stress, led to the subsequent death of large numbers of *Barbus*.

With the decline in lake volume and increases in nutrients in Lake Chilwa, blooms of cyanobacteria (e.g. *Oscillatoria, Anabaena*) occurred, with consequent deoxygenation, causing further stress for the fish, particularly the remaining *Sarotherodon* and *Barbus*. As also occurred in Lake Chad, perennial streams flowing into the lake were refuges for fish migrating out of the lake. This escape appeared to be especially used by fish which normally dwelt in shoreline swamps rather than in open water.

As the lake dried up, the only fish surviving were clariid catfish that sought refuge by burrowing into the mud and undergoing aestivation. With the return of water to the lake in 1969, the clariid catfish were among the first species to repopulate the lake, to be followed two years later by *Barbus*, then by *Sarotherodon*, whose populations took three years to build up to pre-drought levels. Thus, the succession of dominant fish was the inverse of the sequence of their demise in the drought and appeared to have been regulated by the generation of favourable abiotic conditions, the recovery of food resources, and connectivity allowing dispersal from refugia. Not surprisingly, given the severity of the drought, the length of time for recovery was longer (more than three years) than the length of time (one year) that it took for the lake to refill.

Faced with drought, fish in lentic systems may seek refuges, so that populations may survive the drought and subsequently recover. For fish in isolated water bodies, escape to other water bodies that survive the drought may be limited, but in flowing waters, escape is more feasible. Without leaving a lake or pond, some species may survive by aestivating as adults (e.g. clariid catfish) or by producing drought-resistant, diapause eggs (see Chapter 6). Fish in lakes and lagoons during drought may seek refuge in inflowing rivers, as occurred in Lake Chilwa and Lake Chad, or they migrate into the sea, as occurred in the lagoons studied by Mol *et al.* (2000). Fish may seek refuge from the high temperatures associated with drought, such as where groundwater springs enter lentic water bodies (e.g. Hess *et al.*, 1999). A highly valued fish showing this behaviour is striped bass (*Morone saxatilis*), populations of which may be threatened when lengthy droughts deplete groundwater springs (Hess *et al.*, 1999; Baker & Jennings, 2005).

As drought takes hold and reduces habitat and water volume, fish populations are compressed, heightening the pressure of intraspecific interactions such as competition, as well as interspecific interactions such as predation and competition. Kobza *et al.* (2004), in wetlands of the Florida Everglades, observed that small native fish could survive in shallow holes during dry spells but not in droughts. However, in deep holes (depth >1 m) that may persist through drought, predators dominated and preyed upon the small-bodied native fish, augmenting their loss in droughts. Populations of the red shiner (*Cyprinella lutrensis*) in streams flowing into Lake Texoma have been eliminated (Matthews & Marsh-Matthews, 2007). During drought, red shiners move downstream into deep pools. However, in streams flowing into Lake Texoma, this means moving into high predation pressure – a pressure maintained by upstream migration of piscivores from the lake.

Clearly, the knowledge of refuges used by fish living in lakes is fragmentary. We have a poor understanding of the role that refuges may play in allowing lacustrine fish to persist during drought, and in their capacity to recover after droughts.

Changes in abiotic conditions through drought may directly stress fish, causing them to seek refuges or die. Indirectly, drought may cause changes in nutrients and aquatic vegetation, such as algal blooms, that trigger changes affecting particular fish species. Where recovery has been assessed, it appears that in lakes, the recovery of fish populations and assemblages is relatively slow and may result in fish assemblages different from those occurring before a drought.

9.2 Drought and fluvial fish

As for the other groups of aquatic biota, most studies of the effects of drought on fish have been carried out in low-order streams, whereas there have only been a few studies in large rivers. In examining the effects of drought on lotic fish, there have been basically been three types of study:

1 accounts simply of the effects of drought, such as fish kills;
2 studies that have documented the effects of drought and have then reported on recovery after the drought;
3 comprehensive studies that have been the few that document pre-drought conditions, the impacts of drought and the subsequent recovery often with lag effects.

The most interesting studies have been the very few that have covered a number of droughts. Most studies have been short term, but it is good to see that there are some long-term studies on drought and lotic fish. For example, there is the 34-year long study of fish in a small stream in northern England, documenting the effects of drought on population dynamics (Elliott, 2006).

As for most of the studies of drought on freshwater ecosystems, the spatial extent of the studies has been small – a particular site or several sites – whereas the effects of a drought are invariably at a large spatial extent (Matthews & Marsh-Matthews, 2003; Lake *et al.*, 2008). The need to adopt a landscape/waterscape approach to studying the effects of drought accords with the view advocated by Schlosser (1995), Schlosser & Angermeier (1995) and Fausch *et al.* (2002) that, in trying to understand adequately the dynamics of riverine fish populations, investigations need to encompass a cross-scale approach, both in temporal and spatial extents.

As streams become more temporary, the number of fish species declines and, in headwater streams, fish species richness is much lower than those of downstream rivers (Matthews, 1998; Poff *et al.*, 2001). This difference may be partly due to the winnowing effect of low flow events and drought, which affect small streams much more than large rivers. Thus, in many instances, headwater streams may be intermittent, with perennial conditions increasing downstream. This gradient is reflected in the relative tolerances to drought of headwater stream fish in comparison with fish dwelling in perennial stream sections.

As in studies dealing with other biotic groups, it is difficult to determine the nature of the drought in most fish studies. Cases where summer or dry season drought is a normal seasonal event are not necessarily dealt with here but, in some cases, the summer drought appears to be a severe seasonal event (e.g. John, 1964; Ross *et al.*, 1985) or it has been transformed into a supra-seasonal drought (e.g. Bond & Lake, 2005; Dekar & Magoulick, 2007).

Drought initially results in a decline in flow volume, depth, velocity and habitat availability. Further drops in flow lead to a retraction of the stream/river from the normal littoral fringe, which may be important habitat for food, reproduction and for residency. As flows continue to drop, structural habitats such as macrophyte beds, debris dams and coarse wood are lost, and movement across stream sections, such as riffles and runs, may become restricted. The loss of riffles is usually proportionately much greater than the loss of pools (e.g. Hakala & Hartman, 2004). With habitat availability dropping, fish densities rise and fish may disperse to refuges. However, when flow ceases and pools form, dispersal also ceases and fish densities may be high, resulting in an increase in the intensity of biotic interactions. As pool volumes drop, water quality problems can develop, such as diel temperature extremes, low oxygen levels, increased conductivity (salinity), shifts in pH and high levels of dissolved organic carbon. Water quality may decrease to such an extent that fish kills occur. Finally, pools may dry out.

When streams dry, distinct patterns may emerge of sections with flow, sections with pools and ones which are dry. An important point to emerge from the landscape/waterscape ecology of drying is that the spatial distribution for fish as drought develops is critical to their survival and subsequent recovery. Furthermore, pools may be drought refuges for fish, and the spatial distribution and persistence of these pools is critical to their survival and recovery. Similarly, sections with persistent flow can be refuges for rheophilous fish.

While the effects of drought developing and persisting are relatively well documented, the dynamics of recovery by fish after a drought are still relatively unknown. As described in previous chapters, major changes in water quality, aquatic vegetation and invertebrates occur after drought, all of which undoubtedly interact with fish.

9.3   Dealing with the stresses of drought

9.3.1   *Habitat change and behaviour as drought develops*

As drought develops with declining flows, stream fish may alter their behaviour and their habitat use. In an experiment mimicking drought, Sloman *et al.* (2001) found that the dominance hierarchy that brown trout establish in streams broke down, leading to changes in fish distribution with the decline of territoriality. Changes in habitat use in response to drought can be greater than those that the same fish species undergo with normal seasonal variations in hydrology.

In a long-term study (1983–1992) of fish dwelling in a 37 m section of the perennial Coweeta Creek, North Carolina, Grossman *et al.* (1998) found that the fish assemblage consisted of three 'microhabitat guilds: benthic, lower water column and mid water column'. Drought in 1986–88 reduced velocities and the amount of 'erosional substrata' while increasing the amount of 'depositional substrata'. Fish changed their microhabitat use, notably in the mid-water column group, which was species-specific and dependent on habitat availability (Grossman & Ratajczak, 1998; Grossman *et al.*, 1998). The assemblage structure during the drought differed markedly from either the pre-drought or post-drought assemblages.

Rather surprisingly, during the drought, many resident species, especially those in water column 'guilds' (rosyside dace (*Clinostomus funduloides*), northern hogsucker (*Hypentelium nigricans*), warpaint shiner (*Luxilus coccogenis*), rainbow trout (*Oncorhynchus mykiss*), creek chub (*Semotilus atromaculatus*)), increased in abundance. Also, two species not seen before (river chub (*Nocomis micropogon*) and Tennessee shiner (*Notropis leuciodus*)) migrated to the site and three rare species increased in the site. All of these fish appear to have migrated from downstream sites during the drought.

As regards the benthic fish, the abundance of three species (mottled sculpin (*Cottus bairdi*), longnose dace (*Rhinichthys cataractae*), stoneroller (*Campostoma anomalum*)) did not change or declined, whilst the greenside darter (*Etheostoma blennoides*) emigrated from the site (Grossman *et al.*, 1998). Although drought reduced flow, velocities and habitat availability, the marked changes in the water column guilds and small changes in the benthic guild suggest that the drought effects were largely due to the lack of high flow events – sharp floods which cause high fish mortality in this system (Freeman *et al.*, 1988; Grossman *et al.*, 1998).

The movement of fish in streams in response to disturbance becomes evident in studies carried out at the large spatial extent (Schlosser & Angermeier, 1995; Fausch *et al.*, 2002). In such studies, the key migrations may be detected and important refuges identified. However, studies on fish movements as drying sets in are few. The Selwyn River in New Zealand may dry out in its middle reaches (Davey & Kelly, 2007). Movements of fish with drying in the middle reaches have been measured from drying reaches to upstream perennial reaches, and to downstream perennial reaches. During drying, three species, Canterbury galaxias (*Galaxias vulgaris*), upland bully (*Gobiomorphus breviceps*) and brown trout (*Salmo trutta*) migrated upstream and there were no detectable movements of fish downstream. It was assumed that fish at the downstream side of the drying reaches 'became stranded and subsequently died as the stream contracted' (Davey & Kelly,

2007). While the study was carried out during seasonal drying, the results are applicable to the onset of drought.

In two Brazilian streams as drought set in, fish (pike cichlid, *Crenicichla lepidota*) migrated downstream in considerable numbers, such that the density at downstream sites was sharply increased (Lobón-Cerviá *et al.*, 1993). This migration resulted in low fish production at upstream sites during the drought. The movement downstream to more permanent water may, however, be hazardous. Movements by the threatened Arkansas darter (*Etheostoma cragini*) to downstream reaches exposed the species to very strong predation by an introduced predator, the northern pike (*Esox lucius*) (Labbe & Fausch, 2000).

Drought invariably lowers stream volume, so it can create barriers to fish movement. In a small coastal stream in Sweden, Titus and Mosegaard (1992) found that low flows during drought blocked migration from the sea of spawning brown trout and thus limited reproduction. The Cui-ui (*Chamistes cujus*) is a large, long-lived catostomid endemic to Pyramid Lake, Nevada and migrates up the inflowing Truckee River to breed (Scoppettone *et al.*, 2000). In droughts, low flows prevent the fish from migrating and spawning. However, it appears that the fish are adapted to compensate for the non-spawning years by having a long life and also by having a higher fecundity after non-spawning years than after years in which spawning occurred (Scoppettone *et al.*, 2000).

9.3.2   Fish movements and refuges

Movements by fish as drying occurs depend on the landscape/riverscape pattern of drying, the types of refuges that various species use and the distribution and accessibility of these refuges. While most fish, especially in streams with highly variable flows, seek places of persistent water, fish may also seek refuges from extreme temperatures and other taxing stressors, even when streamflow continues. Salmonids, for example, are sensitive to high water temperatures. In a river in Nova Scotia, Canada, during drought and high water temperatures, salmon and trout have been observed to moved into side arms of the river that received spring water and were cooler than the main channel (Huntsman, 1942). Similarly, during drought in Wilfin Beck, an English stream, trout moved into deeper water that was cooler than surface water, even though oxygen levels were sub-optimal (Elliott, 2000).

In streams during drought, pools as refuges allow fish populations and assemblages to resist drought (resistance) and they serve as centres from which colonists move in recovery after the drought (resilience) (Magoulick & Kobza, 2003; see Figure 9.1). They are particularly important for fish dwelling in intermittent streams that undergo large fluctuations in

Figure 9.1 A persistent pool on Middle Creek, Victoria during the 1997–2010 drought. This pool was the only refuge for kilometres, both upstream and downstream, and it harboured an abundant fish fauna consisting of at least 1,153 flathead gudgeons (*Philypnodon grandiceps*), 66 Australian smelt (*Retropinna semoni*) and 11 mountain galaxiids (*Galaxias olidus*), along with 687 yabbie crayfish (*Cherax destructor*). (Photo courtesy of Paul Reich.) (See the colour version of this figure in Plate 9.1.)

flow (e.g. Labbe & Fausch, 2000; Dekar & Magoulick, 2007) and are likely to be impacted by extended supra-seasonal droughts (Bond & Lake, 2005; Magalhães *et al.*, 2007; Perry & Bond, 2009).

The need for refuges in dealing with drought has been recognized for quite some time. Paloumpis (1956, 1957, 1958), in studying drought in an intermittent stream, suggested that recovery of fish after drought (and floods) was dependent on the availability of 'stream havens'. He found that as drought reduced the flow of Squaw Creek, Iowa, USA, fish moved into refuges ('stream havens') that were either 'creekside ponds' or the downstream reaches of the larger Skunk River. Not all 'stream havens' were safe, however; in the winter of the drought, some creekside ponds froze, killing the fish. While the 'stream havens' were described, the extent of recovery after the drought was not observed (Paloumpis, 1958).

Different species of fish differ in refuge use. In an experiment in an artificial stream in New Zealand, Davey *et al.* (2006) found that two species – Canterbury galaxias (*Galaxias vulgaris*) and the upland bully (*Gobiomorphus breviceps*) – differed in their use of refuges. With drying, both species moved upstream, with the bullies migrating from riffles to the deeper runs, while the galaxiids burrowed into the substrate, gaining more protection in cobble substrate than in gravel substrate. If the rate of flow recession is rapid and short-term, the burrowing galaxiids may be advantaged, but if the rate of flow recession is slow and long-term, as in drought, the bullies may survive more effectively. Thus, for fish normally living in the same habitat – riffles – drought may produce differential mortality.

A similar situation, involving brown trout and bullheads (*Cottus gobio*) was observed by Elliott (2006). In drought, with reduced habitat area and water quality (especially high temperatures), trout densities were reduced, with many fish migrating to the thermal refuge of deep pools (Elliott, 2000). On the other hand, densities of the bullheads, which were more tolerant of the drought stress (e.g. high temperatures) increased as they exploited benthic food resources in the shallow areas created by the drought and left vacant by the trout. Fish may survive stream drying in some unusual habitats, such as in Ecuadoran streams where Glodek (1978) observed that deep burrows dug by catfish served as a 'residual habitat' or refuge for fish during periods of low flow and drought.

9.4   The impacts of drought on lotic fish

9.4.1   Tolerance and survival in small streams

Fish dwelling in headwater streams which are periodically exposed to drought are under pressure to develop adaptations to the extreme conditions. That such adaptations are present is suggested by several studies, including that by Matthews & Styron (1981), who found that a headwater cyprinid fish (*Phoxinus oreas*) was more tolerant of sudden changes in oxygen and temperature than were three cyprinids from the mainstem river in the same system. Individuals of the fantail darter (*Etheostoma flabellare*) from headwater streams were more tolerant of low oxygen levels than those individuals from the main stream (Matthews & Styron, 1981).

Laboratory trials found that the critical thermal maxima for the orangethroat darter varied with the thermal regime from which they were collected (Feminella & Matthews, 1984). A critical thermal maximum of $\approx 32.5\,°C$ was found for a population dwelling in a spring with constant temperature of $18\,°C$, whereas, for the population dwelling in the headwater creek and exposed to extreme temperatures with low flow, the critical thermal maximum was $\approx 35\,°C$ (Feminella & Matthews, 1984).

As a group, fish species dwelling in small headwater streams of Missouri, that periodically undergo drought, appear to be well adapted to extremely low oxygen levels and high temperatures (Smale & Rabeni, 1995). For example, the yellow bullhead (*Ameiurus natalis*) had a hyperthermic tolerance of 37.9 °C and a hypoxic tolerance down to 0.49 mg/l (7.3 per cent saturation) (Smale & Rabeni, 1995).

Fish from the pools of the upper Brazos River, Texas, in drought, face extreme temperatures, low oxygen levels and high salinities (Ostrand & Marks, 2000; Ostrand & Wilde, 2001). The species that occurred in the upper river headwater sections, where drying is most severe, were cyprinodontids – the Red River pupfish (*Cyprinodon rubofluviatilis*) and the plains killifish (*Fundulus zebrinus*) (Ostrand & Wilde, 2001). Both of these fish had higher tolerances to high temperatures, low oxygen levels and salinity than fish occurring further downstream.

In examining fish populations at the upper and lower sections of a prairie stream, Oklahoma, Spranza and Stanley (2000) concluded that the upper section was a much harsher environment, supporting more robust fish than individuals of the same species from the lower stream section.

Fish trapped in pools of small streams during drought may be stressed by low oxygen levels associated with the accumulation of dissolved organic matter (DOM) emanating from litter decomposition. Paloumpis (1957) observed, in pools during a drought, that 'the color of the water was black and the bottom mud had an oily odor' – clearly indications of elevated DOM levels. Oxygen levels were very low (0.2 mg/l) and fish were gulping air at the surface, with some dying.

High DOM (DOC) levels and low oxygen levels characteristic of 'blackwater events' (Slack, 1955) can occur during droughts (Paloumpis, 1957; Larimore *et al.*, 1959; McMaster & Bond, 2008). In pools of intermittent streams in south-eastern Australia during drought, McMaster & Bond (2008) recorded low oxygen levels (range 0.4–6.8 mg/l) (4.1–70.1 per cent saturation) and high levels of dissolved organic carbon (DOC) (range 16–50 mg/l). Three native fish species (mountain galaxias (*Galaxias olidus*), southern pygmy perch (*Nannoperca australis*) and western carp gudgeon (*Hypseleotris klunzingeri*)) were unaffected in terms of abundance. In laboratory experiments, the tolerance of the fish to high concentrations of DOC combined with low oxygen concentrations was confirmed (McMaster & Bond, 2008).

Thus, strong evidence suggests that fish dwelling in headwater streams appear to be better adapted to the physiological stresses posed by drought than are species and populations dwelling in high-order perennial streams. However, drought can be so severe that fish in headwater streams may still be depleted, if not locally eliminated (e.g. Ross *et al.*, 1985).

The structure of fish assemblages in drying pools may be structured by the changes in abiotic variables. In dry-season pools in a Texas stream, Capone & Kushlan (1991) found distinct division into three fish assemblages dominated by mosquitofish (*Gambusia affinis*), or by green sunfish (*Lepomis cyanellus*), or by a mixed-species assemblage dominated by golden shiner (*Notemigonus crysoleucas*), black bullhead (*Ictalurus melas*) and mosquitofish. The mosquitofish assemblage dwelt in the least persistent pools, whereas the bullhead assemblage occupied the larger more persistent pools. The key variables determining the assemblages were days with water in the pools, average depth, maximum depth, pool area and bank height (Capone & Kushlan, 1991). All of these variables, with the exception of bank height, were altered by drying. In drought, the composition of fish assemblages would change toward the mosquitofish-dominated assemblage as drying progressed. As in the Florida Everglades, mosquitofish appeared to be very resistant and resilient to the stresses of drying.

In drying stream pools in the Ozark Mountains, Arkansas, USA, the distribution of fish assemblages was determined by location in the catchment. At the level of pools, there were few variables which explained the variability in fish assemblages (Magoulick, 2000). In the pools, fish total density, large central stoneroller (*Campostoma anomalum*) density and small sunfish (*L. cyanellus*) density were all positively correlated with pool depth. Abiotic factors in explaining fish distribution were overridden by regional factors, which were related to large-scale factors such as drought severity. Dekar & Magoulick (2007) found that the variables affecting fish densities differed between a normal year and a year with extreme drying or drought. In the normal year, total fish density was positively correlated with canopy 'openness' and substrate diversity, and was negatively correlated with pool area and maximum depth. However, in the year with extreme drying, total fish density was correlated negatively to substrate diversity. These results indicate that factors structuring fish density change in strength and direction, depending on the severity of the drying.

In a Mediterranean summer drought in the Odelouca Stream, Portugal, a diverse fish assemblage of five native species and six alien species used a variety of pools as refuges (Pires *et al.*, 2010). Species richness increased and overall abundances (fish per catch per unit effort) decreased with increasing pool size. Species richness rose due to the large pools offering more habitat variety, while overall abundance declined due to large numbers of small fish being prey for large predators (Pires *et al.*, 2010). Native fish were favoured in pools with good riparian cover, possibly because of the shading effect of the plant canopy and/or the availability of a prey subsidy of terrestrial insects. Thus, pool characteristics have a strong influence on the nature and abundance of the fish assemblages in these refuges (Pires *et al.*, 2010). These refuges, especially deep pools with overhanging canopies, may also allow the fish assemblage to survive supra-seasonal droughts, although this will vary with drought severity.

With drought setting in, some fish species undergo directed migrations to refuges, such as to permanent water and into pools (e.g. Paloumpis, 1956, 1957, 1958; Magoulick & Kobza, 2003; Davey & Kelly, 2007). Other species appear to have limited mobility and are confined to pools when cease-to-flow occurs. For example, western carp gudgeons (*Hypseleotris* spp.) in small streams in southeastern Australia have a very limited dispersal ability (Perry & Bond, 2009), but they do have a high tolerance of hypoxic conditions and of high concentrations of dissolved organic carbon (McMaster and Bond, 2008). Thus, provided the pools that they occupy do not dry out in drought, they can survive in conditions of very low water quality.

9.4.2 Fish kills

Fish kills can occur in response to drought in intermittent and low-order streams, and even in perennial mainstem rivers, but published reports are relatively few. During severe droughts in 1937 and 1939 in Nova Scotia, Canada, the Moser and the St. Mary Rivers flowed at low levels and air and water temperatures were high – high enough in August ($\approx$29 °C) to kill migrating salmon. Fish that survived aggregated in side channels, into which cool groundwater flowed (Huntsman, 1942).

In shrinking pools of a stream in Ohio, USA, during drought, the combined stress of low oxygen concentrations and high water temperatures caused mass mortality of 12 species of fish (Tramer, 1977). Also in an Ohio stream, Mundahl (1990) observed the death of fish from six species in drought, with water temperatures reaching 38–39.5 °C in an unshaded pool. However, in another pool with shading, fish moved into cooler water (6.5 °C lower) and survived.

In a heat wave during a drought, Matthews *et al.* (1982) observed in shallow pools of a Brier Creek, Oklahoma, USA, that high water temperatures (38–39 °C) had killed orangethroat darters (*Etheostoma spectabile*). This was well above the 'critical thermal maximum' determined for the darter of $\approx$35 °C (Feminella & Matthews, 1984). In small streams with drought, fish may seek refuge in pools but, as drought continues, these pools may dry out, with the loss of the trapped fish.

In large rivers, fish kills are rarer than in small streams, but they can occur in large rivers such as in major tributaries of the Amazon in the recent severe drought (2009) (Figures 9.2a and 9.2b).

On the Canterbury Plain in New Zealand, rivers may cease to flow during drought, creating conditions of low oxygen and high water temperatures in pools, which can kill large numbers of trout (Jellyman, 1989).

Figure 9.2 Fish kills on the Manaquiri River, northwest Brazil, in the severe drought in 2009. (Photos sourced from Reuters.) (See the colour version of this figure in Plate 9.2.)

Fish kills can also occur in rivers that continue to flow through drought. In the middle and flowing reaches of the River Wye in the summer of the 1975–76 drought, there was mass mortality of adult salmon (*Salmo salar*) (Brooker *et al.*, 1977). At the time of the salmon death, large patches of macrophytes were dying and decaying in the river. The decay was enhanced by high water temperatures (up to 27 °C) and resulted in extreme values for dissolved oxygen concentrations, which daily fluctuated from $\approx$18 mg/l (225 per cent saturation) in the day to 1 mg/l ($\approx$13 per cent saturation) at

night. The high water temperatures and low oxygen levels at night synergistically served to kill the salmon (Brooker *et al.*, 1977; see Figure 5.7).

Accounts of drought at the large spatial extent in freshwater ecosystems
are rare. During the 'Dustbowl' drought in the USA, James (1934) compiled
reports from in the mid-west and west of the United States. The accounts
painted a grim picture of large losses of fish populations and fish kills in lakes
and rivers. It was stressed that while most fish kills occurred in summer, low
flows during winter may also kill fish due to 'freezing out'. In 1934, James
(1934) indicated that the severe effects of drought on fish was 'roughly
T-shaped', with the top of the T running across Nebraska and Kansas and
through the Dakotas to Indiana, and the vertical part of the T stretching
down the Mississippi valley to Arkansas – a pattern that fits with the spatial
distribution of extreme drought in 1934 (Cook *et al.*, 2007; see Figure 1.4).

9.4.3 Drying and biotic interactions

Fish in drought-stricken streams not only have to contend with stressful
abiotic variables but also have to contend with adverse biotic interactions.
With reduced habitat space and increased fish densities, new interactions
may be created and existing ones greatly intensified. Though there are only
a small amount of data available, it is likely that with fish occurring in high
densities in key habitats, there would be both intraspecific and interspecific
competition (viz. Zaret & Rand, 1971), which may result in loss of condition
of fish and population reduction.

During a drought in headwater streams in West Virginia, USA, body
condition and population density of brook trout declined (Hakala &
Hartman, 2004), possibly due to reduced habitat space and to reduced
food resources intensifying intra-specific competition. The limitation of
food resources led to loss of fish due to starvation during drought in
intermittent streams in Arizona, USA (John, 1964). In a small stream in
northern England, Elliott (2006) found that in years with severe summer
droughts, trout habitat quality was reduced, while habitat quality and
quantity for bullheads was increased. For trout, drought depleted habitat
space and food resources, and competition for habitat and food favoured
bullheads rather than trout. In normal, non-drought years, trout were the
competitive dominant.

Fish may increasingly be preyed on by terrestrial predators during
drought. Fish in shallow river sections may be eaten by an assortment of
terrestrial predators, including herons and raccoons (Larimore *et al.*, 1959;
Matthews, 1998), garter snakes (John, 1964), killdeer (a plover) and herons
(Tramer, 1977), herons, cormorants, kingfishers and snakes (Lowe-
McConnell,1975), otters (Magalhães *et al.*, 2002), Caspian terns (Antolos

et al., 2005) and snakes (Love *et al.*, 2008). The effects of terrestrial predators during drought are uncertain and have been overlooked (Magoulick, 2000). Magalhães *et al.* (2002) observed that otter predation during summer drought strongly reduced cyprinid densities, especially those in refuge pools.

Caspian terns are proficient predators of juvenile salmonids (*Oncorhynchus* spp.) in the Columbia River, USA. In a drought year (2001), heavy predation by the terns on salmon smolt (young salmon migrating to sea) was favoured by the lower river flows, reduced discharge from hydro-electric plants and the increase in time that migrating smolts spent in the river (Antolos *et al.*, 2005).

Predation of fish by fish in streams during drought may increase and can influence fish abundance and distribution, though firm empirical data are few. As the Arkansas darter retreats downstream with drought, it moves into pools inhabited by an efficient predator, the northern pike (Labbe & Fausch, 2000). In shrinking pools of streams in south-western USA, small fish such as minnows (e.g. *Campostoma*) and topminnows (e.g. *Fundulus*) may be eaten by predators such as largemouth bass (*Micropterus salmoides*), which can withstand the high temperatures and low oxygen levels (Power *et al.*, 1985; Matthews, 1998). Indeed, this predation can be so effective that minnows (e.g. *Campostoma anomalum*) can be excluded from pools containing bass.

Galaxiid fish in the southern hemisphere have been greatly reduced in distribution and abundance by competition and predation from introduced salmonids (McDowall, 2006). In normal years in a stream in Victoria, Australia, the mountain galaxiid (*Galaxias olidus*) is confined to small headwater streams through adverse interactions with brown trout (Closs & Lake, 1995). However, when flow ceased with drought, pools with low water quality formed, leading to conditions toxic for the trout. Consequently, trout were depleted by the severe drought, and when flow returned, the galaxiids moved downstream into their former habitat (Closs & Lake, 1995). Drought thus selectively killed an invader and changed interspecific interactions, benefiting the native species. As Magalhães *et al.* (2007) observed, droughts may serve to reduce, if not eliminate, exotic invaders, as many are not as well adapted to the stresses of low flow and drought. However, in Mediterranean streams in California, a lengthy drought reduced the abundances of the three native fish species and facilitated the establishment of the exotic and hardy green sunfish (*Lepomis cyanellus*) (Bêche *et al.*, 2009).

When, in drought-stricken streams, fish are crowded together in pools, parasitism can increase. As drought progressed, John (1964) noted a heavy infection of the ciliate parasite *Ichthyophthirius*. This parasite causes 'white spot' disease, which can be lethal, and it usually occurs when fish are at high density and stressed. For example, in a Spanish stream during drought, 'white spot' parasitism infected 21 per cent of a redtail barb (*Barbus haasei*) population and significantly reduced both population density and average fish size (Maceda-Veiga *et al.*, 2009).

Samples of the plains killifish *Fundulus zebrinus* from the South Platte River were examined for seven species of parasites (Janovy *et al.*, 1997). The samples were collected from 1982 to 1995, during which there was a drought (1989–1993). With the drought, the level of infection by a parasite (the trematode *Posthodiplostomum minimum*) rose sharply, with a less marked increase by the gill parasite (the trematode *Salsuginus thalkeni*) (Janovy *et al.*, 1997). Five other parasite species did not significantly increase with drought. In a Brazilian stream during a summer drought, the level of infection by the copepod parasite *Lernaea cyprinacea* on fish in pools rose sharply from 21.7 per cent when the stream flowed, to a high of 64.2 per cent in the summer drought (Medeiros & Maltchik, 1999).

Clearly, when fish are confined to habitats such as pools in drought, levels of competition, predation and parasitism can rise. These pressures, combined with those created by the lowering of water quality, serve to act as a strong winnowing force, a powerful environmental filter shaping the viability of fish populations, the composition of the resident fish fauna and the eventual outcomes in post-drought recovery.

9.5   Impacts of drought on fish populations and assemblages and subsequent recovery

The stresses of ramp disturbances such as drought can cause individuals to lose condition, which may lead to failure to breed successfully or to death – and thus, populations are reduced. Alternatively, abiotic stressors and adverse biotic interactions, such as predation, may arise and kill individuals regardless of their condition. The outcome, with time, is that individual populations are depleted and, at a large spatial extent, meta-populations may be fragmented, with some local sub-populations going extinct.

Our knowledge of the effects of drought on stream fish populations primarily comes from studies of salmonids in the northern hemisphere. The studies vary from accounts of loss of condition, and of capacity to breed, to detailed and long-term studies of populations.

In a Welsh upland stream, Cowx *et al.* (1984) found that, in the 1976 drought, the survival of young Atlantic salmon was nil, resulting in the loss of a year class. High water temperatures were deemed to be the cause of the loss. On the other hand, recruitment of the other salmonid, brown trout, was unaffected – due, it was suggested, to their greater tolerance to 'sustained high temperatures' (Cowx *et al.*, 1984).

In Trout Creek, in Colorado, USA, Canton *et al.* (1984) found that brook trout (*Salvelinus fontinalis*) survived drought by moving upstream to flowing water and pools, but the fish lost condition to such an extent that they 'were thin and moribund when handled'. Recovery was, however, rapid, with trout migrating from an upstream lake. Brook trout in small headwater streams in West Virginia during drought in 1999 lost condition and populations declined, especially with a 67 per cent loss in the young of the year, compared with pre-drought levels (Hakala & Hartman, 2004). The loss of condition was suggested to be caused by a loss of invertebrate food resources due to habitat reduction, particularly of riffles. The loss of population, especially young of the year, may have been due to the accumulation of fine silt during the drought in brook trout spawning gravel, reducing the survival of eggs and alevin (hatched trout with yolk sacs) (Hakala & Hartman, 2004). Population density was still low one year after the drought.

Habitat complementation refers to the beneficial state for stream fish populations, whereby habitats required for various critical activities (feeding and spawning), and also for various life history stages, are spatially close together (Schlosser, 1995). In a Rocky Mountains stream with spawning habitat but little adult habitat (i.e. no habitat complementation), drought greatly reduced the population of Bonneville cutthroat trout and spawning ceased, whereas in a second stream with habitat complementation, the trout population survived and recruitment both during and after the drought was successful (White & Rahel, 2008). This is just one example of how the landscape/riverscape approach to stream fish ecology can reveal key sections of rivers that allow species to survive drought, and which serve as centres for post-drought migration and recruitment during recovery from drought.

Sea trout or migratory brown trout (*Salmo trutta*) occurring in a small stream, Black Brows Beck in north-west England, have been subject to a very productive long-term study of 34 years by Elliott & Elliott (2006) and Elliott (2006). The sea trout migrate from the sea to the stream to breed in winter, with spawning occurring in November and December and eggs hatching in spawning redds in February-early March. The alevins remain in the gravel to emerge as fry in early May. The juvenile (parr) stage remains in the stream for about two years, after which they migrate to the sea, to subsequently return as spawning adults in possibly their third, but mainly their fourth year (Elliott & Elliott, 2006).

The relationship between the density of the life stages of fish and recruitment (egg density) was found to fit a Ricker stock-recruitment model

(Elliott, 1985). This model indicates that 'survivor density at different stages in the life cycle was density-dependent on egg density at the start of each year-class' (Elliott *et al.*, 1997). It is remarkable that the relationship even holds between the density of spawning females in their fourth year and their original egg density – a strong indication of density-dependent population regulation in 'normal' years (Elliott & Elliott, 2006).

However, in some drought years, numbers of parr (0+ and 1+) were greatly reduced (Elliott *et al.*, 1997). For example, severe summer droughts in 1976, 1983, 1984, 1989 and 1995, and a severe autumn drought in 1989, reduced both 0+ and especially 1+ parr densities, and subsequently reduced the number of spawning females and eggs from the 1975, 1976, 1982, 1983, 1988 and 1994 year classes (Elliott *et al.*, 1997; Elliott, 2006; see Figure 9.3a). Thus, by reducing juvenile trout (parr) populations, droughts produced a lag effect in reducing the number of spawning females and, consequently, egg production (73–83 per cent reduction) from age classes subjected to drought as parr. Interestingly, while drought exerted strong effects on the trout population, floods had little or no effect (Elliott, 2006).

The reduction of juvenile trout in the beck may have been partly due to high water temperatures with low flow (e.g. Elliott, 2000), but was much more likely to have been due to the effects produced by the loss of habitat in the shallow stream (Elliott, 2006). With the shrinkage of the area of the stream, there would have been a loss of foraging space for the parr. As trout hold territories when feeding (Elliott, 2002), this would have heightened competition between individuals. At the same time, the reduction in flow would have also decreased the supply of food by stream drift. Increased competition for a dwindling supply of food no doubt stressed the parr and increased the risk of death.

In summary, Elliott's work (e.g. Elliott *et al.*, 1997; Bell *et al.*, 2000; Elliott, 2006; Elliott & Elliott, 2006) has produced strong evidence that the sea trout population is, most of the time, regulated in a density-dependent way that acts mainly on the juvenile stages. However, when drought occurs, density-independent mortality of juveniles may occur, causing periodic population reduction.

The loss of habitat for trout created shallow habitat for a hitherto inferior competitor – the bullhead (*Cottus gobio*). These small fish are benthic dwellers and prefer to live in shallow, stony areas with low velocities, such as the areas vacated by the trout parr during the drought (Elliott, 2006). They are mainly nocturnal foragers, in contrast to the diurnal trout, and are more tolerant than trout of high temperatures (incipient lethal levels ≈27°C cf. 25 °C) (Elliott & Elliott, 1995). Furthermore, bullheads breed in spring and the young hatch in June, when the drought is taking effect. With less

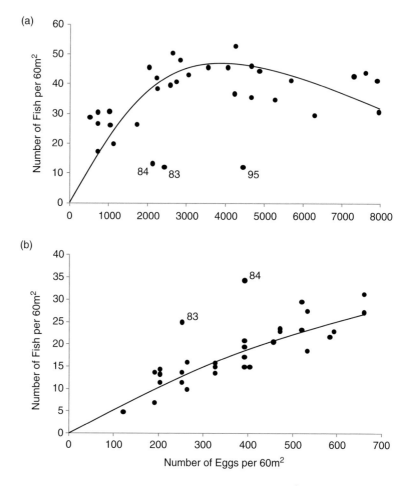

Figure 9.3 (a) Ricker curves relating egg density (eggs m^{-2}) of brown trout (*Salmo trutta*) to the density of trout parr aged 1+ years in Black Brows Beck, UK. Poor survival of parr in 1983, 1984 and 1995 are related to years of summer droughts. (b) Ricker curves relating egg density (eggs m^{-2}) of bullheads (*Cottus gobio*) aged 1+ years. Higher survival than expected occurred in the years 1983 and 1984 with summer droughts. (Redrawn from Figures 2 (c) and 3 (c) in Elliott, 2006.)

interference from trout and ready access to food, bullhead numbers increased in drought years – the inverse of the trout situation of high numbers with less habitable area (Figure 9.3b; Elliott, 2006). Thus, as trout habitat is reduced by drought, habitat favouring juvenile and adult bullheads is created (Figure 9.4).

The bullhead example is one of the very few where drought actually favours a fish species. Other examples are sticklebacks in Mediterranean

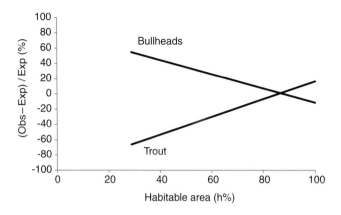

Figure 9.4 A summary of the effects of drought, and hence habitable area, on 1+ bullheads and 1+ trout parr. (Redrawn from Figure 5(b) in Elliott (2006).)

streams (Magalhães *et al.*, 2007), and mountain galaxiids in streams formerly occupied by trout (Closs & Lake, 1995).

The strength of recruitment largely determines the population numbers of trout populations (Elliott & Elliott, 2006; Lobón-Cerviá, 2009b). Mediterranean rivers are noted for their high seasonal variability in flow, with high flows in winter and very low flows in summer (Gasith & Resh, 1999). In Spain, in years of severe summer drought, recruitment of brown trout is very weak (Lobón-Cerviá, 2009a; Nicola *et al.*, 2009), while severe floods appear to have little effect on population numbers (Lobón-Cerviá, 2009a). However, the resilience of the trout to recover from years of very low recruitment is high; this is due, Lobón-Cerviá (2009a) suggests, to the high fecundity of spawning females and high egg survival rates in years with normal summers.

In contrast to the relatively stable trout population of Black Brow Beck (Elliott & Elliott, 2006), trout populations in Spanish streams are largely regulated by density-independent factors coming from the high hydrological variability, with severe droughts being of critical importance (Lobón-Cerviá, 2009a; Nicola *et al.*, 2009). In Spain, trout are at the edges of their southern distribution and, as in other populations of k-selected species at their boundaries of their normal ranges, density-independent factors may exert a stronger and more persistent regulation of population numbers than density-dependent factors (Gaston, 2003).

In a Mediterranean stream in Portugal, the dominant fish are two cyprinids, chub (*Squalius torgalensis*) and nase (*Chondrostoma lusitanicum*) (Magalhães *et al.*, 2002, 2003). Severe summer drought in dry years reduced the chub population, with young fish (0+ , 1+) being especially reduced, but it did not affect the nase population (Magalhães *et al.*, 2003). The decline in chub may be due to chub breeding as summer drying sets in, and young fish being stranded and unable to gain pool refuges. On the other hand, severe spring floods reduced the nase population. Populations of both fish recovered rapidly after losses to the different disturbances. As for trout in Mediterranean streams, both fish species appeared to be regulated by density-independent factors – floods and droughts.

While drought and low flow events are known to negatively affect fish populations, there has been little modelling to determine the viability of fish populations faced with low flow events. The survival of fish populations in intermittent streams is of particular interest, as these populations may be on the edge of survival or local extinction.

The carp gudgeon (*Hypseleotris* spp.) is a small fish with a short lifespan (2–3 years) that dwells in intermittent streams in south-eastern Australia. These streams dry to a series of pools in summer, when the gudgeon breeds (Perry & Bond, 2009). Using field data from the lowland streams where the fish dwells and how the distribution of pools varies each summer, Perry & Bond (2009) built a spatially explicit, individually-based model to assess how populations of the fish survive, given that habitat availability is reduced each summer and greatly reduced when drought occurs. Population numbers were highly variable and were related to annual rainfall, seepage loss from the pools and reduced carrying capacity of pools as they shrank (Perry & Bond, 2009). The population underwent source-sink dynamics, with very successful recruitment in years with good winter rainfall and only a short dry period being sufficient to 'buffer against periods of drought' (Perry & Bond, 2009). Such modelling may apply to other fish species dwelling in stream systems where the advent of severe and long droughts may threaten population viability.

9.6   Assemblage composition and structure and drought

There are many studies of drought impacts on fish assemblages in flowing waters. Some of the changes are difficult to interpret, as sampling at single or a few sites may reveal assemblage changes which are not evident when examined at a larger spatial extent. Changes at sites may reflect patterns of movement rather than losses due to drought. On the other hand, if a stream dries to a series of pools, change in assemblages may be obvious when individual pools are sampled.

9.6.1   Headwater and intermittent streams

The spatial dependence of detecting assemblage changes due to disturbances such as drought is shown in the study by Ross *et al.* (1985) in Brier Creek, a small stream in Oklahoma, that had a severe drought in 1980. Little effect was detected when samples across five sites were aggregated. However, fish from the upstream sites did undergo considerable changes in rank order of species abundances, but these changes only involved a small set of species of the larger assemblages downstream (Ross *et al.*, 1985).

Little effect on assemblage structure of fish was found in downstream sites of another Oklahoma stream (Otter Creek) after a severe drought (1965) (Harrel *et al.*, 1967). However, upstream low-order sites did lose some species and recovery was delayed at these sites. Further work at Brier Creek indicated that a drought in 1998 did not affect the fish assemblages (Matthews & Marsh-Matthews, 2003). However, a drought in 2000 caused considerable changes in assemblage structure, with several common species declining greatly and other species becoming abundant.

In a similar vein, studies of fish assemblages at four river sites in Oklahoma in four drought years did not show any impacts of drought when compared with non-drought years (Matthew & Marsh-Matthews, 2003). However, when single species were examined, some interesting patterns emerged. Three species (red shiner (*Cyprinella lutrensis*), western mosquitofish (*Gambusia affinis*) and bullhead minnow (*Pimephales vigilax*)) all increased in abundance in the drought years – a peak that did not persist into non-drought years (Matthews & Marsh-Matthews, 2003). The cause of the drought peak in the abundances of these three small fish is unknown, though other studies (e.g. Loftus & Ekland, 1994; Ruetz *et al.*, 2005) indicate that mosquitofish are both tolerant to, and recover rapidly from, drought. The lack of substantial changes in fish assemblages in the above studies reflects the likelihood that as the streams are normally subject to harsh conditions, with highly variable flow regimes, the fish have evolved to deal with such conditions. It also illustrates that no two droughts are alike in their characteristics, and their impacts and 'ecological memory' of past droughts may affect the impacts of later droughts.

Droughts in small and intermittent streams can lead to major changes in fish assemblage structure. As mentioned above, severe drought in an intermittent stream led to the elimination of brown trout and, when flow returned, the native galaxiid fish colonized the stream sections formerly occupied by the trout (Closs & Lake, 1995). Depending on the frequency of droughts, the galaxiids may survive so long as there is no trout recolonization from downstream refuges. In an experiment attempting to restore pools in a heavily sedimented stream by adding structures to create scour pools, at first the fish, especially the mountain galaxiid (*Galaxias olidus*), responded positively (Bond & Lake, 2005). However, with the onset of an extended severe drought, the stream dried up and, without any pool refuges in the sedimented sections, three fish species were eliminated. In this case, the disturbance of drought nullified attempts to restore fish populations to sections disturbed by heavy sedimentation from poor catchment management.

Streams in Mediterranean and in wet/dry tropical climatic regions have a predictable and high variability in discharge with regular summer droughts. Thus, as Magalhães *et al.* (2002), describe, the fish assemblages in these streams persist through the summer drought using a variety of refuges, ranging from deep pools to runs. Even within a species, different life stages use different refuges – an example of habitat or refuge complementation (Schlosser, 1995). Thus, young 0+ nase use the runs that persist as refuges, while adult nase take refuge in pools.

A summer drought may persist if winter precipitation is low and become a supra-seasonal drought (e.g. Bravo *et al.*, 2001; Magalhães *et al.*, 2003, 2007). In a Spanish river, the Torgal, drought occurred over three years, with the last (1994–1995) being the driest on record (Magalhães *et al.*, 2003). In the fish assemblage of six native species and two exotic species, drought did not lead to major changes in species richness and overall abundance. However, there were changes in abundances of individual species, with numbers of chub and loach (*Cobitis paludica*) declining, the abundance of sticklebacks (*Gasterosteus gymnurus*) increasing, and both eels (*Anguilla anguilla*) and nase not changing. As Magalhães *et al.* (2007) surmise, current droughts cause only 'relatively small and transient changes' to stream fish assemblages.

Similarly, in a tropical intermittent stream in Brazil, summer droughts caused population reductions in some species but, overall, the drought assemblage had high stability and comprised a suite of distinctive species (Medeiros & Matchik, 2001).

As a general conclusion, it appears that fish assemblages that dwell in streams with high hydrological variability and severe summer or dry season droughts are capable of contending with supra-seasonal droughts. There are four provisos to this conclusion: that the connectivity of the streams is not disrupted by human-installed barriers; that the streams are not polluted; that habitat (refuges) complementation is present for the various species; and that the drought is not of long duration.

9.6.2 Perennial streams

In perennial streams and streams with low flow variability, drought can have marked effects on fish assemblages. In a Colorado montane stream, a severe drought eliminated all three fish species (brook trout and two sucker species) at two downstream sites, while at the upstream site, where flow persisted, the assemblage also persisted (Canton *et al.*, 1984). With the drought breaking, the fish assemblage of three species rapidly recovered, with migration from upstream (Canton *et al.*, 1984).

In streams in an agricultural basin in Illinois, Bayley & Osborne (1993) found that drought eliminated the fish assemblages of first and second order streams, which made up 80 per cent of stream length of the basin. In larger streams, the assemblages persisted through the drought. Within a year after the drought, however, the fish assemblages (species richness and biomass) in the small streams had recovered, and calculations showed that only 17 per cent of the total fish biomass had been lost in the drought. The key to the rapid recovery of the fish was the presence of readily accessible refuges downstream in the persistent stream sections (Bayley & Osborne, 1993).

From 1952 to 1955, a severe drought affected central and south-western USA. Fortunately, before this drought, Larimore *et al.* (1959) had sampled a small stream, Smiths Branch, in Illinois, and had found a diverse fish assemblage of 29 'regularly occurring' species and six 'sporadic species'. Beginning in the autumn of 1953, the stream dried to a few pools and flow was not fully restored until spring in 1954. Fish persisted in the few pools, but these were sampled with rotenone and thus, thanks to the combination of such sampling and mortality due to drought 'the entire fish population was destroyed' (Larimore *et al.*, 1959).

Heavy rains occurred in April 1954 and the stream flowed once more. Within two weeks of flow returning, 21 of the 29 regular species had returned, with many adult fish migrating from the downstream refuges in the Vermilion River. Some fish (bluntnose minnow (*Pimephales notatus*), white sucker (*Catostomus commersoni*) and creek chub (*Semotilus atromaculatus*)) returned rapidly, travelling upstream for about 15 km in two weeks. Other species returned more slowly. For example, longear sunfish (*Lepomis megalotis*) took two years to return, and two species were still missing three years later (1957).

While migration by adults was the major means of recovery, many species reproduced very shortly after their return, and thus recovery was marked by a high proportion of young fish. In terms of assemblage structure, the pre-drought abundance ranking, led by stoneroller (*Campostoma anomalum*), bluntnose minnow, golden redhorse (*Moxostoma erythrurum*) and longear sunfish, was replaced with the return of flow by bluntnose minnow, common shiner (*Notropis cornutus*), stoneroller and hogsucker (*Hypentelium nigricans*). However, as Larimore *et al.* (1959) observed, the recovery of fish populations and assemblage structure in streams 'seems impossible to determine', as 'populations are never constant'.

In summary, it is clear that drought had a strong effect on the fish of Smiths Branch, and recovery for the most part was rapid, though possibly incomplete as some species had not returned three years after flow returned. The importance of downstream refuges in the perennial river was critical to the recovery.

In studies where both invertebrates and fish were studied (e.g. Larimore *et al.*, 1959; Canton *et al.*, 1984; Cowx *et al.*, 1984; Griswold *et al.*, 1982), it does appear that recovery of invertebrate populations and assemblages may be either incomplete or relatively slower in comparison with recovery of fish populations and assemblages. Part of this difference may be due to the directed and strong migration of fish from their drought refuges.

The construction of weirs and barriers across rivers impedes the movement of the fauna, and thus may affect both survival in drought and subsequent recovery. A severe summer drought (1974) dried a channelized section of a mid-west US river, the Little Auglaise (Griswold *et al.*, 1982). An upstream unchannelized section was separated from the channelized section by two weirs. Drought reduced the flow in the unchannelized section and reduced the invertebrate fauna in both sections (Griswold *et al.*, 1982). A year after the drought, the invertebrate fauna had substantially recovered, with densities of oligochaetes, simuliids, baetids and hydropsychids being higher in the channelized section. In the channelized section, the relatively rich fish fauna (30 species) had recovered in 1975, but species richness (12 species) and abundance were much lower in the unchannelized section. The channelized section was rapidly colonized by fish from the downstream larger river, but the weirs limited this recovery pathway at the upstream unchannelized section (Griswold *et al.*, 1982). This situation of weirs impeding recovery from disturbances, especially drought, is probably common, but tends to be unreported.

Some studies have investigated the recovery of fish assemblages only after drought. Two rivers, the Neosho and the Marais des Cygnes in Kansas, USA, became intermittent in the severe drought from 1951 to 1956 (Deacon, 1961). Extensive sampling occurred after flow had returned in 1957, 1958 and 1959. With the return of flow, there was an increase in abundance in these rivers, from 1957 to 1959, of 15 species that preferred permanent flow. This group consisted of species (e.g. long-nosed gar, gravel chub, stoneroller) that preferred permanent flow in large streams/rivers, species such as creek chub and green sunfish that preferred small permanent streams, and species such as stonecat and bluntnose minnows, that prefer runs and riffles (Deacon, 1961).

This increase in species was partly offset by a decline in 18 species that belonged to three separate groups. One group of seven species, which included shad, carp and largemouth bass, had a preference for lentic or slow flowing conditions and probably migrated downstream. Another group of seven species included small-bodied fish that preferred small tributaries, but which had collectively used the main river as a refuge during drought. The third group included channel and flathead catfish that survived in pools during the drought and had great reproductive success when flow returned in 1957. However, in subsequent years, mortality of young catfish was high, which suggests that the catfish are adapted to survive drought and to boom with high recruitment immediately the drought breaks.

In summary, in these rivers, a large number of fish sought refuge downstream with the drought and, upon the return of flow, migrated out of refuges to upstream habitats. Thus, recovery from drought in this study and many others is driven by migration of many species from refuges, and by high recruitment immediately after the drought breaks in some other species.

In some cases, interpretation of the recovery process is difficult, especially when drought has been broken by floods (e.g. Kelsch, 1994; Bravo *et al.*, 2001; Keaton *et al.*, 2005). A six-year drought affecting the Little Missouri River in North Dakota, USA was broken in the summer of 1993 by major flooding, with high turbidity and velocities (Kelsch, 1994). Fish were sampled over a 190 km section of the river, and the fish assemblages were both different and less diverse than previously sampled assemblages. It was concluded that the drought, the flood and the 'rapid transition' between these events possibly produced the depleted and unusual assemblages (Kelsch, 1994).

Creeks in South Carolina, USA, suffered from a severe drought from 1999 to 2003, which was broken by summer floods (Keaton *et al.*, 2005). The post-drought fish assemblage was different from the drought assemblage, with major changes in abundance at both the family and species levels. For example, suckers (Catostomidae) and catfish (Ictaluridae) declined in abundance, while sunfish (Centrarchidae) increased greatly, with a boom in their recruitment immediately after the drought. The changes leading to the post-drought assemblage were attributed to major changes in habitat structure due to the floods rather than simply to the increase in flow after the drought (Keaton *et al.*, 2005). Both of these studies indicate the need for knowing the pre-drought assemblages in order to be able to identify the separate effects of the drought and of the drought-breaking floods, in order to be able to assess recovery from drought.

9.7 Genetics, fluvial fish and drought

Droughts, especially long droughts, impose great stresses on populations of aquatic biota and, no doubt, act as strong selection pressures on them.

Such pressures have served to mould adaptations to the adverse conditions created by drought. These adaptations may be biochemical and physiological, such as the capacity to tolerate high water temperatures and hypoxia, or behavioural, such as the timing of when to change habitat or seek refugia. However, in contrast to the large amount of research with terrestrial organisms on the genetics of adaptations to drought, the effort with aquatic organisms has been minimal.

Droughts can fragment waterways and can thus break up what may be one large panmictic population into a number of sub-populations in isolated pools. With animals such as fish completely dependent on free surface water for dispersal, isolation may be total between the sub-populations, whereas for many macroinvertebrates, especially winged insects, there may be little isolation between the isolated pools.

If the drought is severe, the genetic structure of isolated sub-populations may change in response to the selection pressures imposed by drought (genetic bottleneck) and may also be changed by genetic drift. This appears to be the case for populations of red shiners (*Cyprinella lutrensis*) dwelling in intermittent creeks exposed to seasonal droughts (Rutledge *et al.*, 1990). Populations in some pools may go extinct, while populations in persistent pools may be subject to genetic drift and to strong selection pressures – genetic bottleneck. Correspondingly, as found in *Cyprinella*, drought serves as a bottleneck, restructuring the genomes of the isolated populations, which subsequently colonize the stream when flow returned.

Similarly, drought reduced heterozygosity and sub-divided the genome of the fantailed darter, *Etheostoma flabellare*, dwelling in a small stream, whereas the population in a larger, less drought-prone stream had 'discernible population genetic structure', due presumably to less drought-induced extinctions (Faber & White, 2000).

The flannelmouth sucker *Catostomus latipinnis* is endemic to the Colorado River system. Sampling of mitochondrial genes from fish in the basin revealed that there was surprisingly little genetic variation (Douglas *et al.*, 2003). It was suggested that the species had been through a severe genetic bottleneck due to intense megadrought, such as the megadrought which occurred in south-western USA in the early Holocene (described in Chapter 4). The species now appears to be expanding (Douglas *et al.*, 2003), though movement is severely restricted by the breaking of connectivity by large dams.

In summary, with fish at least, the genetic consequences of drought remain unexplored, though the research that has been done suggests that this area could provide some rewarding insights on past and current evolution.

9.8   Summary and conclusions

Though the studies of drought and lentic fish are few, some interesting points are evident. Only one study, of Lake Chilwa in Africa, provides a complete picture of lentic fish before, during and after a severe drought. Fish in lentic systems appear to be particularly vulnerable to the stresses of drought. As water levels in lakes drop, littoral habitats are lost and loose sediments, which can irritate fish and lower oxygen levels, may be mobilized. Severe water quality stresses, such as high conductivity, high temperatures and low oxygen levels, can develop, and habitat variety and availability may then decline as the volumes are lowered. Increased nutrient concentrations may stimulate algal blooms, notably by cyanobacteria, which deplete oxygen levels. With stresses increasing, fish populations and species richness decline. Refuges in drying lakes may be few, and many fish may migrate to tributaries, and in one particular case to the sea. There are a very few fish that are adapted to deal with the loss of free water and with a lake drying up.

Recovery in fish assemblages in lentic systems after drought is poorly known. It appears that the return of fish populations is dependent on critical habitat requirements (e.g. macrophytes), food resources and, of course, connectivity with the drought refuges. Recovery of fish follows an accumulative succession of species and may take a long period to be complete.

There are many more studies of drought and lotic fish than of lentic fish. Droughts in lotic systems begin as in lentic systems with the loss of shoreline and shallow habitats. As flow volumes drop, water quality can deteriorate and eventually flow may simply stop, creating a streamscape of pools. With low flow or none, adverse conditions such as high temperatures, low oxygen levels and 'blackwater' events can occur and biotic interactions may intensify. Streams can dry up with different spatial configurations, such as upstream sections with no flow and flow continuing downstream. As streams are basically unenclosed linear systems and droughts operate at large spatial extents, a watershed approach to the impacts of drought is preferable to studies based on a specific site or sites. However, the latter are much more common.

As in lentic systems, the impacts of drought are better known than the dynamics of recovery. As droughts set in, fish in streams may change their usual habitats and, with flows reducing, they may move to refuges. Fish kills can occur both in isolated pools and in flowing rivers with high water temperatures. In some cases, the low flows of drought prevent fish from migrating to spawn, or fish may lose condition and forego spawning. The most reported type of refuges consist of pools, which may be used by specific species or by fish assemblages. However, water quality may deteriorate in pools, stressing fish. Heavy predation by both terrestrial and aquatic predators can occur. Levels of parasitism can rise with drought.

Droughts can reduce an intact metapopulation of stream fish to a few scattered sub-populations. As habitat is lost, habitat complementation (different accessible habitats for different life stages) can be weakened, jeopardizing effective recovery. Drought can have direct and lag effects on stream fish populations. Normally, brown trout populations can be under strong density-dependent regulation, whereby recruitment is proportional to adult spawning stock. With drought, as shown in a 34-year-long study Elliott (2006), recruitment may be much less than expected, due to reduced food resources and habitat space. On the other hand, with interactions with trout reduced, the populations of co-existing fish, such as bullheads, can increase. Indeed, in a number of cases, drought favours particular fish species at the expense of other previously common species and invaders may be facilitated at the expense of native species. Brown trout populations at the edge of their range are regulated by density-independent forces, such as droughts, that may diminish populations. However, population recovery is relatively rapid, due to the high fecundity of a low number of spawning females.

Populations of fish in headwater streams are more resistant to the stresses of drought than populations of fish in high-order streams. Where floods break droughts, the resulting changes in habitat structure can greatly alter the trajectory of recovery. In intermittent and perennial headwater streams, upstream migration from downstream refuges is the major driver of recovery. This depends on intact longitudinal connectivity and good dispersal capacity of the fish. Where connectivity has been weakened or broken, the recovery of fish, in particular, compared with other fauna, is delayed.

There is a small amount of genetic evidence that droughts can act as a selection force in moulding fish populations. Recovery after drought in lotic fish populations appears to be relatively faster than recovery in lentic populations. This may be because stream fish populations are normally exposed to flow variability (as opposed to fish in lakes, where variability is small) and also to the continuity of streams, compared with the isolation of many lentic systems.

There are very few studies in which fish are studied along with other major biotic and abiotic variables. The outstanding one is the study of Lake Chilwa (Kalk *et al.*, 1979), with observations on the morphology, water chemistry, phyto- and zooplankton, benthos and fish of Lake Chilwa, before, during and after drought. However, with an increasing interest in long-term ecological research, one hopes that integrated studies of drought in aquatic ecosystems will be forthcoming. In particular, it would be wonderful if such studies could examine the patterns of primary and secondary production and the changes in trophic structure.

10

Estuaries and drought

We have now covered the ecological impacts of drought, from headwater streams to floodplain rivers. The final movement of freshwater from exorheic rivers is into the sea, and this generally occurs via estuaries, which are critical transition zones between freshwater and marine ecosystems. Like rivers, no two estuaries are alike and, thus it is not surprising to find that there are numerous definitions of estuaries (>40; Gillanders, 2007). An estuary may be defined as 'a semi-enclosed coastal body of water with one or more rivers or streams flowing into it, and with a free connection to the open sea' (Pritchard, 1967).

Basically, there are four types of estuaries (Little, 2000; Nybakken & Bertness, 2005):

- coastal plain estuaries, which are formed by the sea inundating river valleys;
- fiords, where glacial valleys are flooded by the sea;
- tectonic estuaries formed by the sea inundating subsided land;
- bar-built or barrier-built estuaries, where offshore sediments have built spits and islands that impede the drainage of rivers into the sea.

Where the mouth of an estuary is restricted, coastal lagoons can occur; these are usually shallow and may be extensive, with unusual gradients in salinity.

Estuaries are transition zones between the saline water of the sea and freshwater from rivers – and also, in many cases, from groundwater. The mixing of sea and fresh waters takes many forms, dependent on the magnitude of the river flow, the tidal amplitude and the morphology of the estuary (Little, 2000; Gillanders, 2007). Where a large river enters an estuary, the river water may move to the sea on top of a layer of dense sea water – a salt wedge. Such estuaries are highly stratified and the amount of mixing is low. In partially mixed estuaries with reduced river

Drought and Aquatic Ecosystems: Effects and Responses, First Edition. P. Sam Lake.

flow and strong tides, there is an increasing gradient of salinity from the surface to the bottom. In some estuaries, the strength of the tide is such that there is total mixing with no stratification.

In all of these types of estuary there is a dynamic gradient of water from fresh to saline. The nature of this gradient varies immensely, being influenced by fluctuations in the freshwater input (floods, droughts) and by variations in tidal strength and wave action (storms). In some localities, where the freshwater input is low and evaporation levels are high, salinity may rise up the river and away from the mouth, creating 'reverse estuaries' (Mikhailov & Isupova, 2008), which may become longer and have sharper salinity gradients in times of drought, especially supra-seasonal droughts. Estuaries consisting of lagoons may be 'reverse estuaries'; for example, the St. Lucia estuary is the largest in Africa and consists of a narrow exit to the sea and three large lagoons or lakes (Forbes & Cyrus, 1993; Cyrus & Vivier, 2006; Taylor et al., 2006). In drought, with the river mouth closed, salinity rises with distance from the mouth (Taylor et al., 2006; Whitfield et al., 2006).

Salinity is a major force regulating the structure and function of the ecological communities in an estuary. In the classical view, where the river enters the estuary, freshwater biota are to be found, some of which may tolerate brackish conditions. As salinity rises down the estuary, diverse communities of euryhaline biota, from plankton to benthos, occur. At the estuary mouth, fully marine stenohaline organisms occur along with euryhaline biota. As salinities fluctuate, mobile species may move, whereas sedentary biota in the estuary are tolerant of the normal salinity changes. However, when disturbances such as floods occur and the estuary fills with freshwater, sedentary biota may suffer high mortality (e.g. Matthews & Constable, 2004).

Estuaries may be permanently open or intermittent and closed for various periods of time. Many estuaries, especially in southern Australia, have sand bars at their entrances which can close the entrance either seasonally, or for long periods such as in droughts. Estuaries are 'sediment traps' (Little, 2000) and the distribution of sediments coming from the river and the sea has a very strong influence on the estuarine biota. In many estuaries, tidal movement is strong. In these tide-dominated systems, the movement and distribution of sediments are under tidal control. Where the tidal influence is mild, wave action and river flow may control distribution of the sediments. Although some estuaries, especially fiords, may have areas of hard surfaces, in most estuaries soft sediments (silt, sand) dominate the bottom.

Estuaries receive organic carbon (OC) compounds (dissolved OC and particulate OC) and nutrients (e.g. phosphorus, nitrogen, silicates) largely from their rivers, though some may come from groundwater, from the sea and from atmospheric deposition. This influx of matter, both particulate and dissolved, has a strong effect in governing estuarine productivity. Floods and high flows may supply nutrients and create conditions for a post-flood period of high productivity for everything from phytoplankton to fish (Gillanders & Kingsford, 2002; Murrell *et al.*, 2007). Retention of the nutrients is essential for such high productivity. During periods of low flows, and especially droughts, nutrient inputs to estuaries can be low and productivity may be checked (e.g. Sigleo & Frick, 2007; Murrell *et al.*, 2007).

Both biodiversity and production can be high in estuaries. The range of different habitats, from mudflats to mangroves and salt marshes, and the dynamic gradients of abiotic conditions, principally salinity, all serve to make estuaries highly productive and commercially valuable ecosystems. While the abiotic forces of salinity, water movements, water quality and substrate availability exert a key role in moulding estuarine ecosystems, it is important to realize that biotic forces, in particular predation, along with competition and parasitism, also exert very strong effects on the estuarine ecology.

Human-imposed low flows of rivers entering estuaries can have strong deleterious effects on estuaries and their biota (Gillanders & Kingsford, 2002). Consistent low flows can cause salinity in estuaries to rise and move upstream. Nutrient, carbon and sediment inputs are lowered, temperatures may rise and oxygen levels may be very low in places. Such conditions also occur with droughts, especially extended supra-seasonal ones.

10.1   Drought and abiotic variables in estuaries

In drought, especially supra-seasonal droughts, freshwater inputs to estuaries are greatly reduced. This particularly applies to surface water inputs, though groundwater inputs may also be curtailed (e.g. Drexler & Ewel, 2001). Droughts may be created and maintained by the loss of floods which may flush estuaries (e.g. Matthews, 2006). As drought is usually associated with high air temperatures, high evaporation rates may occur, reducing the volumes of fresh water. Evaporative losses can be very significant from shallow water and mud flats (e.g. Forbes & Cyrus, 1993). The loss of freshwater inputs and high evaporation serve to affect salinity levels and gradients in drought-affected estuaries. Furthermore, the timing of inflows into estuaries, which stimulate key events such as upstream migration, may be curtailed during drought.

10.1.1 Salinity

In river valley estuaries, distinct salinity gradients normally occur. In drought, with reduced freshwater inflows, saline conditions can occur in reaches of rivers that are normally freshwater:

- In the 1961–67 drought in the eastern USA, abnormally low river flows caused salinity to extend up the Delaware River by 32 km (Anderson & McCall, 1968).
- In the 1975–76 drought in Britain, salinity invaded normally freshwater rivers to such a level that extraction for consumption and irrigation ceased (Davies, 1978).
- In the river Thames, drought in 1989–1990, along with increased water extraction, caused an increase in salinity at the normally freshwater section of the tidal limit (Teddington weir), 110 km up the estuary (Attrill *et al.*, 1996).
- Low river inflows into the Mondego estuary in Portugal in the 2004–05 drought produced marked increases in salinity (up to 20) at sites that were normally freshwater (Marques *et al.*, 2007).
- A severe drought in 1999–2000 in Florida reduced river flows into the Escambia estuary such that, from an average flushing time of 9.5 days, flushing time (freshwater volume divided by the estuarine discharge) rose to more than 20 days and salinity from $\approx$10 to $\approx$25 (Murrell *et al.*, 2007).
- Similarly, in the Satilla River estuary in Georgia, the flushing time rose from 31 to 119 days as drought developed (Blanton *et al.*, 2001).

Thus, in river valley estuaries with drought, flushing times increase, as does salinity, and these changes are powerful determinants of the structure and function of estuarine ecosystems.

In subequatorial rivers in west Africa which drain low rainfall catchments with high evaporation levels, drought can create 'reverse estuaries' for long periods of time (Mikhailov & Isupova, 2008). In an extended drought in the 1980s, for example, the reverse estuary in the Senegal River extended upriver for 330 km, with a salinity reaching 38 – three points above the normal salinity of sea water. Not surprisingly, such conditions, accompanied by low oxygen levels, reduced biomass and the diversity of fish, shrimps and phytoplankton, and also caused both salinization and acidification of catchment soils and groundwater (Mikhailov & Isupova, 2008; Figure 10.1).

Under natural conditions, the inflows from the Sacramento and the San Joaquin River into the tectonically-formed estuary of San Francisco Bay served to create an extended salinity gradient. To determine the dynamics of the salinity gradient, the position of the 2 bottom isohaline (X_2) appears to be

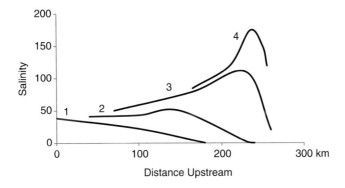

Figure 10.1 The development of a hypersaline reverse estuary in the Casamance River, Senegal, during an extended drought from 1978–1985. (1) The normal situation in June 1968. (2) September 1980. (3) July 1985. (4) June 1986. (Redrawn from Figure 7 (b) in Mikhailov & Isupova, 2008.)

a reliable indicator of habitat conditions and estuarine communities (Jassby *et al.*, 1995). In drought years (ENSO), the X_2 moves away from the sea toward the river mouths. However, the degree to which the movement of X_2 up the estuary is due to drought is debatable, as water extraction and storage for human use make a significant contribution to reductions in river inflows, although in times of drought, extraction may be reduced (Knowles, 2002). This example, and that of the Thames, serve to illustrate that droughts with abnormally low flows may be exacerbated by human water extraction and storage.

In barrier-built estuaries, the changes in salinity may be complicated by the variable morphology of these systems. There may be an entrance to the sea that is open or periodically closed, and behind this there may be a series of coastal lagoons that receive freshwater inflows.

Such an estuary is the St Lucia estuarine system in South Africa. This lake has been studied for a long time, and many aspects of its ecology have been documented in over 300 reports and scientific papers available on this system (Whitfield *et al.*, 2006). The estuary (Figure 10.2) consists of a channel, the Narrows, that periodically discharges into the sea. Above the Narrows there are, in order of distance, three shallow lakes – South Lake, North Lake and False Bay – which have an average depth of about one metre (Forbes & Cyrus, 1993). Four small rivers flow into the two top akes.

As drought sets in and river inflows drop, a normal salinity gradient is present for a short period. However, as evaporation exceeds rainfall, the water levels of the shallow lakes decline such that their levels may be below

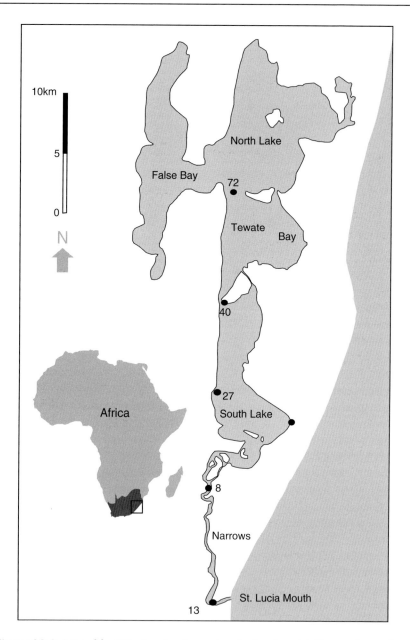

Figure 10.2 Map of the St Lucia estuarine system during drought. The numbers refer to salinities recorded when the system was sampled for fish in December 2004. The drought was severe and occurred from 2002–2009. (Redrawn from Figure 1 in Cyrus & Vivier, 2006.)

mean sea level (Forbes & Cyrus, 1993; Whitfield *et al.*, 2006). This produces hypersaline conditions in North Lake and False Bay and, as water levels drop, more saline water is drawn from South Lake and sea water may be drawn in if the mouth is open (Forbes & Cyrus, 1993). If the mouth closes, as in the drought from 2002 to 2005, high evaporation causes lake levels to decline, salinities to rise and the surface area of the lakes to decline by 25 per cent. These changes produce high salinities (which may be >70; Whitfield *et al.*, 2006), severely restricting diversity to a few species that can tolerate these harsh conditions (Forbes & Cyrus, 1993; Pillay & Perissinotto, 2008; Bate & Smailes, 2008). The 2002–2005 drought was broken by Cyclone Domoina, which caused widespread flooding (Whitfield *et al.*, 2006). In spite of these dramatic oscillations, the system can support a diverse and highly productive biota.

In estuaries in drought, increased retention and low flushing times can cause stratification over a salt wedge and, with organic matter accumulating on the bottom, hypoxia and even anoxia may occur in the bottom waters (e.g. Mackay & Cyrus, 2001; Burkholder *et al.*, 2006; Elsdon *et al.*, 2009). On the other hand, in drought in well-mixed estuaries, with strong tides and with a decline in nutrients due to curtailed streamflow, low oxygen levels may not occur (Attrill & Power, 2000a). In drought in estuaries with reduced freshwater inputs, pH will move toward the range of seawater (pH 7.5–8.4). Suspended solids and turbidity may drop, increasing Secchi disc depths (e.g. Livingston *et al.*, 1997) and water clarity, potentially aiding aquatic photosynthesis. This decline may be a function of reduced inputs from river catchments (Attrill & Power, 2000a), as well as the increased extent of salinity, serving to precipitate suspended particles.

10.1.2 *Nutrients and primary production*

With the decline in river inflows into estuaries during drought, the inputs and concentrations of nutrients required for photosynthesis, principally by phytoplankton, may become limiting. In the Yaquina river estuary in Oregon, USA, most (>94 per cent) of the dissolved nitrate and silica were transported from the catchment into the estuary in wet winter months, whereas inputs were three times lower in drought years (Sigleo & Frick, 2007). Dissolved phosphate loads were low, but were not affected in drought. As explained in Chapter 5, nitrate loads in high flows after drought can be very high, and a pulse of high nitrate entered the Yaquina estuary in the first rains after the drought (Sigleo & Frick, 2007). The effects of such post-drought pulses on estuarine productivity and trophic structure remain to be described.

In the eutrophic Neuse estuary, North Carolina, USA, in a three-year drought, total nitrogen concentrations declined, total phosphorus concentrations declined slightly, a long-term increase in nitrate concentrations was reduced and the concentration of ammonium increased (Burkholder *et al.*, 2006). The increase in the latter may have been due to maintained inputs from 'inadequately controlled, increasing nonpoint sources' (Burkholder *et al.*, 2006). The trends in nutrients were not reflected in phytoplankton chlorophyll-*a* concentrations. In the Thames estuary in the 1989–1992 drought, total nitrogen concentrations did not change, and it was suggested that this was due to the greatly reduced inputs of nitrogen from rivers being balanced by maintained inputs from sewage outlets (Attrill & Power, 2000a).

These two cases illustrate that anthropogenic influences on catchments can affect estuarine nutrient concentrations during drought. If inputs of nutrients from catchments are greatly curtailed due to low inflows during drought, nutrient levels during drought may not be affected by catchment land use. This is indicated in South Australian estuaries during drought, when nutrient levels in estuaries did not reflect differences in catchment land use (rural vs. urban) (Elsdon *et al.*, 2009). However, as pre-drought values and freshwater inflows were not given, this conclusion must be regarded as tentative.

By reducing freshwater inflows to estuaries, drought increases Secchi disk depths (increasing light availability) favouring phytoplankton production, but decreased nutrient loadings and concentrations can inhibit phytoplankton production. In the Escambia estuary, Florida, USA, dissolved inorganic phosphate concentrations rose during drought, while dissolved inorganic nitrogen concentrations were very low and primary productivity was unaffected by the drought (Murrell *et al.*, 2007). When the drought was broken by a flood, primary productivity dropped, largely due to the high turbidity. Following the flood, and with decreased turbidity and increased nutrients, primary productivity boomed. From their observations, Murrell *et al.* (2007) suggested that in drought, primary productivity was limited by nitrogen concentrations, while in times of normal flow, phosphorus was the limiting nutrient.

In large estuarine lagoons, there can be spatial variability in nutrient availability during drought. The Guaiba River flows into the northern end of the large Patos Lagoon in southern Brazil. In a La Niña drought in 1988, high phytoplankton production, largely by diatoms (*Aulacoseira granulata*) occurred in the oligohaline northern part of the lake. Further down the lagoon, with the depletion of silicates, primary production was lower, with cyanobacteria becoming a significant component. Towards the outlet to the sea, with marine salinity levels, primary production was low and limited by low nitrogen and light levels (Odebrecht *et al.*, 2005). Thus, in drought, over the expanse of the lagoon, two nutrients – silicates and nitrogen – served to limit primary production in different parts of the salinity gradient from freshwater to the sea (Odebrecht *et al.*, 2005).

In the St Lucia estuary during the 2002–2007 drought, very high salinities (114–125) were reached in the northern lagoons, and only a few species of diatoms were present (Bate & Smailes, 2008). The few species that were present at the high salinities were not planktonic but were benthic. Indeed, throughout the estuary there did not appear to be a 'true phytoplankton community', and the diatoms collected in the water column were re-suspended benthic species (Bate & Smailes, 2008). Over a lower range of salinities in the estuary (5–44), epiphytic algae on submerged macrophytes appeared not to be regulated by abiotic variables such as salinity, but rather by the availability of biotic factors –the availability of macrophytes and grazing pressure (Gordon et al., 2008). In short, with the very high salinities and greatly reduced macrophyte cover, primary production in the lagoon lakes of this estuary in drought was effectively shut down.

Not only may nutrients be delivered to estuaries from rivers, they can be delivered from offshore upwelling events. Pelorus Sound in New Zealand is a large (50 km long) fiord estuary supporting a valuable mussel industry (Zeldis et al., 2008). In El Niño years, nutrients, especially nitrates, are delivered by nutrient-rich water from offshore upwelling being pushed into the sound by northerly winds and by high river inflows. However, in La Niña drought years, river inflows are decreased and southerly winds prevent the entry of nutrient-rich offshore water. Consequently, in drought (e.g. 1999–2002), concentrations of dissolved inorganic nitrogen, nitrates and ammonia declined, as did particulate nitrogen (seston) and particulate carbon. The decline in nutrients and in seston produced a 25 per cent decrease in mussel yield from the sound (Zeldis et al., 2008). The supply of nutrients from both marine and freshwater sources is unusual, but the trophic consequences of the resulting decline in phytoplankton during drought are documented in many studies.

In some estuaries, while the decline of freshwater inflows in drought is related to the decline in phytoplankton, the actual causes are unclear. In the San Francisco northern bay estuary, salinity rises and the X_2 isohaline moves towards the Sacramento-San Joaquin delta when supra-seasonal droughts occur and produce low river inflows, which are accentuated by water extraction (Jassby et al., 1995). When such El Niño droughts occur, the phytoplankton community changes significantly and biomass declines (Cloern et al., 1983; Lehman & Smith, 1991), but the cause or causes for this are unclear. In contrast to the other cases, where such declines can be related to low nutrient levels, during drought in this estuary they do not appear to be limiting phytoplankton production (Cloern et al., 1983). Conditions created by low inflows and salinity intrusion may have favoured neritic (inshore) diatoms to grow rather than planktonic species (Cloern et al., 1983). A further suggestion is that, in the drought, estuarine benthic grazers and filter feeders (e.g. bivalves, amphipods) were favoured by the increase in salinity and migrated into the top normally brackish area of the estuary, where their feeding was so effective that it reduced the phytoplankton biomass (Nichols, 1985).

10.2   Drought, salinity and estuarine macrophytes

With declining freshwater inputs and rising salinities in drought, aquatic and semi-aquatic macrophytes may be expected to undergo major changes in distribution. In the St Lucia estuary, in response to salinity gradients, aquatic macrophytes come and go. In salinities above about 10, *Potamogeton pectinatus* disappears, while *Ruppia maritima* and *Zostera capensis* can tolerate salinities up to about sea water (Forbes & Cyrus, 1993; Gordon *et al.*, 2008). However, in extended droughts in areas with high salinities ($\approx$50), *Ruppia* is the survivor, while *Zostera* disappears (Forbes & Cyrus, 1993; Gordon *et al.*, 2008). In this reverse estuary, with severe drought, the macrophytes disappear from the shallow hypersaline lagoons.

Similar results in terms of the effects of elevated salinity during a severe La Niña drought (1999–2001) were found by Cho & Poirrier, (2005) in Lake Pontchartrain, an estuary near New Orleans, USA. With drought, salinity rose ($\approx$3 to $\approx$9) and water clarity increased. With this change, *Ruppia maritima* expanded, while freshwater species either declined in area (e.g. *Vallisneria americana, Myriophyllum spicatum*) or simply disappeared (e.g. *Potamogeton perfoliatus, Najas guadalupensis*) (Cho & Poirrier, 2005). Within two years of the drought ending and salinity and water clarity declining, *Ruppia* had contracted and the freshwater species, excluding *Potamogeton*, had returned. The switch from *Potamogeton* to *Ruppia* in estuaries with increasing salinity may not occur necessarily due to drought, but perhaps to other causes such as restriction of freshwater inputs (e.g. Shili *et al.*, 2007).

Fringing and marsh vegetation in estuaries may be affected by the high salinities created by lack of freshwater inputs. In the St Lucia estuary, very high salinities from drought in the upper lagoon killed the reed *Phragmites australis* (Forbes & Cyrus, 1993). In the severe 1999–2001 la Niña drought mentioned above, in the Barartaria estuary, near New Orleans, salinities rose and altered the salt marsh vegetation (Visser *et al.*, 2002). The changes were more severe in the marsh vegetation normally exposed to low salinities, in particular the 'oligohaline wiregrass' vegetation type dominated by *Spartina patens* (Visser *et al.*, 2002). In this vegetation assemblage, with the change, *Spartina* remained the dominant, but freshwater species disappeared. During the drought, the area of brackish and saline marshes expanded at the expense of the fresh and intermediate marches (Visser *et al.*, 2002). Presumably, as suggested by other studies, recovery will involve a return to the pre-drought proportional areas. *Spartina* species vary in their salinity tolerance, and thus different species may occur along salinity gradients.

In the same 1999–2002 drought in the eastern USA, the salinity along the Altamaha River estuary rose, increasing markedly in freshwater and brackish regions (White & Alber, 2009). Consequently, *Spartina cynosuroides*, with a salinity tolerance up to ≈14, retreated 3 km or so up the estuary. *Spartina alterniflora*, with a much greater salinity tolerance, invaded this new habitat area and remained there after the drought. With the return of normal freshwater flows, *S. cynosuroides* returned to its pre-drought position, but in the areas it had vacated in the drought, it now occurred alongside *S. alterniflora* (White & Alber, 2009). Thus, in this case, drought conditions facilitated the expansion of an invader.

As indicated above, *Spartina* salt marshes may be changed by the increased salinity created by drought. Drought has thus been regarded as a major trigger for a number of stresses (e.g. salinity, soil acidification), producing the large-scale dying off of *Spartina alterniflora* in salt marshes in the southern USA (e.g. McKee *et al.*, 2004). However, it appears that, while this dying off is associated with drought, grazing by snails (*Littoraria irrorata*) is also a major contributing force (Silliman *et al.*, 2005). The snail feeds on pathogenic fungi associated with *Spartina* (Silliman & Newell, 2003). *Spartina* plants stressed by drought are damaged by snail grazing, and fungi subsequently thrive in the grazing wounds and are consumed by the snails. Thus, the snails facilitate the spread and growth of an important food source for them – a facultative mutualism (Silliman & Newell, 2003). In these patches of die-off, snail populations progressively build up and become transformed into 'consumer fronts', destroying *Spartina* (Silliman *et al.*, 2005). Once triggered by drought conditions, these fronts may continue after the drought has broken. Thus, with drought creating stressful conditions, grazing and fungal infection synergistically act to destroy *Spartina*. No doubt in many other environments, drought acts to create stressful conditions that allow particular biota to thrive and magnify the drought's effects.

10.3 Estuarine invertebrates and drought

Many invertebrates, being mobile, can move as drought effects set in. Hence, mesohaline and oligohaline taxa may move into brackish or freshwater areas if access is available, and fully marine stenohaline taxa may move into estuaries; they have two refuges, depending on their physiology. Sedentary taxa such as oysters, other bivalves and even meiobenthos must either tolerate the conditions or die.

As freshwater inflow drops, marine water pushes further up estuaries. This is reflected in the distribution of zooplankton. For example, in a Portuguese estuary in severe drought, saline water pushed upstream into the zone that was previously fresh water. The freshwater community was replaced by one with marine affinities, being dominated by the copepod *Acartia tonsa* (Marques *et al.*, 2007). No doubt such upstream intrusions by marine zooplankton in drought occur elsewhere and remain unreported.

Meiobenthos, being small and interstitial, have limited mobility and thus may be substantially affected by drought. This was certainly true for the meiobenthos of the St Lucia estuary in a severe drought. In the shallow highly saline waters of the North Lake-False Bay region, the meiobenthos was reduced in abundance and species, with only hardy ostracods and nematodes remaining (Pillay & Perissinotto, 2009). In the lower seaward region of the estuary, a normal marine meiobenthic fauna persisted unaffected by the drought (Pillay & Perissinotto, 2009).

The macrobenthos of the St Lucia estuary showed a more drastic pattern as the hypersaline and shallow North Lake and False Bay area was 'devoid of macrofauna during any sampling season' during drought (Pillay & Perissinotto, 2008). With much of this area drying up, drought (2002–2009) created a fragmented scape that hindered movements of adults and their larvae. In the southern part of the estuary, a diverse and abundant community occurred. This drought was extreme, with the drying out and fragmentation of large amounts of habitat, high water temperatures and hypersalinity all affecting the biota, from diatoms to macrobenthos and fish.

Most bivalve molluscs are sedentary, as they are either fixed to hard surfaces or live in sediments. With fluctuations in salinity generated by either floods or droughts, they are thus very susceptible to stress if not death. In high salinities generated by drought (1950–1957) in a Texan estuary, reefs of the oyster *Crassostrea virginica* were replaced by the oyster *Ostrea equestris* and the mussel *Brachidontes exustus* (Hoese, 1960). With the return of freshwater inflows and low salinities in 1957, the *Ostrea* reefs were replaced by the more normal *C. virginica* and *B. recurvus* community.

High salinities favour many parasites and pathogens of oysters, and thus low salinities with occasional short floods are regarded as beneficial, since they reduce the parasite-pathogen load (Copeland, 1966). Extended floods with long periods of low salinity are, however, harmful (Buzan *et al.*, 2009). Furthermore, in drought, as suggested by Wilber (1992), marine predators may move into estuaries and increase the predation pressure on larvae and spat. Drought may, however, benefit some estuarine bivalves. The small, sediment-dwelling clam *Soletellina alba* suffers mass mortalities when winter floods flush the Hopkins estuary in southern Australia (Matthews, 2006), but when droughts occur and winter floods are diminished, causing moderate salinities ($\approx$10–20) to prevail, clam populations thrive.

Decreased freshwater inputs to large river estuaries means that salinity moves upstream and, if drought is extended, major faunal changes occur. In a severe drought (1979–1981) in the Hawkesbury River estuary in Australia, species richness rose along the estuary, especially at the upper-most sites, where two polychaete annelids and a bivalve mollusc dominated the benthos during, but not after, the drought (Jones, 1990).

In drought in the Thames at its upper tidal limit, normal freshwater conditions gave way to brackish conditions (salinity 3–5) (Attrill *et al.*, 1996). Consequently, at this site, freshwater fauna such as Caenidae (Ephemeroptera), Leptoceridae (Trichoptera) and Hydracarina were deplet-ed, while brackish-tolerant taxa such as the exotic snail *Potampyrgus antipodarum* (previously *P. jenkinsi*) and the amphipod *Gammarus zaddachi* were abundant (Attrill *et al.*, 1996). Further down the Thames estuary, at West Thurrock, Essex, rather than changes in salinity during drought being significant, temperature and dissolved oxygen were important variables affecting the fauna (Attrill & Power, 2000b). The abundance of the crab *Carcinus maenas* rose, whereas abundances of the amphipods *Gammarus* spp. and the shrimps *Palaemon longirostris* and *Crangon crangon* declined sharply in summer.

This study illustrates that changes in just a single variable (salinity) during drought may not necessarily be influential in affecting estuarine biota. The decline in *Gammarus* and *Crangon*, in particular, may have changed the trophic structure of the estuary, as both taxa comprise a major part of the diet of important fish species (Attrill & Power, 2000b). The commercial catch of shrimps and prawns in estuaries has been reported to decline during drought (Copeland, 1966), possibly due to a drop in primary production because of reduced inputs of organic matter and nutrients.

As mentioned above, in a Texas estuary with high salinities from drought, there were major changes in the taxa associated with oyster reefs, including the oyster species themselves (Hoese, 1960). With drought in the Texan estuary, Mesquite Bay, large populations of stenohaline marine and polyha-line estuarine species occurred including a thriving infaunal community dominated by the clams *Mercenaria campechiensis* and *Chione cancellata* and large colonies of marine sponges (Hoese, 1960). The drought was broken by a large flood that produced an extended period of low salinity in the estuary. The marine infaunal community 'suffered complete mortality', as did the *Ostrea equestris* populations (Hoese, 1960). Similar to the findings of

Matthews (2006) for bivalves in an Australian estuary, floods induced mass mortalities of bivalves in the Texas estuary, illustrating the point that drought-breaking floods may be highly damaging to stream and estuary dwelling biota.

10.4 Drought and estuarine fish

Faced with changes in estuaries due to drought, marine fish may retreat to the sea and freshwater species may move into the inflow systems. However, as happens in barrier estuaries, the lack of inflows – and hence outflow – from the estuary may mean that the mouth of the estuary closes (e.g. Hastie & Smith, 2006). In the St Lucia estuary, with the mouth closing and high evaporation, very harsh aquatic conditions occur, along with major losses of habitat (Whitfield *et al.*, 2006; Taylor *et al.*, 2006). In barrier estuaries with mouth closure in drought, fish populations may be drastically reduced, as escape to the sea is not possible and access to upstream freshwater areas may be restricted.

In Portugal, in the Mondego estuary, during drought there were increases in salinities and summer water temperatures, and changes occurred in the proportions of 'ecological guilds' in terms of species richness and abundance (Martinho *et al.*, 2007; Baptista *et al.*, 2010). Not surprisingly, freshwater species disappeared, along with catadromous fish species (fish that migrate from freshwater to the sea to breed). There was a marked increase in abundance of four species of 'marine adventitious' (Martinho *et al.*, 2007) or 'marine straggler' (Baptista *et al.*, 2010) fish, although the abundance (but not the species richness) of 'estuarine residents', declined but rapidly recovered within a year after the drought (Martinho *et al.*, 2007).

Drought in this estuary was characterized by increases in five species, three of which are 'marine stragglers'. During drought, fish production was estimated to have dropped by 15 to 45 per cent, with major decreases in European sea bass and two estuarine gobies (Dolbeth *et al.*, 2008). While there was substantial recovery of 'estuarine residents' after the drought, overall estuarine fish production did not recover rapidly.

A quite different outcome from drought for fish occurred in the very large barrier lagoon, the Patos lagoon estuary in southern Brazil (Garcia *et al.*, 2001). During a La Niña drought in 1995–96, freshwater inflows to the lagoon dropped and salinity in the lagoon, while fluctuating, rose by ≈15. Consequently, the abundance of estuarine-resident fish – especially the silverside (*Atherinella brasiliensis*) – and of estuarine-dependent fish – especially the mullet (*Mugil platanus*) – rose greatly. Freshwater vagrants

virtually disappeared. Thus, in a La Niña event in this estuary, fish populations and production rose. The productivity of the estuary rose because
nutrient-rich marine water enhancing primary production was drawn into
the estuary during the drought (Garcia et al., 2001).

In summary, in the Portuguese river estuary, the estuary fish community
composition underwent some changes with drought, but fish production
dropped considerably. In the Brazilian lagoon estuary, with drought there
were considerable changes in fish community composition, but instead of
fish production dropping, it more than doubled as assessed by changes in
catch per unit effort. The differences may reflect the fact that the Portuguese
estuary, being much smaller, was sampled along its length, while sampling
in the immense Patos lagoon (280 km long, maximum width of 70 km) was
confined to near the ocean mouth. Both studies show that droughts, even
relatively short ones, can exert strong effects of estuarine fish populations
and production.

By altering the variability of salinity (temporally and spatially) in both
river and barrier estuaries, drought alters the fish biomass and, hence,
commercial fishing operations. A study of commercial catch per unit effort
for fish in nine estuaries in eastern Australia (Gillson et al., 2009) found that
with drought in 2003–2006, there were significant declines in the catches of
four species out of five. The exception was yellowfin bream (*Acanthopagrus
australis*), which can live in brackish and fresh water, and which moved
downstream with the drought into the estuary and thus into the fishery
(Gillson et al., 2009).

Recruitment of larval bass (*Morone saxatilis*) drastically declined in the San
Francisco Bay estuary in the 1987–1992 drought (Bennett et al., 1995). It
was suggested that lack of food (zooplankton) was a major cause for this loss.
However, Bennett et al. (1995) found strong histological evidence that the
larvae had been exposed to toxic agents. They hypothesized that the larvae
had been exposed to agricultural pesticides which may have been concentrated because of the very low freshwater flows into the bay. This, along with
other drought-induced effects such as reduced food, acted synergistically to
reduce the larval bass populations – a further example of the multiple stresses
that drought, a decline in freshwater inflows, can create.

Reverse estuaries in which, as one moves upstream from the mouth,
salinity rises, are found in areas where freshwater inflows may be low or
highly variable, and where evaporation rates are high. Such estuaries were
described by Mikhailov & Isupova (2008) in rivers of west Africa, with the
most remarkable reverse estuary occurring in the Casamance River in
Senegal (see Figure 10.1). The saline region of the estuary extended
upstream for 230 km during a severe drought (Albaret, 1987; Mikhailov
& Isupova, 2008).

The reverse salinity gradient resulted in a highly unusual distribution of fish. Upstream from the mouth, for about 60 km, normal marine salinities prevailed and the fish fauna was a rich one (30–45 species), consisting of normal estuarine residents with some oceanic species (Albaret, 1987). From there, for the next 150 km, salinity was high (50–83), without much seasonal variability. Species richness dropped to 6–7 abundant species, including the tilapia *Sarotherodon melanotheron*, the clupeid *Ethmalosa fimbriata*, the mullet *Mugil bananensis* and ariid catfish (Albaret, 1987).

Above this point (207 km from the mouth), salinity dropped (1–8) early on in the drought but, as the drought wore on, the salinity in this section rose to levels of ≈64. Not surprisingly, this zone of variable-rising-to-high salinities harboured few fish, principally only three species – two cichlids, *S. melanotheron* and *Tilapia guineensis* and a clariid catfish, *Clarias anguillaris*, with *S. melanotheron* occurring in abundance. In terms of tolerance of salinity, oxygen levels and temperature, *S. melanotheron* is a remarkable fish and, in the severe conditions of the drought-stricken Casamance, it is the last fish swimming. That this species is actually favoured by drought is probably too grand a claim, but it is interesting to note that in salinities stressful to other species, 'seul *S. melanotheron* prolifère' (Albaret, 1987).

The St Lucia estuarine system is probably the best studied estuary for the effects of drought on the physico-chemical environment and on the biota – algae, macrophytes, plankton, invertebrates, fish, birds, reptiles and mammals (e.g. Forbes & Cyrus, 1993; Taylor *et al.*, 2006; Whitfield *et al.*, 2006). The effects of long droughts on the estuary depend largely on whether the mouth of the estuary is closed, as is normal in drought, or whether through management intervention it is kept open (Taylor *et al.*, 2006). In a recent drought, (2002–2005) the mouth remained closed (Cyrus & Vivier, 2006). The estuary dried and water levels dropped to such an extent that the water body became fragmented into four distinct areas, with hypersaline conditions (≈70) in the North Lake and False Bay (Cyrus & Vivier, 2006; see Figure 10.2). This fragmentation meant that fish could not move to refugia containing lower salinities.

Near the mouth of the estuary, low salinities prevailed (8–15) and fish species richness was high (17–24). The number of fish species dropped (9–12) in the South Lake, with a salinity of 27–29. At Fanies Island at the bottom of the North Lake, salinity was elevated (40), with 12 species present and the dominant fish being the cichlid *Oreochromis mossambicus*. Finally, at Hells Gate, about 40 km from the mouth, hypersaline conditions prevailed, with a salinity of ≈70, with numerous dead fish and with *O. mossambicus* being the last fish swimming (Cyrus & Vivier, 2006). Clearly, in this drought, salinity and fragmentation governed the distribution of the fish in the estuary. Salinity may have acted as a powerful osmotic regulator of the fish, but as Whitfield *et al.* (2006) point out, salinity also had powerful effects on primary production, such that it and food resources in general declined with rising salinity. In hypersaline conditions, food resources for *O. mossambicus* were reduced to microphytobenthos and detritus.

It is worth pointing out that across the range from low to hyper-salinity in St Lucia estuary, the cichlid *O. mossambicus* not only survived but reproduced. Similarly, in the reverse hypersaline estuary of the Casamance River estuary, Senegal, across the wide salinity range, another cichlid (*Saratherodon melanotheron*) survived and reproduced (Albaret, 1987).

In the 2002–2005 period of drought, the surface area of fish habitat was reduced to 25 per cent of its original area and a large amount of this area was hypersaline (Whitfield *et al.*, 2006). Normally, the St Lucia estuary occupies a large area ($\approx$35,000 ha), with salinities not exceeding sea water levels. Accordingly, the estuary is a major nursery area for many fish species. Thus, a major impact of the drought may be to cause a substantial decline in fish abundance, not just at the local level, but more significantly at the regional level (Whitfield *et al.*, 2006). As estuaries, besides harbouring resident species, are also major areas of fish recruitment, the effects of droughts may have major implications for fish stocks at the large regional scale.

In open estuaries, the effects of drought on fish are definitely more predictable than they are for barrier estuaries. With the latter, the question of whether the system remains open or becomes closed in drought is critical. In barrier estuaries, the morphology of the basin containing the lakes or lagoons is crucial to the responses of the fish because, in severe cases (e.g. St Lucia), fragmentation of the water body can occur preventing fish from moving into less stressful habitats.

10.5   Drought and changes in faunal biomass and trophic organization

As detailed above, only a limited amount of information exists on the effects of drought on the ecology of estuaries. Most studies have concentrated on particular biotic components, be they phytoplankton or molluscs or particular fish species. A much wider appreciation of the effects of drought on estuarine communities comes from the extensive and long-term work on the St Lucia estuarine system in South Africa. However, this work does not readily provide an understanding of how the trophic organization of an estuarine ecosystem may be changed by drought. Such an understanding comes from the studies by Livingston (1997, 2000) and Livingston *et al.* (1997) on the effects of drought on abiotic variables and biota of the Apalachicola Bay estuary in Florida.

The Apalachicola estuary is a barrier island estuary with the unregulated, pollution-free Apalachicola River providing most of the freshwater inflow to the Apalachicola Bay system, which consists of East Bay, Apalachicola Bay, St Vincent Sound and St George Sound. East Bay, a shallow (≈ 2 m) bay, was the focus for a 9.5–year study that produced before-drought (1975–1980), during drought (1980–1982) and after-drought (1982–1984) data (Livingston, 2002; Livingston *et al.*, 1997). The strengths of this study are the focus on trophic organization and the capacity to partition variability due to season from changes due to drought.

Basically, in comparison with the non-drought situation, the drought gave rise to low freshwater inflows, higher salinities (≈ 15), greatly increased Secchi disc depth and reduced colour (Figure 10.2). The increases in light availability and of higher oxygen anomalies strongly indicate that, in the first year of the drought, there were high levels of primary production, notably by benthic microalgae (Livingston *et al.*, 1997; Livingston, 2002). However, in the second year of the drought, primary production dropped, possibly due to low nutrient levels as a function of low river inflows. Similarly, detritus inputs – both dissolved and particulate – declined in the drought.

The non-drought infaunal macroinvertebrate component of the bay largely consisted of polychaetes (e.g. *Streblospio benedicti*, *Mediomastus ambiseta*) and clams (e.g. *Mactra fragilis*, *Macoma mitchelli*), whereas the epibenthic macroinvertebrates largely consisted of shrimps, prawns and crabs. The fish of the bay consisted of planktivorous and benthic feeding primary carnivores, secondary carnivores and tertiary piscivorous carnivores such as sea bass and gars (Livingston, 2002). During the drought, species richness of the infauna initially rose, then tailed away as the drought ended, while species richness of both the epibenthic macroinvertebrates and the fish dropped, especially near the end of the drought. Both infaunal and epibenthic macroinvertebrate biomasses rose in the drought, notably at the end of the first drought year, while fish biomass peaked in the winter at the end of the first drought year and then declined to very low levels (Livingston, 1997, 2000).

Biomass of macroinvertebrates and fish were divided into five trophic groups: herbivores, omnivores, and primary (feeding on herbivores and omnivores), secondary (feeding on primary carnivores and omnivores) and tertiary carnivores (feeding on omnivores, primary and secondary carnivores) (Figure 10.3). In the drought, herbivore biomass (mainly bivalves) rose to an unprecedented high level in the first year, before tailing off in the second year, while omnivore biomass rose sharply, coincident with the rise in herbivore biomass, and then declined. Primary carnivore biomass reached high biomass levels toward the end of the first year, followed by hitherto unprecedented levels toward the end of the drought. Secondary carnivore

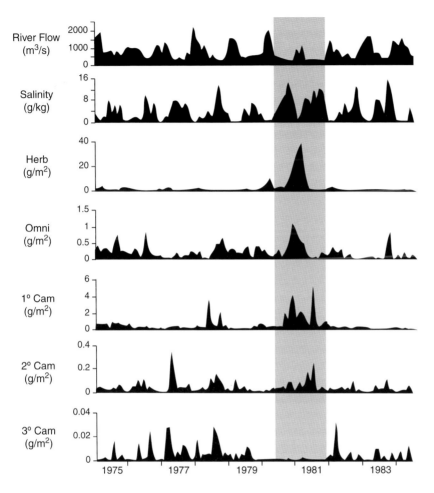

Figure 10.3 Monthly totals for river flow, estuarine salinity, and biomasses of herbivores, omnivores, and primary, secondary and tertiary consumers in the estuary of East Bay, Florida from 1975 to 1984. (Redrawn from Figure 7 in Livingston *et al.*, 1997.)

biomass increased toward the end of the drought, while tertiary carnivore biomass dropped sharply and these fish virtually disappeared from the estuary.

The herbivores clearly capitalized on the increase in primary productivity, with a noticeable drop in the second year, due possibly to the decline in primary production and/or predation by the primary carnivores. The omnivore biomass may have benefited from the organic matter produced by the boom in primary production which subsequently tailed off near the end of the drought. The increase in primary carnivore biomass after the herbivore-omnivore biomass peak may have been due to increased predation on herbivore-omnivore. However, with herbivore-omnivore biomass declining, primary carnivore production was also checked. The peak in secondary carnivore production occurred slightly after the primary carnivore peak, suggesting a surge in production originating with the herbivore-omnivore production pulse. The loss of tertiary carnivores suggests that this group was more affected by stress from abiotic variables (e.g. salinity) than by any check in prey availability.

With the end of the drought, there were major and rather sudden changes (over about six months) in the biomasses of the trophic groups. Herbivore biomass dropped greatly, and the biomasses of omnivores and the carnivore groups sharply decreased, except for the tertiary carnivores (Livingston et al., 1997; Livingston, 2002). The causes of these marked changes are unclear. The tertiary carnivores increased their biomass and, presumably, returned from their refuges in freshwater (e.g. gars, smallmouth bass) or the sea (spotted sea trout, southern flounder). However, they returned to an estuary with relatively low biomass levels of suitable prey.

What is remarkable in this study is the rapidity in which the effects of drought showed up, and the rapidity with which the biota of the estuary returned to pre-drought levels, with the possible exception of the tertiary carnivores. It is also clear from the river flow data that the drought did not break with a devastating flood, as has occurred in other estuaries (e.g. Hoese, 1960). The rapid changes with and after drought may be due to mobile estuarine organisms using the sea and the freshwater river as refuges from which they could rapidly return. The return of sedentary organisms may be a result of the return of propagules from unaffected parts of the large embayment system.

10.6   Summary

Clearly, the effects of drought on estuaries and their biota depend on the nature of the estuary (whether open or closed), the inputs of freshwater, the changes in chemical conditions (salinity, nutrients) and the nature of the pre-drought biota, especially if the biota have been affected by human activities (e.g. pollution, harvesting).

In drought, with lowered freshwater inputs and increased evaporation, salinity in estuaries rises and can invade normally freshwater areas and, in some instances, harsh conditions of 'reverse estuaries' can occur. If salt wedges develop, then anoxia in bottom water can develop. As a transition zone between the marine and freshwater domains, most estuaries are marked by a salinity gradient of considerable length, and normally under tidal influence. With drought, the gradient can become sharp in salinity and abrupt in distance.

Because drought reduces inflows, low nutrient concentrations in estuaries can limit phytoplankton production in spite of increased water clarity due to reduced suspended sediments from freshwater inflows. In some urbanized estuaries, the reduction in nutrients from freshwater inflows can be counteracted by nutrients from sewage and reduced dilution. As described in Chapter 5, when droughts break, there can be pulses of nutrients released from catchments downstream. These pulses end up, presumably, in estuaries, but their effects remain unexplored.

Freshwater and brackish aquatic macrophytes are depleted in drought due to increases in salinity. In some cases, depending on the duration of the drought, euryhaline macrophytes may replace the low salinity species. Drought can trigger the dying off of extensive areas of *Spartina* salt marsh, but the mechanism for this is complicated. As drought stresses *Spartina*, concurrently it is grazed by *Littorina* snails and the snails feed on pathogenic fungi that grow in grazing wounds. Thus, a positive feedback is generated, with snails grazing on fungi facilitated by snail grazing, which consequently kills the *Spartina*.

Mobile freshwater benthic invertebrates may migrate upstream to refuges with drought, while marine species, be they zooplankton or benthos, can migrate into estuaries and thrive in times of drought. Indeed, in some cases with the incursion of euryhaline and marine species, estuaries can become highly productive. On the other hand, sedentary and interstitial invertebrates are at risk in drought, as they cannot migrate and thus may be killed by the increased salinities.

Fish, being mobile, undergo quite rapid changes with drought, with freshwater and brackish species withdrawing from the estuary and some marine species invading. Of course, the movements of fish depend on whether the estuary is open or closed. In closed estuaries, salinities can rise due to evaporation and low freshwater inputs and reverse estuarine conditions can occur. In the latter situation, few species survive. In reverse estuaries, an unusual gradient may occur, with marine species moving in from the sea; however, further upstream, very high salinities occur and few species persist. Droughts may disrupt reproduction, especially in those species that breed in estuaries and go to sea as adults. In general, fish production in estuaries drops with drought, as does the production of shrimps and prawns.

All of the above effects suggest that not only does drought alter species assemblages, but that it produces major changes in the trophic structure and production of estuaries. This is borne out by the study of Livingston

(1997, 2002) and Livingston *et al.* (1997) on the effects of drought on the macrobiota of the Apalachicola Bay estuary in Florida. Basically, as this drought set in, water clarity increased greatly, producing a great increase in primary production, which in turn stimulated a rise in herbivores, followed by omnivores. This increase then tailed off in the second year of the drought, perhaps due to the depletion of incoming nutrients or to the ravages of predation. Certainly, the latter was important, as the biomasses of both primary and secondary carnivores started to rise to high levels with the drop in herbivores and omnivores. To summarise, in this case, drought induced major changes in the trophic structure of the estuary. More striking was the rapid return to 'normal' trophic conditions within a year of the breaking of the drought.

11

Human-induced exacerbation of drought effects on aquatic ecosystems

Human needs and socio-economic activities have for long been threatened and damaged by drought. It is obviously a much feared catastrophe and, in Christian tradition, for example, drought has been described as coming to those societies which have turned their back on God. As described in Chapter 4, drought has been wholly or partly responsible for damaging, if not demolishing, past societies. In modern times, droughts still diminish agricultural productivity, force people to migrate, reduce industrial activities and greatly deplete water resources for urban and rural societies and their economic activities. The effects of drought on human social and economic activities have produced a very substantial literature and given rise to a considerable research effort. In contrast, the literature and research efforts on the effects of drought on both natural and human-influenced ecosystems have been much less, especially as regards aquatic ecosystems.

Previous chapters have described the effects of drought on aquatic ecosystems. This chapter deals with how human activities have exacerbated these effects – an area that has not received much scientific consideration or research activity.

Increasingly, the onset of droughts have become able to be predicted, allowing some measures to reduce the damage, but their intensity and the duration are much harder to predict. Many other catastrophes (e.g. floods, cyclones, earthquakes, etc.) may also be hard to predict but, being pulse disturbances, they are invariably short-term events with discrete ends (Bryant, 2005). Droughts, on the other hand, are ramp disturbances with intensities and durations that are hard to forecast, and it is this property which makes droughts so dangerous and feared. Their duration makes supra-seasonal droughts disturbances that may be either ameliorated or exacerbated by human activities as they occur. However, this can be

Drought and Aquatic Ecosystems: Effects and Responses, First Edition. P. Sam Lake.
© 2011 P. Sam Lake. Published 2011 by Blackwell Publishing Ltd.

substantially reduced if we come to understand how human activities can exacerbate the effects of drought.

In examining the ways that human activities have exacerbated, and possibly even produced, droughts, I have distinguished between human activities on catchments which have effects on downslope aquatic ecosystems and human activities that exacerbate drought in the water bodies themselves, be they wetlands, lakes or rivers. The former may be seen as indirect catchment-mediated effects, while the latter are direct effects.

11.1 Human activities on catchments and drought

11.1.1 Changes in land use and land cover which influence regional climates

Large-scale changes in land use can influence climate (e.g. Foley *et al.*, 2003; Brovkin *et al.*, 2006; Lawrence & Chase, 2010). In tropical and temperate biomes, it appears that deforestation can lead to warming (Feddema *et al.*, 2005; Lawrence & Chase, 2010), whereas in boreal biomes, deforestation produces cooling (Betts *et al.*, 2001; Bala *et al.*, 2007) through an increase in surface albedo.

The changes in tropical and temperate biomes are largely due to reduced evapotranspiration rather than an increased surface albedo (Brovkin *et al.*, 2006; Lawrence & Chase, 2010), giving rise at a regional level to reduced precipitation and runoff. In warmer parts of the world, deforestation and land clearing lead to reductions in evapotranspiration and precipitation, which can produce an increase in water table levels (McAlpine *et al.*, 2007, 2009; Lawrence & Chase, 2010). It appears that in such warmer regions, there is a positive feedback between vegetation and precipitation (e.g. Los *et al.*, 2006; Shiel & Murdiyarso, 2009) and between soil moisture and precipitation (Koster *et al.*, 2004). Thus, converting terrain covered with forests and woodlands into grazed pastures and cropping land may decrease precipitation – which, in turn, intensifies and prolongs droughts.

As described in Chapter 8, Lake Chad in the Sahel is drying, due to ongoing drought and to water extraction from its inflowing streams. The Sahel region of Africa occurs south of the Sahara Desert between 10°N to 20°N and 18°W to 20°E. This region has been locked in a megadrought since 1973 (Dai *et al.*, 2004a; Held *et al.*, 2005; Lenton *et al.*, 2008) and has suffered a 40 per cent loss in rainfall, which has created widespread famine. Up to a million people have died and lakes, wetlands and rivers have dropped in volume, with many having dried out. It is possible that there has been a regime shift to a 'dry Sahel' climatic regime from a prior 'wet Sahel' regime (Foley *et al.*, 2003). Indeed, this whole region has been identified as containing a 'tipping element' where a subsystem 'of the Earth system ... can be switched ... into a qualitatively different state by small perturbations' (Lenton et al., 2008).

There has been extensive land clearance and loss of vegetation cover in the region, especially in the Sahel (e.g. Leblanc et al., 2008), which may increase surface runoff when rain occurs and, in the long term, increase groundwater volumes. On the other hand, though, with a substantial reduction in vegetation, the weakening of the feedback between vegetation and rainfall may serve to prolong the drought and strengthen the 'dry Sahel' regime (Foley et al., 2003; Wang et al., 2004; Wang, 2004; Los et al., 2006).

Vegetation loss is only one contributor to the ongoing drought, however, as the drought may have been produced by the interplay between 'global ocean forcing, human-induced land use/land cover changes, and the regional climatic feedback due to vegetation dynamics' (Wang et al., 2004). It is worth mentioning that there is uncertainty about the future dynamics of rainfall in the Sahara/Sahel region, with models predicting climate change will weaken the West African monsoon, producing either increased or decreased rainfall (Lenton et al., 2008).

Australia is the most arid inhabited continent, with a highly variable climate. Like the Sahel/Sahara region, the climate functions as a 'tightly coupled ocean-atmosphere-land surface system' (McAlpine et al., 2009) with a similar legacy of extensive land clearing leading to changes in the types and spatial extent of vegetation cover. Land clearing and conversion to grazing and/or cropland have been very extensive in south-western Western Australia and in eastern Australia, with ≈15 per cent of the continent being cleared and/or greatly modified. From 1780 (just prior to European settlement) to 1980, the areas of forest and woodland decreased by 80 million hectares, which were mostly converted to grasslands and croplands (Gordon et al., 2003). In the economically important Murray-Darling Basin, 12–20 billion trees have been removed (Hatton & Nulsen, 1999). With the loss of wooded vegetation cover across Australia, it has been estimated that there has been a ≈10 per cent drop in water vapour flows or ≈340 km^3 (Gordon et al., 2003), a loss that depletes precipitation and could exacerbate droughts.

Recent modelling of droughts in Australia has concentrated on increased temperatures and reduced rainfall, which have been regarded as being due to global climate change. However, as indicated by McAlpine et al. (2007, 2009), land use/land cover change (LUCC) could be significantly interacting with global warming in increasing the severity of droughts, especially in south-eastern Australia. In south-western Australia, clearance of land for grazing and wheat production may explain the increased temperatures and reduced precipitation. Removal of native vegetation reduces surface roughness, increasing horizontal wind velocities and thus allowing moisture-laden air from the ocean to be pushed further inland and lost to south-western Western Australia (Pitman *et al.*, 2004). Loss of riparian vegetation raises stream temperatures and evaporation; thus, extensive riparian restoration is seen as a measure to counteract stream temperature increases due to climate change (Davies, 2010). Large-scale reforestation of the extensively cleared regions is posited as a policy option in contending with climate change.

Human-generated, large-scale deforestation can lead to an increase in groundwater levels. In Australia, this rise has been accompanied by large-scale dryland salinity problems in eastern and western Australia, with 5.7 million hectares of land and 24 of 79 river basins affected (National Land and Water Resources Audit, 2001; Gordon *et al.*, 2003, 2007). Thus, wetlands and streams in areas of dryland salinity also face increasing salinity when their volumes are depleted by drought.

For example, in the Wimmera River, Victoria, a river in a region of dryland salinity, the effects of drought on the aquatic fauna were exacerbated by high salinities (Lind *et al.*, 2006). Similarly, in areas of dryland salinity in wetlands and lakes – especially closed systems – high salinities may be reached in drought, taxing the biota (Williams, 2001).

A possible adaptive response to reduced precipitation and increasing dryness/droughts comes from the ideas of Makarieva & Gorshkov (2007) and Makarieva *et al.* (2009), which are summarized by Shiel & Murdiyarso (2009). Basically, forests, with their high transpiration fluxes, generate precipitation, and the consequential lowering of atmospheric pressure may draw moisture-laden air from the ocean. In transects from the coast to inland regions, precipitation may not decline along those that are forested, whereas in transects that run from coastal forest to deforested inland areas, precipitation drops sharply with distance from the coast (Makarieva *et al.*, 2009; Shiel & Murdiyarso, 2009). Such gradients of decline in precipitation where there is loss of forest or woodland are very obvious in much of Australia. If the hypothesis of a 'forest pump of atmospheric moisture' proves sound, then a powerful form of enhancing precipitation, restoring landscapes and ameliorating the effects of droughts and global climate change is extensive reforestation of inland areas that were previously cleared (McAlpine *et al.*, 2009; Shiel & Murdiyarso, 2009; Makarieva *et al.*, 2009). However, it must be noted that the Biotic Pump Theory may not be supported by 'basic physical principles' (Meesters *et al.*, 2009).

Removing deep-rooted native vegetation and replacing it with shallow-rooted pasture grasses, or ploughing the land for shallow-rooted crops (e.g. wheat), undoubtedly reduces evapotranspiration and increases air temperatures (Glantz, 1994, 2000). The exposure of soils through ploughing or overgrazing, coupled with the increased winds that occur in droughts, causes the likelihood of dust storms to increase sharply. As described by Worster 1979), in the Dustbowl drought in the mid-USA in the 1930s, immense and lengthy dust storms were generated. In the 'Millennium' drought (1997–2010) in south-eastern Australia, inland dust storms swept hundreds of kilometres to the east coast, blanketing the city of Sydney. The loss of valuable topsoil in dust storms during droughts is bad enough, but raised dust and dust storms may also reduce solar radiation on the surface beneath them and reduce both evaporation and precipitation (Miller & Tegen, 1998), potentially increasing drought severity.

Modelling of the Dustbowl drought on the basis of sea surface temperature variations does not satisfactorily predict the large spatial extent occupied by that drought (Seager *et al.*, 2008; Cook *et al.*, 2008). However, its spatial extent, especially its northward extension, can be satisfactorily accounted for when data are included on atmospheric dust loads, which reduced precipitation (Cook *et al.*, 2008).

Overall, the loss of vegetation – forests, woodlands and deep-rooted grasslands – can exacerbate droughts by reducing evapotranspiration and consequently lowering precipitation and surface water availability. The full extent of this influence on droughts awaits clarification, but recognition of the atmosphere-vegetation interactions could offer a policy for ameliorating the effects of climatic drying and of droughts – namely, reforestation.

11.1.2   Local effects on droughts: Accumulation and mobilization of pollutants

As described earlier during droughts, materials may accumulate in catchments and be changed by chemical processes, principally oxidation. Many of these materials may be produced or greatly augmented by human activities. A vivid example of this is the deposition of sulphur released from the burning of coal onto wetlands in eastern Canada (see Chapters 5 and 8). The sulphur may accumulate in the reduced form in wetland sediments (e.g. Dillon *et al.*, 1997, 2003). However, in drought, such wetlands can dry out and, the reduced sulphur becomes sulphates. With drought-breaking rain, these sulphates can generate sulphuric acid, which can acidify both standing and running waters (Dillon *et al.*, 2003; Laudon *et al.*, 2004). Due to the acidic conditions in catchment wetlands, heavy metals may be mobilized and also released when rains break a drought (e.g. Adkinson *et al.*, 2008).

In exposing sediments of rivers and wetlands to the air, drought can generate acidic sulphate conditions. This is a particular problem along the Murray River system in south-east Australia (Hall *et al.*, 2006; Lamontagne *et al.*, 2006), where, during the 1997–2010 drought, large areas of acid sulphate soils in wetlands, lake beds and river channels have been exposed. When the drought breaks, there is a danger of creating acidic conditions, along with elevated concentrations of heavy metals, in water bodies of the Murray River (e.g. Simpson *et al.*, 2010).

Sulphur occurs in swampy soils of coastal floodplains and can be oxidized to sulphates by exposure through land development, especially with wetland drainage schemes. In eastern Australia, during droughts, sulphate concentrations build up and are released as acidic pulses into estuaries when the droughts break, resulting in fish kills (Cook *et al.*, 2000; Heath, 2009). It is both past human development and current land management practices that have created and maintained this problem in coastal floodplains (White *et al.*, 2007).

Nitrogen, as nitrates produced from the burning of fossil fuels in high compression engines, can be aerially deposited on catchments. Along with nitrogen added as fertilizers and in animal faeces, it can accumulate in soils during drought (Burt *et al.*, 1988; Reynolds & Edwards, 1995). With the breaking of a drought, elevated loads of nitrates may be exported downstream and into groundwater aquifers, affecting freshwater ecosystems either in the short or long term (see Chapter 5). This threat appears to be unstudied.

In the above situations, chemicals accumulate in the catchment during drought and are then swept into downstream or downslope aquatic ecosystems when the drought breaks. The post-drought effects of the chemical pulses may be to delay the recovery from drought and/or, in the case of lakes, to setback restoration.

11.1.3   Groundwater and drought

As described in Chapter 2, far less is known about groundwater droughts than about surface water droughts. Two important points are that the onset of groundwater droughts lags behind the onset of surface water droughts and that the recovery from surface water droughts occurs long before the recovery of groundwater droughts. In the Murray-Darling Basin of Australia, surface water drought began in 2001 with loss of surface water resources, with some groundwater losses at the same time (Leblanc *et al.*, 2009). By early 2007, surface water resources had reached a record low, but in 2008, in the northern part of the basin, they increased with a return to normal rainfall. On the other hand, groundwater resources continued to decline in 2008, even though the surface water drought was easing (Leblanc *et al.*, 2009).

Surface water is, in most situations but not all, linked with groundwater storages. Groundwater aquifers may be unconfined or confined. Unconfined aquifers are usually recharged by precipitation and may discharge to surface waters. Many streams are spring-fed and heavily dependent on groundwater, especially in karstic catchments. Base flow and maintenance of hyporheic zones in streams is largely controlled by groundwater levels, as are water levels in many wetlands (Sophocleous, 2002; Glennon, 2002; Evans, 2007). Many lakes and wetlands are groundwater-dependent and may be sites of groundwater recharge. In drought, groundwater levels may drop and, along with evaporation, may cause losses in the volumes of lakes and wetlands. For example, in the Corangamite catchment, a lake region in south-eastern Australia, the causes of water loss in the recent drought varied from evaporation alone to evaporation combined with reduced groundwater discharge (Tweed et al., 2009).

Human-induced reductions in recharge and excessive harvesting of groundwater resources may alter the state of surface water systems and strongly influence their responses to drought (Glennon, 2002). Indeed, in many irrigation areas where groundwater is the major form of supply, groundwater extraction far exceeds recharge – a situation that is unsustainable and has been called 'groundwater mining' (Scanlon et al., 2006).

Humans can reduce the recharge of aquifers by creating large amounts of impervious surfaces, such as in urban areas, by reducing the flooding on floodplains by building flood control structures (levees) and by controlling water flows from dams (Sophocleous, 2002). Aquifers can be reduced by excessive water extraction (extraction > recharge), and excessive groundwater depletion, mainly for irrigated agriculture, has occurred in many parts of the world (e.g. north Africa, the Middle East, south and central Asia, north China, north America and Australia; Glennon, 2002; Konikow & Kendy, 2005; Pearce, 2007). Excessive extraction of groundwater lowers water tables and, as in the case of riparian hydrologic drought (Groffman et al., 2003), water tables may move below the beds of streams and wetlands (Sophocleous, 2002; Evans, 2007). Where the streams are connected with active water movement between groundwater and channel water, gaining streams may be converted into losing streams. Wetlands may lose water such that their volumes decrease, and streams may lose water so that flow is greatly reduced (Danielson & Qazi, 1972; Glennon, 2002; Sophocleous, 2002; Evans, 2007).

The process of streamflow and wetland depletion varies with the levels of groundwater extraction and with the distance that the extraction points are from the stream channel or wetland (Evans, 2007). Wells near a water body may draw water from the stream/wetland in a matter of days, while wells kilometres from the water body, but using the same aquifer as the water body, may take years to significantly deplete the water in the stream (Glennon, 2002; Evans, 2007).

In extreme cases, perennial streams, wetlands and lakes may become temporary and some may even disappear. A particularly dramatic example comes from Florida, where, due to excessive groundwater pumping, lakes went dry, only to be refilled with groundwater pumped from other aquifers (Glennon, 2002). Also in the southern USA, the basin of the Apalachicola-Chattahoochee-Flint River system is subject to very heavy groundwater extraction (Glennon, 2002). Such extraction undoubtedly serves to strengthen the periodic droughts (e.g. 1999–2002) that affect this river system and its impoundments, along with the highly productive Apalachicola estuary (e.g. Livingston et al., 1997).

Quite simply, due to excessive groundwater extraction in their catchments, water bodies with declining water tables are subject to extended and artificial hydrological drought. Loss of groundwater connected with water bodies also increases their vulnerability to the stresses of seasonal and supra-seasonal droughts.

Particularly sensitive to damage from groundwater extraction are streams flowing in karstic systems – chalk or limestone. In Britain, groundwater extraction from chalk systems has exacerbated hydrological drought, as indicated by changes in water levels and the invertebrate fauna (e.g. Wright & Berrie, 1987; Armitage & Petts, 1992; Bickerton et al., 1993; Wood & Petts, 1994; Castella et al., 1995; Wood, 1998). To maintain flows in rivers for both human use and environmental protection in times of drought, groundwater pumping may be used. For example, in England, the Shropshire Groundwater Scheme pumps water from an aquifer into the River Severn (Shepley et al., 2009). However, the scheme needs to be very precisely managed, as such extraction threatens small groundwater streams dependent on the aquifer (Shepley et al., 2009). The risk to groundwater-dependent water bodies of unsustainable extraction has been recognized in Britain, and steps have been put in motion to limit new and present groundwater abstractions, based on comprehensive assessment of groundwater resources at the groundwater catchment level (Burgess, 2002). Hopefully, such a policy will prevent groundwater extraction from exacerbating surface water supra-seasonal droughts.

11.1.4   Catchment condition and drought

In land that has been cleared for grazing, soil compaction, direct exposure to solar radiation and reduced vegetation cover due to grazing (which may be increased in drought) can produce impermeable soil surfaces. Such soils may also be poor retainers of moisture due to the loss of organic matter by clearing and grazing. When droughts break with heavy rain on sloping land surfaces, high surface runoff can produce soil erosion. Sediments mobilized through both sheet and gulley erosion may be transported downstream, to be deposited in areas with low stream power. These deposited sediments drastically reduce habitat availability and diversity by smothering stream bottoms. Pools and backwaters, which may be drought refuges, can be filled in and lost in these systems choked with sand slugs (Davis & Finlayson, 2000).

The clearing and grazing of land invariably results in the loss of riparian vegetation. Intact riparian zones, with trees and shrubs, are key interceptors of catchment-derived sediments and processors of nutrients. The high amounts of nutrients and sediments which may come from catchments when droughts break may be retained and processed by intact riparian zones, thus protecting streams and wetlands. Intact riparian zones supply particulate and dissolved organic matter which supports, wholly or partly, the trophic structure of many running waters.

Furthermore, the depletion of the riparian vegetation can result in a loss of shading of streams and other water bodies. As droughts are invariably marked by having extreme high temperatures, the loss of vegetated riparian zones, leading to there being little or no shading, can give rise to high water temperatures, stressing the aquatic biota (Rutherford *et al.*, 2004). Maintaining and restoring intact riparian zones is thus a key measure to mitigate the effects of drought, both during the drought and afterwards when droughts break (Davies, 2010).

In many agricultural regions of the world, especially in dryland regions, water may be stored on catchments in constructed farm dams or ponds. The numbers of such farm dams have been steadily increasing, (e.g. O'Connor, 2001; Schreider *et al.*, 2002; Callow & Smettem, 2009). Each farm dam captures the runoff from a small area of the catchment, but the cumulative effect of many dams in a catchment is to decrease the total runoff and streamflow (Finlayson *et al.*, 2008; Callow & Smettem, 2009). In South Australia, for example, farm dams have reduced the annual streamflow of the upper Marne River catchment by 18 per cent (Savadamuthu, 2002) and that of the Onkaparinga River catchment up to 20 per cent (Teoh, 2002).

The effects of farm dam storage on streamflow are especially marked in dry periods and droughts, rather than in wet periods. These reductions in streamflow can have damaging ecological effects. For example, in a southern African stream, loss of water to farm dams significantly reduced streamflow in drought to such an extent that riparian trees were stressed and many died (O'Connor, 2001). The volumes of water in farm dams after dry periods or droughts are usually very low, so, when drought breaks in a catchment, capturing considerable volumes of water in the refilling of farm dams may delay the recovery from drought in streams (Finlayson *et al.*, 2008). Thus, regulation of the volumes of water stored in farm dams on catchments can be a strategic measure to mitigate the effects of drought.

Urbanization of catchments can increase the area of impervious surfaces, producing rapid runoff, especially where guttering and drains are well developed (Walsh *et al.*, 2005). During droughts, impervious surfaces such as roads and parking areas accumulate nutrients, salts, heavy metals and organic pollutants which, upon the drought breaking, can be delivered directly in storm runoff to urban streams and wetlands (Hatt *et al.*, 2004; Walsh *et al.*, 2005). Through the rapid drainage of precipitation from these impervious surfaces, urban streams can have sharp storm hydrographs, which have great erosional power and can cause channel incision in unlined channels.

In urban areas with high levels of impervious surfaces, groundwater recharge is limited, and thus groundwater levels drop and may move below the riparian zones, especially in incised streams (Hardison *et al.*, 2009). A result of both rapid drainage and low groundwater recharge is that most of the flow in urban streams is usually low and akin to hydrological drought. Furthermore, the situation where the riparian zone is permanently perched above the groundwater level, and thus is much drier than usual, has been termed 'riparian hydrologic drought' (Groffman *et al.*, 2003). This long-term riparian drought reduces the vegetation and the processing of the riparian zone, as well as contributing to the impacts of supra-seasonal droughts when they occur.

Many streams in urbanized catchments receive treated sewage and industrial effluents and in normal times these are diluted by the streamflow, reducing their potentially damaging effects. However, in drought, the volume of natural streamflow decreases, and sewage and effluent volumes may thus remain constant or be only partly diminished. The threat of insufficient sewage dilution may be partly offset by the lack of storm flows from impervious surfaces and reduced non-point pollution from agricultural areas but nevertheless, during drought, declines in water quality due to increases in effluent pollutants, in pathogenic bacteria and in low oxygen concentrations have been recorded in rivers (e.g. Slack, 1977; Davies, 1978; Chessman & Robinson, 1987; Andersen *et al.*, 2004; Zwolsman & van Bokhoven, 2007; Van Vliet & Zwolsman, 2008; Canobbio *et al.*, 2009) and in estuaries (e.g. Anderson & McCall, 1968; Attrill & Power, 2000a, 2000b).

The declines in water quality appear to be related to the nature and volumes of the effluents, the quality and volumes of the receiving water, water temperatures and, in the case of estuaries, the residence time. Hence, when streamflow is reduced in droughts, treatment of wastes from point sources must be upgraded to lower their polluting potential. Increasingly, with more efficient sewage and wastewater treatment, the lowering of water quality by effluents in droughts can be diminished (e.g. Wilbers *et al.*, 2009).

In impoundments and lakes which have human activities such as agriculture and urbanization on their catchments, water quality problems may arise during periods of low water levels due to drought. Thus, in Lake Eymir, a eutrophic lake in Turkey, drought increased the level of eutrophication, marked by increased concentrations of cyanobacteria (Beklioğlu & Tan, 2008). In impoundments used to supply drinking water, the risk of eutrophication and cyanobacteria blooms during drought is a major cause of concern, for example in Brazil (e.g. Bouvy *et al.*, 2003) and in the USA (Touchette *et al.*, 2007b). During drought, with very low dam levels, anoxic conditions and subsequent water column mixing may occur. These elevate phosphorus concentrations and trigger cyanobacteria blooms which, travelling with the released water, create potentially toxic conditions for considerable distances downstream. Such a situation occurred in Hume Dam on the Murray River in south-eastern Australia during the 1997–2010 drought (Baldwin *et al.*, 2010). Toxic algal conditions occurred for about 150 km downstream the dam in the summer of 2007–2008 (Baldwin *et al.*, 2010) and re-occurred for a greater distance ($\approx$680 km) during the summer of 2009–2010 (Ker, 2010).

At both the regional and local spatial extent, there are human-induced changes that can exacerbate the damaging effects of droughts on aquatic ecosystems. Some, such as the lack of sufficient dilution of wastes in waterways, have been acknowledged for some time, but the majority of threats ranging from the large-scale climatic effects of vegetation clearance to the hydrological effects of groundwater extraction have only been recognized recently. For any catchment with human settlement, it will be a rare situation in which only one of these human-induced changes is acting; usually it is the case that the changes are many, are cumulative and interact in ways which may be difficult to disentangle. In a catchment, the climatic effects of land clearance may be a panoply in which are embedded local effects due to such problems as impervious surfaces, loss of riparian vegetation and groundwater extraction.

In many localities, the goal of catchment management appears to be to restore and/or maintain ecologically sustainable aquatic ecosystems. However, in many others, this is clearly not a goal, and either ecosystem degradation is accepted as the necessary price of human development or, with no interest at all, unplanned neglect drifts on.

If the goal is to protect and restore catchments and aquatic ecosystems, there are paths to follow in dealing with drought. To strengthen the resilience of a catchment to drought, an audit of the strength and location of the human-induced changes and the forces causing them would be a good step to guide planning. In an ideal situation, this could be followed by targeted efforts of mitigation. As it stands now, throughout the world, humans have created catchment conditions which may be economically beneficial in the short-term, but which will produce a legacy of more severe droughts and more severe effects when they do occur. Proactive measures include: increasing catchment tree cover (particularly along riparian zones); capturing and storing stormwater in urban catchments; reducing fertilizer and pesticide use; and avoiding catchment overgrazing.

11.2 Human-induced exacerbation of drought effects within water bodies

Human activities on water bodies themselves may clearly exacerbate the effects of drought. The boundary between these effects and those occurring on catchments is a blurred one. For example:

- groundwater mining can reduce streamflow and wetland water volume;
- removal of riparian vegetation can lower stream water quality;
- non-point pollutants from catchment land use can create algal blooms;
- salt from dryland salinity can salinize wetlands

Yet, in terms of management, there are explicit human interventions in water bodies that do exacerbate droughts and their effects.

The range of water bodies to consider extends from ponds to wetlands to lakes, and from small streams to rivers to estuaries. Across this wide range, the effects of drought differ and the impacts of human activity will differ, such as between damming a low-order stream and damming and regulating the flow of a large river. Unfortunately, little has been reported on how drought may be exacerbated by human activities across the wide spectra of different types of water bodies.

11.2.1 Dams and impoundments

Starting from the 1920s, worldwide there has been a massive program of building large dams. This reached its peak in the 1970s and has since tailed off in most parts of the world (Chao *et al.*, 2008), with the notable exception of China. It is estimated that large dams can now impound about

$10,800\,km^3$ of water, including estimations of seepage, or $8,300\,km^3$, based on the storage capacity of 29,484 dams (Chao et al., 2008). Thus, large dams store about 19 per cent of the global total of annual river flow. These figures do not include the numerous small dams and weirs.

Globally, the total human-generated withdrawal of water from all sources is estimated to be about $4,000\,km^3$ annually, about one-tenth of the world's renewable water resources (Döll et al., 2009). Of this amount, the consumptive water use is estimated to be between $1300-1400\,km^3$ per year, 90 per cent of which goes to irrigation (Döll et al., 2009).

Along with impoundment and water extraction, the damming of rivers has changed the flow regime and volumes of rivers – very starkly, in many cases. Thus, due to water extraction and diversion, many rivers have had significant reductions in their volumes. In some cases, the dams have been used to divert entire rivers, leaving only a small volume to flow downstream. For example, in the Kosciusko region of south-east Australia, until recently the Snowy River below Jindabyne Dam only received one per cent of its annual flow, and similarly the Tantangara Dam on the upper Murrumbidgee River also released only one per cent of the annual flow. Below such dams, the rivers are in a permanent state of human-generated drought, with reduced habitat space and biodiversity, terrestrialization of the river channel, siltation of coarse sediments and increased variability of water temperatures.

Most large dams are operated to regulate river flow, reducing floods and, more importantly from an economic perspective, delivering water to consumers – principally irrigated agriculture. The effects of river regulation and alteration of flow regimes are many and are well known (e.g. Poff et al., 1997; Poff & Zimmerman, 2010; Bunn & Arthington, 2002; Postel & Richter, 2003; Nilsson et al., 2005).

In relation to the potential exacerbation of drought, river regulation can have major effects. The obvious effect of dramatically reducing flows below dams because of wholesale diversion has already been mentioned. The effects of drastically limiting releases can resemble those generated by progressive withdrawals of water downstream of a dam, resulting incrementally in a diminished flow volume. Such withdrawals are primarily made to meet irrigation demands, and they tend to fluctuate with the timing of demand (e.g. Wilber et al., 1996; Miller et al., 2007). Indeed, in times of natural drought, urban and rural pressures to extract water from water bodies can increase sharply.

In a review, Dewson et al. (2007) concluded that the effects of artificially decreased flow on habitat and stream invertebrates appear to be rather inconclusive, though this may be due to there being marked differences of the degree and duration of low flow periods in different studies examined, and to the aggregation of responses to both natural and artificial low flow episodes.

Miller *et al.* (2007), in a study of an Oregon river with progressive downstream extractions of water for irrigation, found that water extraction exacerbated the effects of drought on macroinvertebrates due both to reduced flows and to marked increases in water temperatures. In south-eastern Australia, Finn *et al.* (2009) compared the responses of two neighbouring rivers, with one river having natural flows and the other subject to 'artificial drought' created by water extraction in the period of seasonal low flow. With 'artificial drought', conductivity rose and periphyton growth increased, though this growth appeared to be limited by low nutrient levels, due to diminished runoff and consequential low nutrient inputs. The response of the invertebrates was marked changes in species composition, rather than changes in species richness and abundance. The 'artificial drought' produced an invertebrate fauna comprising taxa known to be very tolerant of low flow conditions. More interestingly was the suggestion that the fauna produced by the low flows was shaped more by the long-term history of artificial droughts rather than the particular drought event in which the streams were sampled (Finn *et al.*, 2009). Thus, a history of low flows created by heavy water extraction may mould a novel fauna quite different from the natural one of the region.

So great can the extraction of water be from large rivers that the discharge of water into the sea can be reduced to a mere trickle, thus creating drought conditions at their mouths and in their estuaries. Vivid examples of this are the lower Murray River in South Australia, where, due to no flows from upstream, the mouth to the sea is now closed for 40 per cent of the time (Brookes *et al.*, 2009), and the Colorado River in the USA, which now only occasionally flows through its delta to reach the sea (Adler, 2007). In both cases, permanent estuaries have been converted into temporary ones.

In estuaries prone to closing and without adequate freshwater inputs, hostile hypersaline conditions may set in, as found in the St Lucia estuary in South Africa (Whitfield *et al.*, 2006) and the Coorong arm of the Murray River estuary (Brookes *et al.*, 2009)

River regulation can greatly change the temporal nature of the flow regime. Thus, in south-east Australia, the normal high river flows in winter and spring are curtailed as water is stored, while in summer, large volumes of water are released to downstream irrigators. Thus, 'anti-droughts' occur when normal seasonal low flow periods are replaced by high flows (McMahon & Finlayson, 2003), even in periods of severe drought to meet irrigation demands.

Dams can control floods; indeed this is a major management aim of many dams. Hence, in many cases the floods that normally inundate floodplains are eliminated or greatly reduced, and only very large and rare floods inundate floodplains. Lack of floodplain flooding, often combined with levee construction and wetland drainage, weakens or eliminates the booms in biotic production and diversity and the stimulation of key ecosystem processes incorporated into the concept of the flood pulse (Junk *et al.*, 1989; Tockner *et al.*, 2000, 2008; Lake *et al.*, 2006). Flooding stimulates ecological processes in a rich variety of habitats, and it generates vital avenues of lateral connectivity between the river channel and the floodplain and between the floodplain habitats – lagoons, wetlands, runners. This connectivity is critical to the movements of nutrients and carbon (POM, DOM) and of biota, be they seeds, invertebrates, fish or tortoises.

With flooding, lagoons and wetlands are filled, nutrients are mobilized, groundwater may be recharged, primary and secondary production boom, many species of invertebrates, fish and birds breed, and floodplain forests flourish (Tockner *et al.*, 2000, 2008; Lake *et al.*, 2006). Indeed, provided the flooding occurs, riverine floodplains are one of the most productive ecosystems in the world (Tockner *et al.*, 2008) – a classic case of pulse production. With no or truncated flooding, the active floodplain is simply inadequate in area and/or too briefly flooded to generate a legitimate flood pulse.

For example, elimination of the flood pulse by dams on the Paraná River, Brazil, and the consequential loss of floodplain lagoons, greatly reduced populations of the important fish *Prochilodus lineatus*, which depends on access to floodplain lagoons as part of its life cycle (Gubiani *et al.*, 2007). In south-east Australia, lack of flooding due to river regulation has caused a large wetland complex – the Lowbidgee floodplain on the Murrumbidgee River, which was once rich in biodiversity – to be essentially destroyed (Kingsford and Thomas, 2004; Khan *et al.*, 2009). Around the world there are many examples of floodplain systems impoverished by river regulation and floodplain development. Thus, one can say that because floods are vital to the ecological integrity of floodplains, their elimination or great reduction is a stark example of severe and human-generated drought, or even megadrought.

11.2.2 *Water extraction*

Many cities and agricultural areas occur along rivers and on the shores of lakes, and many of these draw water for domestic, industrial and agricultural use from waterways. Water is extracted from impoundments, lakes and flowing waters, and this can result in considerable decreases in water levels and volumes in standing and flowing waters. Thus, prior to drought, water bodies may have lost significant amounts of water. Furthermore, when droughts occur, direct water extraction from streams and wetlands may increase, lowering volumes to artificially low levels (Gasith & Resh, 1999; Dewson *et al.*, 2007). In the 1997–2010 drought in south-eastern Australia, in small streams, refuge pools were pumped out by landholders and wetlands were depleted to provide water for stock.

11.2.3   *The critical importance of connectivity*

Intrinsically linked with flow regulation by dams are the obvious impacts of fragmentation, whereby the axes of connectivity that are vital to the natural integrity of rivers are abolished or greatly reduced (Ward, 1989a; McCully, 2001; Nilsson *et al.*, 2005). The three vital axes of connectivity are longitudinal (upstream-downstream), lateral (channel-riparian zone-floodplain) and vertical (surface water-hyporheic zone-groundwater (Ward, 1989a). Dams obviously sever longitudinal connectivity, limiting the downstream movements of water, sediments, nutrients and biota as well as the upstream movements of migrating biota.

Even without droughts, the severing of longitudinal connectivity has major impacts on ecological processes. Some biota may, under normal conditions, be capable of migrating over dams – for example, in tropical streams, shrimp may migrate through small dams (Benstead *et al.*, 1999; Pringle *et al.*, 2000). However, when drought and water extraction produce very low flows, these dams are an absolute barrier, preventing juvenile shrimps from migrating upstream and replenishing upstream adult populations. In the 'bottleneck' below the dams in drought, predation greatly reduces the populations (Benstead *et al.*, 1999; Crook *et al.*, 2009). Droughts and water extraction produce very low flows such that dams become absolute barriers and fishways are ineffective (Pelicice & Agostinho, 2008). After drought, by acting as barriers to both upstream and downstream movements, dams curtail the movements of biota, especially fish, from drought refuges to favourable habitats and to sites vital for reproduction.

River flow regulation, combined with structures such as levees, have isolated rivers from their floodplains. Such strictures can create permanent drought conditions on the floodplains, and generate an environment in which riparian trees lose condition and die (e.g. Doody & Overton, 2009; Horner *et al.*, 2009) and floodplain wetlands are reduced, if not eliminated (Kingsford & Thomas, 2004; Pearce, 2007; Khan *et al.*, 2009). Worldwide, floodplain wetlands have been dried out and converted to agricultural land producing crops and pastures.

Hydrological connectivity between a river channel and the hyporheic zone and the water table can be reduced through the clogging of interstices by fine sediment (colmation). Such colmation may occur in the shallow, low velocity flows found below many dams as well as in streams whose channels have been filled with sediments from catchment erosion. In times of drought, colmation reduces the effectiveness of the hyporheic zone as a refuge for invertebrates.

Fragmentation of rivers by dams and the consequential severing or weakening of axes of connectivity – longitudinal, lateral and vertical – have served to exacerbate the effects of drought on aquatic and riparian biota. The loss of lateral connectivity through flow regulation and flood mitigation measures (e.g. levees) have cut the vital links in nutrients, food resources and biota between a river and its floodplain, have greatly curtailed floodplain productivity and have prevented the breeding of floodplain-dependent species. These conditions may be viewed as akin to permanent drought.

Severing longitudinal connectivity not only fragments populations but prevents migrations essential for successful breeding, for moving to refugia before and during droughts and for moving to favourable habitat when droughts break.

11.2.4   Habitat availability and refuges

The building of dams and weirs, divorcing floodplains from their channels, diverting water away from rivers and lakes, draining wetlands, and many other forms of deliberate human intervention, have all greatly reduced habitats for many aquatic, semi-aquatic and riparian biota. Habitat loss and the fragmentation of species populations have reduced populations and made many such populations increasingly vulnerable to the hazard of drought. In addition to these impacts, there are the deleterious effects of introduced species through predation, competition and parasites/pathogens.

For example, in a severe drought (2008), a hitherto unknown protozoan parasite, *Ichthyophthirius multifilis*, was detected in a population of redtail barb (*Barbus haasi*) in a Spanish stream. The new parasite infected 21 per cent of the population and reduced both fish density and average fish size (Maceda-Viega *et al.*, 2009). The virulence of the parasite may have been due to the fish being stressed by the low water availability and high water temperatures created by the drought.

Not only have forces such as channelization and the construction of barriers greatly reduced habitat availability, but in many cases refugia have also been destroyed. In the widespread practice of 'river improvement', coarse wood (logs) or snag loadings were greatly reduced in stream channels. Not only is coarse wood valuable habitat, but it may offer biota refuge during drought. For example, in a drought in the Flint River basin in Georgia,

USA, depressions underneath coarse wood acted as an effective refuge for mussels (Golladay *et al.*, 2004). In two Australian streams heavily sedimented by massive amounts of sand, pools were created by placing log structures into the stream channel. These structures, with associated pools, served as refuges for fish species in the early part of the 1997–2010 drought. Eventually, however, surface water in the streams disappeared and the fish were lost (Bond & Lake, 2005).

As mentioned before, a major effect of drought in small low-order streams is to greatly reduce or stop flow and for pools to form – pools that may be valuable refuges. Unfortunately, as water availability in the landscape declines, such pools become targets for water extraction or for the watering of domestic stock. In larger rivers, during drought, particular pools in which fish populations become concentrated may be subject to heavy fishing pressure. Decreases in the volumes of floodplain lagoons due to drought can produce a concentration of fish populations that may be heavily exploited by an increased fishing effort (e.g. Merron *et al.*, 1993; Laë, 1995).

Climate change is causing sea levels to rise, which could marginally reduce brackish and freshwater habitat area in open estuaries. With more sea water entering estuaries with high tides, salt wedges may lengthen and the boundary between fresh and sea water could become more abrupt. In times of low freshwater inputs, especially in droughts, estuaries, instead of having a gradient between fresh and sea water, may become fully marine environments.

11.2.5 Invasive species

The biodiversity of freshwater ecosystems is declining at a higher rate than for either terrestrial or marine ecosystems, with a major threat, especially to lentic ecosystems, being 'biotic exchange' –the impacts of invasive species (Sala *et al.*, 2000). The decline of native species at the expense of invasive species, from population reductions to local extinctions, occurs through a variety of processes. An unfortunate outcome of the success of invading species has been not only the reductions in populations of native biota but marked fragmentation of their distribution. This is clearly illustrated by the current distribution of many galaxiid fish populations in streams in Australia and New Zealand, where the galaxiids are confined to sections of headwater streams, above barriers to trout and their predatory impact (McDowall, 2006).

Invasions, by reducing and fragmenting populations of indigenous species, by altering habitat availability and by altering trophic structure, have reduced the capacity of native species to resist and recover from supra-seasonal droughts in many cases. Indeed, in some cases, it appears that droughts may favour introduced alien species, as exemplified by mosquito fish in Australia and *Tamarix* in the south-western USA. Furthermore, species such as these may not only have their invasion facilitated by drought, but may be able to persist and thrive after the drought ends.

In both depleting native species and in being favoured, invasive species in aquatic ecosystems can exacerbate the ecological impacts of drought.

11.3 Climate change and drought

Climate change refers to overall effects accompanying global warming created by the greenhouse effect in the atmosphere, due to a steady build-up of greenhouse gases – principally carbon dioxide and methane. The principal force generating the greenhouse gas build-up is the burning of fossil fuels by humans. In the pre-industrial era, the level of carbon dioxide concentration in the world's atmosphere hovered around 280 ppm, but this concentration had risen rather sharply to 379 ppm by 2005, and that of methane had risen from a pre-industrial level of 715 ppb to 1774 ppb in the same period (IPCC, 2007). The greenhouse warming may be moderated by aerosol concentrations, which, due to pollution abatement programs, have declined greatly in recent years (e.g. Cox *et al.*, 2008).

Nevertheless, these greenhouse gases have caused the earth's temperature to rise by 0.5 °C between 1910 and 2000, and it is estimated that, as greenhouse gas concentrations reach double their pre-industrial levels, there will be a net rise in the range of 2–4.5 °C, with 3 °C being the most likely estimate. Driven by the thermal expansion of the sea and the melting of glaciers and ice sheets, it is projected that the temperature rise will also be manifested in sea level rises which will range from 0.1 to 0.2 m (IPCC, 2007).

Increases in surface temperatures due to global climate change are expected to increase evapotranspiration rates and change precipitation levels, with differences between evapotranspiration and precipitation being dependent on the region. With these increased temperatures speeding up the hydrological cycle, extreme events – principally floods and droughts – are expected to increase in frequency and strength. Also, overall changes in surface water runoff are expected to be reflected in changes in surface water resources and in groundwater, albeit with lags. Thus, water levels in wetlands and lakes may rise or fall and streamflow may increase or decrease, depending on the region (Milly *et al.*, 2005; Kundzewicz *et al.*, 2007).

With climate change, runoff is expected to increase in the high latitude regions of North America and Eurasia, in the La Plata basin of South America, in eastern equatorial Africa and on equatorial islands of the eastern Pacific (Milly *et al.*, 2005). Regions with decreasing runoff include southern Africa, southern Europe, the Middle East and western North America (Milly *et al.*, 2005). Overall, modelling suggests that the area of decreased runoff will increase with time (Milly *et al.*, 2005, Kundzewicz *et al.*, 2007).

Decreased runoff affects both lotic and lentic ecosystems, or at least those systems not subject to significant human interference. Drying of a region could occur through extended and constant loss of moisture – a drying phase. However, this is not usually the case; extended drying usually occurs with a series of droughts interspersed with wet periods.

In some cases, as indicated in Chapter 4, regional drying occurs through a long period of constant drought – a megadrought. Along with the runoff and streamflow declines, droughts are expected to increase in duration, severity and spatial extent with climate change (Dai, 2010; Dai *et al.*, 2004b). Modelling by Burke *et al.* (2006) predicts that the number of moderate drought events may not alter with global climate change, but the number of severe and extreme droughts will double by the end of the 21st century. Consequently, the proportion of the land surface in drought annually is projected to sharply increase from ≈ 1 per cent at present to ≈ 30 per cent.

Severe drying (drought) is expected to increase over the 21st century in Amazonia, the United States, northern Africa, southern Europe and western Eurasia (Burke *et al.*, 2006). Furthermore, regional analyses predict that droughts will increase in southern Europe (Lehner *et al.*, 2006), Amazonia (Cox *et al.*, 2008), southwestern North America (Seager *et al.*, 2007), west Asia and Middle East (Kim & Byun, 2009) and Australia (Mpelasoka *et al.*, 2008). With the exception of Amazonia, the regions are in the mid-latitudes, which may be affected as the climate changes by the movement polewards of 'the descending branch of Hadley Cells' (Seager *et al.*, 2007). However, it does appear that the projected changes in droughts are not going to be driven by significant changes in the strength or frequency of El Niño events (Coelho & Goddard, 2009).

While climate change unleashes an array of many disturbances (e.g. higher temperatures, sea level rise, floods, droughts), it is intricately inter-meshed with the formidable array of ongoing human-created and non-climatic generators of disturbances such as pollution, land use change, water extraction, urbanization and invasive species. Thus, with more severe droughts in the future, it may be difficult in many cases to distinguish the effects of climate change from those exerted by the pre-existing and ongoing array of forces of human interventions that have exacerbated, if not created, drought conditions.

For example, changes in land cover, principally deforestation, have increased surface temperatures and evaporation to the extent that droughts in south-eastern Australia are strengthened both in duration and severity (McAlpine *et al.*, 2007; Deo *et al.*, 2009). Water extractions from many rivers are high, and when these extraction levels are combined with declines in flow due to climate change, severe reductions (up to 75 per cent) of fish species in riverine assemblages are forecast (Xenopoulos *et al.*, 2005). Planned and implemented reductions in water extraction may offset the losses due to climate change.

Human extraction and diversion of water from standing and flowing waters have greatly altered water availability (both surface and ground) and hydrological seasonality. With rising human populations, the pressures to extract water from natural and man-made water bodies will increase. Hence, both present and future pressures to meet ever-growing demands will tend to blur climate change signals. Overall, the hydrological changes from climate change, especially regional drying with droughts, will undoubtedly increase the stresses on human populations and economies (Alcamo *et al.*, 2007; Palmer *et al.*, 2008). Particularly affected will be populations dependent on dammed rivers in drought-prone regions (Palmer *et al.*, 2008) such as the Middle East (e.g. Kura River), north-eastern Brazil (e.g. Parnaiba River), central Africa (e.g. Lake Chad region) and southeastern Australia (e.g. Murray-Darling Basin).

Linked with climate change, there are effects that will undoubtedly interact with those of drought. With the overall rise in global surface temperatures, evaporation levels will rise, as will the temperatures of freshwater ecosystems. High evaporation levels (possibly accompanied by lower precipitation) independent of drought events may lower the volumes and levels of lentic systems, be they small ponds or large lakes. Many shallow wetlands, both temporary and perennial, may simply dry out even without droughts (e.g. Covich *et al.*, 1997; Nielsen & Brock, 2009). In the process of wetlands drying, salinities can rise, altering biotic communities and hindering possible future restoration efforts (Nielsen & Brock, 2009). Rising temperatures in lakes may increase thermal stability and the intensity of stratification, which in turn reduce mixing (Adrian *et al.*, 2009). Consequently, hypoxic, if not anoxic, conditions may occur or last for longer in the hypolimnion of lakes, limiting the hypolimnion as a thermal refuge for fish and allowing phosphorus to be re-mobilized from sediments. The latter outcome will serve to increase the risk of cyanobacteria blooms (Johnk *et al.*, 2008; Williamson *et al.*, 2009).

Over the long term, and influenced by drought events, temperatures have also been rising in flowing waters (Webb, 1996). For example, temperatures in the upper Danube since 1901 have risen by 1.4–1.7 °C (Webb & Nobilis, 2007), the temperatures of French rivers rose by about 2.6 °C from 1979 to 2003 (Daufresne & Boët, 2007), temperatures of Welsh streams rose by 1.4

to 1.7 °C from 1981 to 2005 (Durance & Ormerod, 2007) and, in New South Wales, Australia, stream temperatures are estimated to be rising by 0.12 °C per annum (Chessman, 2009). It is indicated that the increases in water temperatures have changed both invertebrate (Durance & Ormerod, 2007; Chessman, 2009) and fish assemblages (Daufresne & Boët, 2007), with cold water species and families declining. Warm water species may be favoured, provided that migration pathways are not blocked by such barriers as dams (e.g. Daufresne & Boët, 2007).

Human activities have played a major role in exacerbating droughts by changing land cover and in extracting and diverting large volumes of surface and groundwater, and also in largely inducing climate change. However, it may still be possible to predict the effects that climate change-induced droughts may have on aquatic systems. Given that climate change will lead to more severe supra-seasonal droughts, the effects on both lentic and lotic systems could be extremely stressful, taxing not only the biota's resistance capacity but also impairing its resilience. Climate change is predicted to increase the frequency and strength of extreme events, and this poses a particular problem for droughts as the likelihood increases that droughts may be broken by extreme floods. The coupling of these two extreme events can be highly damaging to aquatic ecosystems and very difficult to manage effectively.

Severe droughts deleteriously change the physical, chemical and biological components of freshwater ecosystems and these impacts can be expected to increase with climate change. For example, with higher temperatures, water quality deterioration in drought may increase. In standing waters, thermal stratification and hypolimnial anoxia may intensify and increase outbreaks of cyanobacteria blooms.

With the increased temperatures in severe droughts, conditions for the fauna of small streams may lead to local extinctions with poor recoveries after droughts. Indeed, fragmentation due to drought and climate change can become a major problem for the biota of flowing waters. In flowing waters and open lakes with incoming and outgoing streams, climate-change droughts could favour the expansion of warm water species at the expense of cold water ones.

In lakes and wetland complexes, especially shallow systems, harsh conditions may arise (such as described for Lake Chad and Lake Chilwa), with low volumes, higher ionic concentrations and temperatures and low oxygen levels leading to fish kills and the demise of invertebrate populations. In closed lakes, loss of populations and species in drought may mean long-term loss of diversity, due to isolation of the lake restricting the immigration of new species.

In summary, the inexorable changes due to climate change (e.g. increased water temperatures), the already degraded state worldwide of many aquatic ecosystems due to human activities, combined with the predicted increase in extreme events (floods and droughts) will further degrade, if not destroy, many aquatic ecosystems. It must also be emphasized that human populations and their activities are now, and will – especially in poorer nations – be badly affected by the trio of major forces outlined above (Alcamo *et al.*, 2007; Palmer *et al.*, 2008; Kundzewicz *et al.*, 2008). Indeed, climate change effects on water resources will be one of the major stresses on natural and human populations, communities and ecosystems. However, while it seems that climate change will very difficult to arrest, let alone reverse, we can initiate actions now to repair and restore aquatic ecosystems and increase their capacity to contend with droughts.

11.3.1 Mitigation and adaptation

Faced with increasing human demands for fresh water, the degraded state of many freshwater ecosystems and the steadily building threat of climate change, measures should be set in train to mitigate the impacts of climate change and to adapt to the pressures of climate change. In many cases, addressing these initiatives means accepting that many freshwater ecosystems are in a degraded condition and are being (mis)managed in an unsustainable way. There are, of course, a few freshwater ecosystems in remote regions that have not been changed by past and current human intervention, but these nevertheless will now be affected by climate change. In particular, we are dealing with the likelihood that the strength and duration of droughts will increase due to climate change.

Basically, in dealing with climate change, there are two courses of action – reactive and proactive (Palmer *et al.*, 2008). Reactive management means implementing measures in response to current impacts. Proactive management means the implementation of measures in response to current impacts – and much more importantly, it means acting in anticipation of future threats and impacts. In the case of droughts, management measures should be designed to aid the biota of aquatic ecosystems to resist, where feasible, the impacts of drought, and more importantly to recover after drought has broken. Reactive measures are usually short-term (Palmer *et al.*, 2008, 2009) and, in the case of drought would involve such activities as the emergency release of water from dams, the pumping of groundwater to maintain streams and wetlands, facilitating the movement of aquatic fauna to refuges, the identification and protection of refuges and the destratification of lakes/reservoirs (Bond *et al.*, 2008; Palmer *et al.*, 2009).

Although these measures may significantly reduce the damage inflicted by droughts on aquatic ecosystems, their effectiveness may be compromised and they may impede, for funding reasons, the development and implementation of a long-term proactive strategy. Proactive measures are now, and will be in most cases, linked with restoration. They operate on the basis that, in restoring an ecosystem to an ecologically sustainable state, both resistance and resilience for contending with drought are strengthened. In the long term, the costs of implementing proactive measures for drought, while initially high, may be less than the costs incurred by doing nothing and then acting reactively when drought hits (Bond *et al.*, 2008).

Faced with the threat of climate change, there is clearly a need for a feasible strategy for managing both standing and flowing water ecosystems at the appropriate spatial extent (Palmer *et al.*, 2008, 2009). Palmer *et al.* (2009) have suggested what they call a 'place-based approach', with the obvious place being the catchment level. Manageable catchments would need to be local enough to not be overwhelmed by complexity, but large enough to allow flexibility in planning and implementation.

In terms of contending with increasing dryness and drought, climate change predictions based on scaled-down climate models with predictions of human population and land use trends could be used to compile future scenarios of changes in temperature, precipitation, runoff and streamflow or water volumes in lentic systems. Although very provisional, the outcome of such an exercise would allow identification of vulnerable ecosystems, of communities, species, populations and habitats worth protecting and of ecosystems and habitats that, although they are degraded, would be worthwhile to restore. In dealing with drought, it is important that there is sufficient monitoring to detect changes in surface water volumes and in groundwater levels.

Based on the modelled climatic scenarios and the distribution of natural and human-affected ecosystems, restoration and protection measures should be planned and implemented. Large-scale restoration is invariably a proactive measure, as effective restoration takes time (years to decades) to come to fruition. Restoration measures that increase the resistance and resilience of rivers, lakes and wetlands to drought include:

- the restoration of riparian zones;
- the reinstatement of lost avenues of connectivity (e.g. river channel to floodplain);
- changes in water management to install 'natural' flow regimes, or at least key flow events with environmental flows;
- changes in catchment land use to protect wetlands and to curtail the leakage of chemicals and sediments from catchments.

Currently, many floodplain systems are divorced from their rivers because of barriers, levees and flow regulation, and are thus locked into permanent, artificial drought. Therefore, a powerful form of restoration would be the return of regular floods to these floodplains, which would allow the recharge of groundwater, the reinvigoration of riparian vegetation and the build-up of floodplain-dependent fauna – all of which would, in turn, increase resilience to drought. As the full effects of such measures are speculative, monitoring of key projects with adaptive management is essential to improve the science and practice of restoration ecology.

Besides restoration, there is also a requirement to identify and protect particular species, communities and habitats. This may involve such activities as removing barriers in streams to allow stream animals to migrate, removing harmful invaders, preventing dry wetlands from being ploughed and/or grazed and replicating habitats for isolated populations or species (Palmer *et al.*, 2009).

Overall, at the catchment level, and among stakeholders, there is the prime need to produce a water management plan that returns and then maintains ecological integrity to the aquatic ecosystems of the catchment in the face of droughts and of meeting essential human needs.

The challenge of dealing with, and adapting to, climate change and forthcoming extreme events is a daunting and very challenging one. Currently, in terms of aquatic ecosystems, their management in drought is at present poor to marginal at best, and yet we now know enough to improve management substantially. The difficulty lies in convincing management authorities and governments to make proactive measures rather than, as they are tending to do currently, acting reactively with emergency measures.

12

Conclusions

Drought has long been recognized and feared by human societies, and it has played a major role in deciding the fate of human societies and their economies. It could be argued that humans, in building economies which are basically dependent on having reliable water supplies, have developed social and economic systems which are particularly vulnerable to drought. As expected, given this vulnerability, there is a voluminous literature on the social and economic impacts of drought, and there has been considerable research on ways to mitigate drought impacts and to decrease the vulnerability of human enterprises to drought. By contrast, the literature on the impacts of drought on natural ecosystems, especially aquatic ecosystems, is meagre, fragmentary and geographically scattered in terms of its spread. Thus, this book has attempted to review the impacts of drought on the biota and ecological processes of aquatic ecosystems and to assess the vulnerability of these ecosystems to the extreme forces of drought.

One would expect that the biota of natural ecosystems may be shielded from the effects of drought to some extent, as they have evolved mechanisms over millions of years to contend successfully with droughts. However, this evolved tolerance must be tempered by the fact that particular droughts, in terms of severity and duration, can be very taxing to biota. Furthermore, in many instances, the forces of drought are accompanied by other natural disturbances, such as extreme temperatures and wildfire. With the expansion of human population, many of the disturbances that now accompany drought are human-induced, such as loss of habitats and refuges, along with excessive water extraction and imposition of barriers, all of which can exacerbate the effects of natural droughts (see Chapter 11). Thus, in many droughts, both natural and human-generated disturbances act in concert and may together overwhelm natural ecosystems.

Drought and Aquatic Ecosystems: Effects and Responses, First Edition. P. Sam Lake.
© 2011 P. Sam Lake. Published 2011 by Blackwell Publishing Ltd.

12.1   Large-scale, long-term ramp disturbances

As a disturbance, drought is unusual in that it rises from a deficiency of water, as opposed to most other disturbances that arise from excesses. Drought is particularly damaging because it occurs across a large spatial extent, across large regions and can last for a long time. Compared with most other disturbances, which are usually pulses, droughts occurs as ramps, emerging slowly and then steadily expanding in strength and in area (see Chapters 1 & 2).

While the focus of research on drought has been on its development and persistence, the ways in which droughts end have been neglected. How droughts end may make a big difference to the strength of recovery. For example, droughts that end with the pulse of a large flood may be more damaging than those that end with a ramp of steady precipitation. This speculation awaits exploration.

In recent years, stimulated undoubtedly by concerns about global climate change, research directed at understanding the climatology and meteorology of droughts has increased sharply. Evidence is steadily emerging that most droughts arise from oceanic dipole oscillations in sea surface temperatures and air pressure; this understanding, coupled with real-time data on sea surface temperatures and winds, has undoubtedly increased the capacity to predict droughts. At the same time, however, it has become clear that many droughts occur not from the operation of a single dipole oscillation but from the more complicated combination of two or even more dipole oscillations. This makes predicting and understanding the origins and persistence of droughts a difficult challenge, but one which needs continued and concerted research effort. This effort will also need to increase knowledge on the effects that global climate change is having and will continue to have on drought frequency and severity. In this regard, one of the very few benefits to emerge from the concern about climate change has been the increased interest and support for research on climate, especially on extreme events, and the effects of changing climate on biota and ecosystems.

12.2   Meteorological, hydrological and groundwater droughts – a sequence in time and severity

Meteorological droughts lead on to hydrological droughts, with surface waters being obviously affected. In time, groundwater droughts may develop. These lag in their beginning behind surface water droughts, and they may persist long after surface water droughts have ceased. Groundwater drought is clearly a major hazard in many areas of the world, yet it remains neglected in terms of research effort. This deficiency should be addressed urgently, especially as, in many parts of the world, groundwater extraction by humans is now way beyond sustainable levels.

Meteorological droughts may be reliably and readily measured in terms of indices such as the Rainfall Deciles Index and the Standardized Precipitation Index. The same does not apply to hydrological droughts. The rather complicated Palmer Drought Severity Index, based on soil moisture levels, is used to indicate agricultural drought and may be modified to indicate hydrological drought (see Chapter 2). However, this is unsatisfactory, as it relies on extensive data on soil moisture levels at the regional level. Hydrological drought is due to a deficiency of water in the landscape that causes standing and flowing waters to lose water beyond normal levels. This deficiency is reflected in lake volumes and streamflows, but there is still no universally accepted index to indicate hydrological drought in surface waters and in groundwater. Such an index needs to be standardized in terms of local conditions and capable of distinguishing between normal low flows, seasonal droughts and the more unpredictable and lengthy supra-seasonal droughts. These requirements may necessitate the development of other reliable indices.

The distinction between normal, seasonal droughts and abnormal supra-seasonal droughts is important to make, because many accounts on the ecology of drought, especially from areas with Mediterranean climates, are simply concerned with seasonal droughts. A standardized hydrological drought index would greatly improve our understanding of the effects of drought, as it would allow meaningful comparisons to be made from place to place and from drought to drought. At present, in making ecological comparisons between different localities and/or droughts, the severity of the effects may be the only guide.

12.3 Recognizing the importance of past droughts

Over the Holocene period (the last 10,000 years) droughts – especially megadroughts (those lasting ten or more years) – have occurred, with the evidence of such droughts coming mainly from lake sediments. In many places, the mid-Holocene was marked by lengthy and severe megadroughts (see Chapter 4). For droughts in more recent times (i.e. the past 1,000 years), tree rings can be used to detect droughts with finer temporal resolution than lake sediments. Both of these sources of evidence strongly indicate that many past droughts have been severe and lengthy, and that many human societies are ill-equipped to deal with them (see Chapter 4).

Currently, a detailed appreciation of past droughts, especially over the last 1,000 years, is only available for North America. An international research effort to develop the drought history of other regions of the world would therefore be very worthwhile, not only to the discipline of palaeoclimatology, but also in terms of providing a forewarning of what sort of droughts could be expected in any one region. Such a research effort may also be able to assist in making predictions of the effects that global climate change might exert on future droughts. Mapping and measuring megadroughts of the past could provide an estimate of future droughts and could thus be used to assess how contemporary societies could contend – successfully or not – with the threat of megadroughts in the future.

12.4 Ecological effects of drought

12.4.1 Disconnections and variable effects

Across landscapes, there are waterscapes consisting of disconnected and connected water bodies (e.g. lakes, ponds, rivers). Connectivity may be between different surface waters and between surface waters and groundwater. In any one catchment, there can be a great variety of temporary and permanent water bodies differing in many factors, including levels of connectivity, size, catchment condition, morphology, chemical composition and their biota. As a large-scale disturbance, drought can affect water bodies across catchments and across regions to greatly varying extents.

The differing impacts of drought and the varying responses to it that undoubtedly occur over catchments or regions have not been investigated at the large spatial extent, and thus we have a poor understanding of the integrated effects of drought across a variety of water bodies. To gain such an integrated understanding may be seen as an impossible task but, with a rapidly increasing interest in long-term ecological research, and with the rapid advances being made in remote sensing, such a task may become feasible. It may even become possible for the effects of drought on components of both terrestrial and aquatic ecosystems to be assessed and integrated across large areas. Such an integrated approach to the study of droughts would be great value to managers, as it may identify features of land- or waterscapes that are at a higher risk of damage by drought than others. Such knowledge could then be used to build resistance and/or resilience to the effects of drought.

Droughts set in by decreasing the amount of water in catchments through the lack of precipitation and increases in evapotranspiration. There is a sharp decline in inputs of chemicals – especially nutrients, sediments and organic matter – from catchments to their water bodies. Within catchments, chemicals such as nitrates and dissolved organic carbon can accumulate with drought.

The vegetation of riparian zones during drought typically becomes stressed and declines, especially in long droughts, when the water table may recede. This decline may impair the recovery processes in water bodies after drought. Knowledge of biogeochemical processes in catchments (and riparian zones) during and after droughts is limited and fragmentary.

12.4.2   Abiotic and biotic effects

As with other disturbances, droughts exert a variety of different effects on both abiotic variables and on aquatic biota and on their ecosystems. The changes in abiotic variables, such as changes in water quality, may be a major means by which the biota are affected and to which the different biota have evolved different adaptive responses. The main target of research on abiotic variables during drought has focused on the changes in such variables as temperature, oxygen concentrations and salinity (electrical conductivity). When droughts break, high loadings of many chemicals (e.g. nitrate, dissolved organic matter (DOC), heavy metals) may be delivered from catchments to aquatic ecosystems (see Chapter 5). These high loadings of inorganic chemicals and organic matter into both lotic and lentic systems may impair or stimulate recovery, but this phenomenon remains unstudied.

Linked with the recognition of catchment inputs influencing the responses of aquatic biota to drought is the realization that for both catchments and their water bodies, the antecedent conditions prevailing before drought exert a very strong influence on the impacts of drought. For example, if a catchment has a lingering water deficit due to past droughts, the effects of a further drought will be more severe than if the catchment and its water bodies were at normal water levels. As mentioned in Chapter 11, the antecedent conditions of catchments and water bodies created by human activities can exert a strong influence on the impacts of subsequent droughts. This influence is an important and unstudied area of research. A prime reason for this knowledge gap is simply that most studies of drought in aquatic systems have been begun with droughts developing or already developed, and they have not included catchment factors.

As drought sets in and water is steadily lost from a water body, distinct thresholds with major ecological effects may occur. In lentic systems (lakes, ponds), a critical threshold is the loss of connection between the littoral zone and the water. A further threshold evident in shallow lakes is that the loss of volume can lead to the emergence of separate water basins. Other thresholds dependent on the nature of the standing water body include the cessation of surface and groundwater inflows and outflows. In most flowing waters, the major thresholds are the loss of the littoral zone and of longitudinal connectivity.

In both systems, a further threshold is the drying from free surface water to dry sediment. The crossing of these thresholds undoubtedly has major ecological effects and yet, in most cases, and especially in lentic systems, these effects have been poorly described both in terms of abiotic and biotic changes.

Of particular note are the effects of drought on floodplain systems. Longitudinal connectivity may be maintained throughout a drought, albeit with greatly reduced flow. With drought, lateral connectivity between the river channel and the floodplain can be severed, and consequently the wetlands of the floodplain are not replenished by floods and progressively dry out. This leads to losses of biota, of biotic production and of the subsidies both between a river channel and its floodplain, and between the floodplain and its hinterland.

Human activities causing river regulation and water extraction have meant that many floodplain systems are now locked into permanent hydrological drought, and it has taken a long time for the full implications of this sad state to be realized by water resource managers. To restore floodplain ecosystems that have been in prolonged and human-created droughts, an integrated approach is required which involves much more than allowing occasional floods on selected parts of floodplains.

Neglected until recently has been the effects of drought on estuaries. The type of estuary – whether open or closed – can have a major bearing on the effects of drought. In closed estuaries, with low or no freshwater inputs and with high evaporation, very high salinities can eventuate. Salinities may also rise in open estuaries, but not to the same extent – and, with open passage to the sea, many species may escape to refuges. As in freshwater systems, the breaking of a drought can lead to high loadings of nutrients, with consequential changes in phytoplankton and attached algae. Many estuaries are centres of human settlement, and human activities such as pollution often exacerbate the effects of drought.

The responses of the biota to drought, as for other disturbances, have been divided into those that involve resistance (the capacity to withstand) and those which involve resilience (the capacity to recover after losses). This division is useful, although for many biota, both resistance and resilience are present in varying degrees. Many organisms – microbial, plant and animal – have mechanisms to resist drought, and the presence and strength of these mechanisms vary immensely.

For example, in the case of variable resistance, freshwater fish mostly require free water with normal water quality, but some can survive by air breathing in foul, deoxygenated waters and a very few can survive in cocoons in dry mud. Similarly, and linked with resilience, there are many mechanisms for populations of biota to be depleted by drought and yet allow some to survive and recover through migration or reproduction or both. Many fish, especially in wetlands, appear to be able to detect the drying out of their habitat, and they consequently move to areas where water may persist (refugia). With the breaking of the drought, the survivors and their offspring may return and repopulate their former habitat.

There has been considerable research on the effects of drought on biota in terms of resistance or lack of it, but little on the mechanisms of resilience. A clear reason for this lack is that many studies are short-term and do not follow recovery after drought.

Key to both resistance and resilience of biota is the presence or use of refuges. In the past 20 years or so, in the study of freshwater ecology there have been many reports on various forms of refuges to disturbance, notably in relation for floods and, to a lesser extent, for droughts. For droughts, refuges involving particular life histories and traits, patterns of habitat changes, and migration pathways have been described and categorized. However, due to the abovementioned lack of recovery studies, the value of refuges for the recovery of many organisms after drought remains rather uncertain.

Linked with the availability of refuges for recovery, especially in flowing waters, is the critical need for recovery also to restore connectivity in order to allow migration from refuges to the newly available habitats. In managing to identify and protect drought refuges in rivers, the key to successful recovery lies not just in maintaining the refuges, but in ensuring that connectivity after the drought is maintained and/or restored.

12.5 Recovery from drought – a neglected field

In terms of the capacity to survive drought and to effectively recover, there are wide differences between groups of plants and animals. A consistent finding in lentic water bodies, notably shallow lakes and lagoons, is that, as drought sets in, the normal phytoplankton may be supplanted by blooms of cyanobacteria, and the normally crustacean-dominated zooplankton may become dominated by rotifers.

In flowing waters, biofilm bacteria and algae are diminished by drought, yet their recovery with re-wetting can be rapid. In the invertebrates, there are wide differences among groups but as, a generalization in comparison with other groups (e.g. oligochaetes, crustaceans, molluscs), insects appear to be more effective in surviving droughts. No doubt this tolerance is a function of the great diversity and high mobility of insects, although some

groups (e.g., Plecoptera, Ephemeroptera, Odonata) are much more susceptible to depletion by droughts than others such as the Coleoptera, Hemiptera
and Diptera.

An interesting finding in recovery from drought in both flowing and
standing waters is the short-term high abundance of otherwise rare species
which briefly flourish and then disappear – possibly examples of disturbance-
dependent fugitive species. In standing waters, the recovery by fish after
droughts can take time and there is generally a succession of species, with
small-bodied, short-generational species coming before large-bodied, long-
generational fish. In flowing waters, with the restoration of connectivity
after drought, the recovery of fish populations can be relatively rapid.

Perhaps with further research on population densities and assemblage
structure, consistent findings on drought tolerances and capacities to
recover will emerge, although such findings will need to be tempered by
the particular geographical and ecological settings of the affected water
bodies. For effective conservation of biota, it is imperative that the drought
tolerance and recovery capacities of species of high conservation value be
determined, so that strategies for them to survive drought can be devised
and implemented.

As a broad generalization, supra-seasonal droughts exert much stronger
effects on the biota of permanent water bodies than that of temporary
water bodies. This is not surprising, as the fauna and flora of temporary
water bodies have presumably evolved adaptations to contend with seasonal
drying and wetting and even the very occasional and unpredictable
presence of water, interspersed with long dry periods. The biota of temporary
water bodies have many adaptations to contend with drought, such
as seeking and using refuges and having specialized physiological adaptations to tolerate adverse aquatic physico-chemical conditions. However,
if temporary or permanent streams or lakes dry down to remnant pools
and then to dry sediments, the loss of biota, notably fauna, can be
quite considerable.

The differences in contending with drought are illustrated by the fauna of
floodplains. To survive drought, most fish have to migrate to permanent
water (e.g. a river channel), whereas both the aquatic flora and most
invertebrates have resistant mechanisms to persist in the dry sediments
of the floodplain. Although reliable evidence on recovery (resilience) is
rather limited, it does appear that recovery in permanent systems, such as
shallow lakes and rivers, can be lengthy and incomplete, in that some species
remain missing. This difference is very evident when comparing the high
resilience to drought of the biota of streams in Mediterranean-climate areas,
which seasonally have low flows or even dry out, with the highly variable
resilience of biota in perennial streams. Again, as stressed above, a vital gap

in our knowledge of the effects of drought lies in the lack of extensive studies of ecological recovery from drought.

From studies up to the present time, an understanding of the effects of drought on biota is emerging for a wide variety of aquatic ecosystems (lotic and lentic, permanent and temporary, freshwater or saline or estuarine). However, there have been very few studies of ecosystem properties and processes. Ecosystem properties that have been barely touched upon include trophic structure and food webs, and ecosystem processes in need of further study include primary and secondary production, decomposition and nitrogen dynamics, to mention just a few. Clearly, the ebbing and flowing, the contraction and expansion of ecosystem processes and structure with drought remains a large knowledge gap – a gap of major importance, given the predictions of climate change.

12.6   The future: studying drought and human interactions

Drought is a very difficult phenomenon to study. To understand drought in a holistic sense at a locality requires coordinated monitoring of meteorology, hydrology (surface and groundwater), biogeochemistry and ecology. Rarely has such a coordinated effort been executed. Consequently at present, in practical terms, coming to understand droughts ecologically means evaluating studies with quite different aims and variables across different droughts (usually poorly defined) or even, in some cases, different studies with different variables for the same drought.

Being large-scale disturbances of long duration, and also being relatively unpredictable, droughts do not fit into the normal agendas of researchers. Short-term, small-scale experiments may provide insights into such aspects as physiological and behavioural adaptations to drought, but they do not allow an appreciation of drought as a ramp disturbance acting on ecosystems. To understand drought adequately requires long-term research with adequate funding. If this is possible, cross-disciplinary, long-term monitoring at selected localities could provide a holistic appreciation of droughts from their initiation, their peak and their cessation with ecosystem recovery.

Humans and their activities have directly and indirectly affected many, if not most, inland water bodies in the world. In altering catchments by changing land cover and land use, humans have indirectly affected water bodies by altering the inputs from catchments into water bodies. By actions such as pollution, river regulation, barrier construction, channelization, water extraction and introducing alien species, humans have been the major force of degradation of many aquatic ecosystems. It seems that, as human activities have spread, the scale of human-generated disturbances has also inexorably increased from small local ones to global-scale distur-bances such as climate change. Many of these disturbances, both local and regional, have, as described in Chapter 11, served to exacerbate the damaging effects of drought on aquatic ecosystems. This strengthening of drought effects is widespread and has, in many instances, greatly increased the extent of damage due to drought. In other instances, such as floodplain ecosystems, by long-term prevention of seasonal flooding, humans have actually created long-term hydrological drought.

Awareness of how humans exacerbate the effects of drought is increasing but, as yet, steps to reduce this exacerbation have been very limited. Where steps have been made to reduce the effects of drought, they are invariably reactive measures to ameliorate a crisis, such as the threatened extinction of a species. A more sensible approach to ameliorating the human-induced exacerbation of drought would be to take a comprehensively planned and implemented proactive strategy of restoring or rehabilitating damaged catchments and aquatic ecosystems. Such a strategy rests upon the concept that if an ecosystem is restored, the capacity for that ecosystem to contend with the stresses of drought (and other disturbances) would be increased. This concept clearly warrants a major research effort in order to develop effective restoration measures for increasing the resistance/resilience of the biota of damaged ecosystems. Again, as in the case of drought research, the need is for adequately resourced long-term interven-tion and research.

With global climate change, it is likely that, in many areas, droughts may increase in frequency and duration. This threat heightens the need for proactive, restorative measures to be implemented now. Droughts have always damaged human societies and they remain a very threatening disturbance which we poorly understand, and with which we contend quite inadequately.

References

Abbott, M.B., Binford, M.W., Brenner, M. & Kelts, K.R. (1997) A 3500 ^{14}C yr high-resolution record of water-level changes in Lake Titicaca, Bolivia/Peru. *Quaternary Research* **47**, 169–180.

Acosta, C.A. & Perry, S.A. (2001) Impact of hydropattern disturbance on crayfish population dynamics in the seasonal wetlands of Everglades National Park, USA. *Aquatic Conservation: Marine and Freshwater Ecosystems* **11**, 45–57.

Acuña, V., Muñoz, I., Giorgi, A., Omella, M., Sabater, F. & Sabater, S. (2005) Drought and post-drought recovery cycles in an intermittent Mediterranean stream: structural and functional aspects. *Journal of the North American Benthological Society* **24**, 919–933.

Acuña, V., Giorgi, A., Muñoz, I., Sabater, F. & Sabater, S. (2007) Meteorological and riparian influences on organic matter dynamics in a forested Mediterranean stream. *Journal of the North American Benthological Society* **26**, 54–69.

Adams, S.B. & Warren, M.L. (2005) Recolonization by warmwater fishes and crayfishes after severe drought in upper Coastal Plain Hill streams. *Transactions of the American Fisheries Society* **134**, 1173–1192.

Adamson, J.K., Scott, W.A., Rowland, A.P. & Beard, G.R. (2001) Ionic concentrations in a blanket peat bog in northern England and correlations with deposition and climate variables. *European Journal of Soil Science* **52**, 69–79.

Adkinson, A., Watmough, S.A. & Dillon, P.J. (2008) Drought-induced metal release from a wetland at Plastic Lake, central Ontario. *Canadian Journal of Fisheries and Aquatic Sciences* **65**, 834–845.

Adler, R. (2007) *Restoring Colorado River Ecoystems. A Troubled Sense of Immensity*. Island Press, Washington, DC.

Adrian, R., O'Reilly, C.M., Zagarese H., Baines, S.B., Hessen, D.O., Keller, W., Livingstone, D.M., Sommaruga, R., Straile D., Van Donk, E., Weyhenmeyer, G.A. & Winder, M. (2009) Lakes as sentinels of climate change. *Limnology and Oceanography* **54**, 2283–2297.

Agnew, C.T., Clifford, N.J. & Haylett, S. (2000) Identifying and alleviating low flows in regulated rivers: the case of the Rivers Bulbourne and Gade, Hertfordshire UK. *Regulated Rivers: Research & Management* **16**, 245–266.

Albaret, J.J. (1987) Les peuplements de poissons de la Casamance (Sénégal) en période de sécheresse. *Revue de Hydrobiologie Tropicale* **20**, 291–310.

Alcamo, J., Florke, M. & Märker, M. (2007) Future long-term changes in global water resources driven by socio-economic and climatic changes. *Hydrological Sciences Journal* **52**, 247–275.

Alin, S.R. & Cohen, A.S. (2003) Lake-level history of Lake Tanganyika, East Africa, for the past 2500 years based on ostracode-inferred water-depth reconstruction. *Palaeogeography, Palaeoclimatology, Palaeoecology* **199**, 31–49.

Allan, J.D. & Castillo, M.M. (2007) *Stream ecology. Structure and Function of Running Waters.* (2nd Edition) Springer, Dordrecht, The Netherlands.

Allan, R.J. & Haylock, M.R. (1993) Circulation features associated with the winter rainfall decrease in southwestern Australia. *Journal of Climate* **6**, 1356–1367.

Allan, R.J., Lindesay, J. & Parker, D. (1996) *El Niño, Southern Oscillation and Climatic Variability.* CSIRO Publishing, Collingwood, Victoria.

Amlin, N.M. & Rood, S.B. (2003) Drought stress and recovery of riparian cottonwoods due to water table alteration along Willow Creek, Alberta *Trees* **17**, 351–358.

Andersen, C.B., Lewis, G.P. & Sargent, K.A. (2004) Influence of wastewater-treatment effluent on concentrations and fluxes of solutes in the Bush River, South Carolina, during extreme drought conditions. *Environmental Geosciences* **11**, 28–41.

Anderson, P.W. & McCall, J.E. (1968) Effect of drought on stream quality in New Jersey. *Journal of the Sanitary Engineering Division Proceedings of the American Society of Civil Engineers.* **94** (SA5)779–788.

Angeler, D.G. & Rodrigo, M.A. (2004) Ramp disturbance-ramp response: a simple model for wetland disturbance ecology. *Marine and Freshwater Research* **55**, 33–37.

Ångström, A. (1935) Teleconnections of climate changes in present time. *Geografisker Annaler* **17**, 242–258.

Anlauf, A. (1990) Cyst formation of *Tubifex tubifex* (Müller) – an adaptation to food deficiency and drought. *Hydrobiologia* **190**, 79–82.

Anonymous, (2008) Opportunistic observations. *Codwatch* **27**, 4.

Antolos, M., Roby, D.D., Lyons, D.E., Collis, K., Evans, A.E., Hawbecker, M. & Ryan, B.A. (2005) Caspian tern predation on juvenile salmonids in the mid-Columbia River. *Transactions of the American Fisheries Society* **134**, 466–480.

Arab, A., Lek, S., Lounaci, A. & Park, Y.S. (2004) Spatial and temporal patterns of benthic invertebrate communities in an intermittent river (North Africa). *Annales de Limnologie* **40**, 317–327.

Arfi, R., Bouvy, M., Cecchi, P., Corbin, D. & Pagano, M. (2003) Environmental conditions and phytoplankton assemblages in two shallow reservoirs of Ivory Coast (West Africa). *Archiv für Hydrobiologie* **156**, 511–534.

Armitage, P.D. & Petts, G.E. (1992) Biotic score and prediction to assess the effects of groundwater abstraction on river macroinvertebrates for conservation purposes. *Aquatic Conservation* **2**, 1–17.

Arnott, S.E. & Yan, N.D. (2002) The influence of drought and re-acidification on zooplankton emergence from resting stages. *Ecological Applications* **12**, 138–153.

Arnott, S.E., Yan, N., Keller, W. & Nicholls, K. (2001) The influence of drought-induced acidification on the recovery of plankton in Swan Lake (Canada). *Ecological Applications* **11**, 747–763.

Artigas, J., Romaní, A.M., Gaudes, A., Muñoz, I. & Sabater, S. (2009) Organic matter availability structures microbial biomass and activity in a Mediterranean stream. *Freshwater Biology* **54**, 2025–2036.

Aspbury, A.S. & Juliano, S.A. (1998) Negative effects of habitat drying and prior exploitation on the detritus resource in an ephemeral aquatic habitat. *Oecologia* **115**, 137–148.

Australian Bureau of Statistics (2004) Impact of the drought on Australian production in 2002–2003. *Yearbook Australia 2004*. Australian Bureau of Statistics, Canberra, ACT.

Attrill, M.J. & Power, M. (2000a) Modelling the effect of drought on estuarine water quality. *Water Research* **34**, 1584–1594.

Attrill, M.J. & Power, M. (2000b) Effects on invertebrate populations of drought-induced changes in estuarine water quality. *Marine Ecology Progress Series* **203**, 133–143.

Attrill, M.J., Rundle, S.D. & Thomas, R.M. (1996) The influence of drought-induced low freshwater flow on an upper-estuarine macroinvertebrate community. *Water Research* **30**, 261–268.

Axtell, R.L., Epstein, J.M., Dean, J.S., Gumerman, G.J., Swedlund, A.C., Jarburger, J., Chakravarty, S., Hammond, S., Parker, J. & Parker, M. (2002) Population growth and collapse in a multiagent model of the Kayenta Anasazi in Long House Valley. *Proceedings of the National academy of Sciences* **99** (Suppl. 3) 7275–7279.

Babbitt, K.J. & Tanner, G.W. (2000) Use of temporary wetlands by anurans in a hydrologically modified landscape. *Wetlands* **20**, 313–322.

Baber, M.J., Childers, D.L., Babbitt, K.J. & Anderson D.H. (2002) Controls on fish distribution and abundance in temporary wetlands. *Canadian Journal of Fisheries and Aquatic Sciences* **59**, 1441–1450.

Baker, T.L. & Jennings, C.A. (2005) Striped bass survival in Lake Blackshear, Georgia during drought conditions: implications for restoration efforts in Gulf of Mexico drainages. *Environmental Biology of Fishes* **72**, 73–84.

Bala, G., Caldeira, K., Wickett, M., Phillips, T.J., Lobel, D.B., Delire, C. & Mirin, A. (2007) Combined climate and carbon-cycle effects of large-scale deforestation. *Proceedings of the National Academy of Sciences* **104**, 6550–6555.

Balcombe, S.R. & Arthington, A.H. (2009) Temporal changes in fish abundance in response to hydrological variability in a dryland floodplain river. *Marine and Freshwater Research* **60**, 146–159.

Baldwin, D.S. & Mitchell, A.M. (2000) The effects of drying and re-flooding on the sediment and soil nutrient dynamics of lowland river-floodplain systems: A synthesis. *Regulated Rivers: Research & Management* **16**, 457–467.

Baldwin, D.S., Rees, G.N., Mitchell, A.M. & Watson, G. (2005) Spatial and temporal variability of nitrogen dynamics in an upland stream before and after a drought. *Marine and Freshwater Research* **56**, 457–464.

Baldwin, D.S., Hall, K. C., Rees, G.N. & Richardson, A.J. (2007) Development of a protocol for recognizing sulfidic sediments (potential acid sulfate soils) in freshwater wetlands. *Ecological Management and Restoration*. **8**, 56–60.

Baldwin, D.S., Gigney, H., Wilson, J.S., Watson, G. & Boulding, A.N. (2008) Drivers of water quality in a large water storage reservoir during a period of extreme drawdown. *Water Research* **42**, 4711–4724.

Baldwin, D.S., Wilson, J., Gigney, H. & Boulding, A. (2010) Influence of extreme drawdown on water quality downstream of a large water storage reservoir. *River Research and Applications* **26**, 194–206.

Ballinger, A. & Lake, P.S. (2006) Energy and nutrient fluxes from rivers and streams into terrestrial foodwebs. *Marine and Freshwater Research* **57**, 15–28.

Balogh, C., Muskó, I.B., Tóth, L.G. & Nagy, L. (2008) Quantitative trends of zebra mussels in Lake Balaton (Hungary) in 2003–2005 at different water levels. *Hydrobiologia* **613**, 57–69.

Baptista, J., Martinho, F., Dolbeth, M., Viegas, I., Cabral, H. & Pardal, M. (2010) Effects of freshwater flow on the fish assemblage of the Mondego estuary (Portugal): Comparison between drought and non-drought years. *Marine and Freshwater Research* **61**, 490–501.

Bärlocher, F., Mackay, R.J. & Wiggins, G.B. (1978) Detritus processing in a temporary vernal pool in southern Ontario. *Archiv für Hydrobologie* **81**, 269–295.

Barrat-Segretain, M-H. & Cellot, B. (2007) Response of invasive macrophyte species to drawdown: the case of *Elodea* sp. *Aquatic Botany* **87**, 255–261.

Barros, A.P. and Bowden, G.J. (2008) Toward long-lead operational forecasts of drought: an experimental study in the Murray-Darling River basin. *Journal of Hydrology* **357**, 349–367.

Bate, G. C. & Smailes, P.A. (2008) The response of the diatom flora of St Lucia Lake and estuary, South Africa, to a severe drought. *African Journal of Aquatic Science* **33**, 1–15.

Battin, T.J., Kaplan, L.A., Findlay, S., Hopkinson, C.S., Marti, E., Packman, A.I., Newbold, J.D. & Sabater, F. (2008) Biophysical controls on organic carbon fluxes in fluvial networks. *Nature Geoscience* **1**, 95–100.

Baumgärtner, D., Mörtl, M. & Rothhaupt, K-O. (2008) Effects of water-depth and water-level fluctuations on the macroinvertebrate community structure in the littoral zone of Lake Constance. *Hydrobiologia* **613**, 97–107.

Bayley, P.B. & Osborne, L.L. (1993) Natural rehabilitation of stream fish populations in an Illinois catchment. *Freshwater Biology* **29**, 295–300.

Bayley, S.E., Schindler, D.W., Parker, B.R., Stainton, M.P. & Beaty, K.G. (1992) Effects of forest fire and drought on acidity of a base-poor boreal forest stream: similarities between climatic warming and acidic precipitation. *Biogeochemistry* **17**, 191–204.

Bêche, L.A. & Resh, V.H. (2007) Short-term climatic trends affect the temporal variability of macroinvertebrates in California 'Mediterranean' streams. *Freshwater Biology* **52**, 2317–2339.

Bêche, L.A., McElravy, E.P. & Resh, V.H. (2006) Long-term seasonal variation in the biological traits of benthic-macroinvertebrates in two Mediterranean-climate streams in California, U.S.A. *Freshwater Biology* **51**, 56–75.

Bêche, L.A., Connors, P.G., Resh, V.H. & Merenlender, A.M. (2009) Resilience of fishes and invertebrates to prolonged drought in two California streams. *Ecography* **32**, 778–788.

Beebee, T.J.C. & Griffiths, R.A. (2005) The amphibian decline crisis: A watershed for conservation biology? *Biological Conservation* **125**, 271–285.

Beissinger, S.R. (1995) Modeling extinction in periodic environments: Everglades water levels and snail kite population viability. *Ecological Applications* **5**, 618–631.

Beissinger, S.R. & Takegawa, J.E. (1983) Habitat use and dispersal by snail kites in Florida during drought conditions. *Florida Field Naturalist* **11**, 89–106.

Beklioğlu, M. (2007) Role of hydrology, nutrients and fish in interaction with global climate change in effecting ecology of shallow lakes in Turkey. Proceedings of the International Congress on River Basin Management. pp. 583–595.

Beklioğlu, M. & Tan, C.O. (2008) Restoration of a shallow Mediterranean lake by biomanipulation complicated by drought. *Fundamental and Applied Limnology* **171**, 105–118.

Bell, V.A., Elliott, J.M. & Moore, R.J. (2000) Modelling the effects of drought on the population of brown trout in Black Brows Beck. *Ecological Modelling* **127**, 141–159.

Bender, E.A., Case, T.J. & Gilpin, M.E. (1984) Perturbation experiments in community ecology; theory and practice. *Ecology* **65**, 1–13.

Bénech, V., Durand, J-R. & Quensière, J. (1983) Fish communities of Lake Chad and associated rivers and floodplains. In: Carmouze, J.-P., Durand, J.-R.& Lévêque, C. (eds.) *Lake Chad. Ecology and Productivity of a Shallow Tropical Ecosystem.* pp. 293–356. Dr W. Junk, The Hague.

Benenati, P.L., Shannon, J.P. & Blinn, D.W. (1998) Desiccation and recolonization of phytobenthos in a regulated desert river: Colorado River at Lees Ferry, Arizona. *Regulated Rivers: Research and Management* **14**, 519–532.

Bennett, K.D., Tzedakis, P.C. & Willis, K.J. (1991) Quaternary refugia of north European trees. *Journal of Biogeography* **18**, 103–115.

Bennett, W.A., Ostrach, D.J. & Hinton, D.E. (1995) Larval striped bass condition in a drought-stricken estuary: Evaluating pelagic food-web limitation. *Ecological Applications* **5**, 680–692.

Bennetts, R.E., Colloy, M.W. & Rodgers, J.A. (1994) The snail kite in the Florida Everglades: A food specialist in a changing environment. In: Davis, S.M. & Ogden, J.C. (eds.) *Everglades. The Ecosystem and its Restoration.* pp. 507–532. St. Lucie Press, Delray Beach, FL.

Benson, L., Kashgarian, M., Rye, R., Lund, S., Paillet, F., Smoot, J., Kester, C., Mensing, S., Meko, D. & Lindström, S. (2002) Holocene multidecadal and multicentennial droughts affecting northern California and Nevada. *Quaternary Science Reviews* **21**, 659–682.

Benson, L., Petersen, K. & Stein, J. (2006) Anasazi (Pre-Columbian native-American) migrations during the middle-12th and late-13th centuries – were they drought induced? *Climatic Change* **83**, 187–213.

Benson, L.V., Berry, M.S., Jolie, E.A., Spangler, J.D., Stahle, D.W. & Hattori, E.M. (2007) Possible impacts of early-11th-, middle-12th-, and late-13th-century droughts on western Native Americans and the Mississippi Cahokians. *Quaternary Science Reviews* **26**, 336–350.

Bentley, P.J. (1966) Adaptations of Amphibia to arid environments. *Science* **152**, 619–623.

Benstead, J.P., March, J.G., Pringle, C.M. & Scatena, F.N. (1999) Effects of water abstraction and damming on migratory stream biota. *Ecological Applications* **9**, 656–688.

Berrie, A.D. (1992) The chalk stream environment. *Hydrobiologia* **248**, 3–9.

Betts, R.A., Falloon, P.D., Goldewijk, K.K. & Ramankutty, N. (2001) Biogeophysical effects of land use on climate: model simulations of radiative forcing and large-scale temperature change. *Agricultural and Forest Meteorology* **142**, 216–233.

Bickerton, M., Petts, G., Armitage, P. & Castella, E. (1993) Assessing the ecological effects of groundwater abstraction on chalk streams: three examples from eastern England. *Regulated Rivers: Research & Management* **8**, 121–134.

Binford, M.W., Kolata, A.L., Janusek, J.W., Seddon, M.T., Abbott, M. & Curtis, J.H. (1997) Climate variation and the rise and fall of an Andean civilization. *Quaternary Research* **47**, 235–248.

Bjerknes, J. (1969) Atmospheric teleconnections from the equatorial Pacific. *Monthly Weather Review* **97**, 163–172.

Blair, W.F. (1957) Changes in vertebrate populations under conditions of drought. *Cold Spring Harbor Symposia on Quantitative Biology* **22**, 273–275.

Blanton, J., Alber, M. & Sheldon, J. (2001) Salinity response in the Satilla River estuary to seasonal changes in freshwater discharge. In: Hatcher, K.J. (ed.) *Proceedings of the 2001 Georgia Water Resources Conference*. Institute of Ecology, University of Georgia, Athens, GA.

Bonacci, O. (1993) Hydrological identification of drought. *Hydrological Processes* **7**, 249–262.

Bonada, N., Rieradevall, M., Prat, N. & Resh, V.H. (2006) Benthic macroinvertebrate assemblages and macrohabitat connectivity in Mediterranean-climate streams of northern California. *Journal of the North American Benthological Society* **25**, 32–43.

Bonada, N., Rieradevall, M. & Prat, N. (2007) Macroinvertebrate community structure and biological traits related to flow permanence in a Mediterranean river network. *Hydrobiologia* **589**, 91–106.

Bond, G., Kromer, B., Beer, J., Muscheler, R., Evans, M.N., Showers, W., Hoffman, S., Lotti-Bond, R., Hajdas, I. & Bonani, G. (2001) Persistent solar influence on North Atlantic climate during the Holocene. *Science* **294**, 2130–2136.

Bond, N.R. & Lake, P.S. (2005) Ecological restoration and large-scale ecological disturbance: the effects of drought on the response by fish to a habitat restoration experiment. *Restoration Ecology* **13**, 39–48.

Bond, N.R. Lake, P.S. & Arthington, A.H. (2008) The impacts of drought on freshwater ecosystems: an Australian perspective. *Hydrobiologia* **600**, 3–16.

Bonis, A. & Grillas, P. (2002) Deposition, germination and spatio-temporal patterns of charophyte propagule banks: a review. *Aquatic Botany* **72**, 235–248.

Boqiang, Q., Weiping, H., Guang, G., Liancong, L. & Jinshan, Z. (2004) Dynamics of sediment resuspension and the conceptual schema of nutrient release in the large shallow Lake Taihu. *Chinese Science Bulletin* **49**, 54–64.

Borges, P.A.F. & Train, S. (2009) Phytoplankton diversity in the upper Paraná River floodplain during two years of drought (2000 and 2001). *Brazilian Journal of Biology* **69**, supplement 2, 637–647.

Botterill, L.C. (2003) Government responses to drought in Australia. In: Botterill, L.C. & Fisher, M. (eds.) *Beyond Drought. People, Policy and Perspectives*, pp. 49–66. CSIRO Publishing, Collingwood. Victoria, Australia.

Boulton, A.J. (1989) Over-summering refuges of aquatic macroinvertebrates in two intermittent streams in central Victoria. *Transactions of the Royal Society of South Australia* **113**, 23–34.

Boulton, A.J. (2003) Parallels and contrasts in the effects of drought on stream macroinvertebrate assemblages. *Freshwater Biology* **48**, 1173–1185.

Boulton, A.J. & Brock, M.A. (1999) *Australian Freshwater Ecology, Processes and Management*. Gleneagles Publishing, Adelaide, Australia.

Boulton, A.J. & Lake, P.S. (1990) The ecology of two intermittent streams in Victoria, Australia. I. Multivariate analyses of physicochemical features. *Freshwater Biology* **24**, 123–141.

Boulton, A.J. & Lake, P.S. (1992a) The ecology of two intermittent streams in Victoria, Australia. II. Comparisons of faunal composition between habitats, rivers and years. *Freshwater Biology* **27**, 99–121.

Boulton, A.J. & Lake, P.S. (1992b) The ecology of two intermittent streams in Victoria, Australia. III. Temporal changes in faunal composition. *Freshwater Biology* **27**, 123–138.

Boulton, A.J. & Lake, P.S. (1992c) Benthic organic matter and detritivorous macroinvertebrates in two intermittent streams in south-eastern Australia. *Hydrobiologia* **241**, 107–118.

Boulton, A.J. & Lake, P.S. (2008) Effects of drought on stream insects and its ecological consequences. In: Lancaster J. & Briers R.A. (eds.) *Aquatic Insects: Challenges to Populations*, pp. 81–102. CAB International, Wallingford, UK.

Boulton, A.J. & Lloyd, L.N. (1992) Flooding frequency and invertebrate emergence from dry floodplain sediments. *Regulated Rivers: Research and Management* **7**, 137–151.

Boulton, A.J. & Stanley, E.H. (1995) Hyporheic processes during flooding and drying in a Sonoran Desert stream. II. Faunal dynamics. *Archiv für Hydrobiologie* **134**, 27–52.

Boulton, A.J., Stanley, E.H., Fisher, S.G. & Lake, P.S. (1992a) Over-summering strategies of macroinvertebrates in intermittent streams in Australia and Arizona. In: Robarts, R.D. & Bothwell, M.L. (eds.) *Aquatic Ecosystems in Semi-arid Regions: Implications for Resource Management* pp. 227–237. National Hydrological Research Institute Symposium Series 7, Environment Canada, Saskatoon.

Boulton, A.J., Peterson, C.G., Grimm, N.B. & Fisher, S.G. (1992b) Stability of an aquatic macroinvertebrate community in a multiyear hydrologic disturbance regime. *Ecology* **73**, 2192–2207.

Bouma, M.J., Kovats, R.S., Goubet, S.A., Cox, J.S.H. & Haines, A. (1997) Global assessment of El Niño's disaster burden. *The Lancet* **350**, 1435–1438.

Bouvy, M., Molica, R.J.R., De Oliveira, S., Marinho, M. & Beker, B. (1999) Dynamics of a toxic cyanobacterial bloom (*Cylindrospermopsis raciborskii*) in a shallow reservoir in the semi-arid region of Northeast Brazil. *Aquatic Microbial Ecology* **20**, 285–297.

Bouvy, M., Falcão, D., Marinho, M., Pagano, M. & Moura, A. (2000) Occurrence of *Cylindrospermopsis* (Cyanobacteria) in 39 Brazilian tropical reservoirs during the 1998 drought. *Aquatic Microbial Ecology* **23**, 13–17.

Bouvy, M., Pagano, M. & Trousesellier, M. (2001) Effects of a cyanobacterial bloom (*Cylindrospermopsis raciborskii*) on bacteria and zooplankton communities in Ingazeira reservoir (northeast Brazil). *Aquatic Microbial Ecology* **25**, 215–227.

Bouvy, M., Nascimento, S.M., Molica, R.J.R., Ferreira, A., Huzar, V. & Azevedo S.M.F.O. (2003) Limnological features in Tapacurá reservoir (northeast Brazil) during a severe drought. *Hydrobiologia* **493**, 115–130.

Bowman, M.F., Somers, K.M., Reid, R.A. & Scott, L.D. (2006) Temporal response of stream benthic macroinvertebrate communities to the synergistic effects of anthropogenic acidification and natural drought events. *Freshwater Biology* **51**, 768–782.

Boyle, J.F. (2001) Inorganic geochemical methods in paleolimnology. In: Last, W.M. & Smol, J.P. (eds.) *Tracking Environmental Change using Lake Sediments. Volume 2: Physical and Geochemical Methods*, pp. 83–141. Kluwer Academic Publishers, Dordrecht, The Netherlands.

Bradshaw, W.E. & Holzapfel, C.M. (1988) Drought and the organization of tree-hole mosquito communities. *Oecologia* **74**, 507–514.

Bradley, R.S., Hughes, M.K. & Diaz, H.F. (2003) Climate in Medieval time. *Science* **302**, 404–405.

Bravo, R., Soriguer, M.C., Villar, N. & Hernando, J.A. (2001) The dynamics of fish populations in the Palancar stream, a small tributary of the river Guadalquivir, Spain. *Acta Oecologia* **22**, 9–20.

Brenner, M., Rosenmeier, M.F., Hodell, D.A. & Curtis, J.H. (2002) Paleolimnology of the Maya lowlands. Long-term perspectives on interactions among climate, environment, and humans. *Ancient Mesoamerica* **13**, 141–157.

Briggs, S.V. & Maher, M.T. (1985) Limnological studies of waterfowl habitat in south-western New South Wales. II. Aquatic macrophyte productivity. *Australian Journal of Marine and Freshwater Research* **36**, 707–716.

Briggs, S.V., Maher, T. & Carpenter, S.M. (1985) Limnological studies of waterfowl habitat in south-western New South Wales. 1. Water chemistry. *Australian Journal of Marine and Freshwater Research* **36**, 59–67.

Brochet, P. (1977) La sécheresse 1976 en France: Aspects climatologiques et conse-quences. *Hydrological Sciences Bulletin* **23**, 393–411.

Brock, M.A. (1998) Are temporary wetlands resilient? Evidence from seed banks of Australian and South African wetlands. In: McComb, A.J. & Davis, J.A. (eds.) *Wetlands for the Future*. pp. 193–206. Gleneagles Press, Adelaide, Australia.

Brock, M.A., Nielsen D.L., Shiel R.J., Green J.D. & Langley J.D. (2003) Drought and aquatic community resilience: the role of eggs and seeds in sediments of temporary wetlands. *Freshwater Biology* **48**, 1207–1218.

Brooker, M.P., Morris D.L. & Hemsworth R.J. (1977) Mass mortalities of adult salmon, *Salmo salar*, in the R. Wye, 1976. *Journal of Applied Ecology* **14**, 409–417.

Brookes, J.D., Lamontagne, S., Aldridge, K.T., Berger, S., Bissatt, A., Bucater, L., Cheshire, A.C., Cook, P.L.M., Deegan, B.M., Dittman, S., Fairweather, P.G., Fernandes, M.B., Ford, P.W., Geddes, M.C., Gillanders, B.M., Grigg, N.S., Haese, R.R., Krull, E., Langley, R.A., Lester, R.E., Loo, M., Munro, A.R., Noell, C.J., Nayar, S., Paton, D.C., Revill, A.T., Rogers, D.J., Roston, A., Sharma, S.K., Short, D.A., Tanner, I.E., Webster, I.T., Well-man, N.R. & Ye, Q. (2009). *An ecosystem assessment framework to guide management of the Coorong*. Water for a Healthy Country National Research Flagship, CSIRO, Canberra, ACT, Australia.

Brovkin, V., Claussen, M., Driesschaert, E., Fichefet, T., Kicklighter, D., Loutre, M.F., Matthews, H.D., Ramankutty, N., Schaeffer, M. & Sokolov, A. (2006) Biogeophysical effects of historical land cover changes simulated by six Earth system models of intermediate complexity. *Climate Dynamics* **26**, 587–600.

Brown, E.T. & Johnson, T.C. (2005) Coherence between tropical East African and South American records of the Little Ice Age. *Geochemistry, Geophysics, Geosystems* **6**, doi: 10.1029/2005GC00059.

Bruton, M.N. (1979) The survival of habitat desiccation by air breathing clariid catfishes. *Environmental Biology of Fishes* **4**, 273–280.

Bryant, E.A. (2005). *Natural Hazards*. 2nd edn. Cambridge University Press, Cambridge, UK.

Buckley, B.M., Anchukaitis, K.J., Penny, D., Fletcher, R., Cook, E.R., Sano, M., Nam, L.C., Wichienkeeo, A., Minh, T.T. & Hong, T.M. (2010) Climate as a contributing factor in the demise of Angkor. *Proceedings of the National Academy of Sciences* **107**, 6748–6752.

Bunn, S.E. & Arthington, A.H. (2002) Basic principles and ecological consequences of altered flow regimes for aquatic biodiversity. *Environmental Management* **30**, 492–507.

Bunn, S.E., Thoms, M.C., Hamilton, S.K. & Capon, S.J. (2006) Flow variability in dryland rivers: boom, bust and the bits in between. *River Research and Applications* **22**, 619–635.

Burbidge, A.A. & Kuchling, G. (1994). Western swamp tortoise recovery plan. Western Australia Wildlife Management Program Report 11, Department of Conservation and Land Management, Como, Western Australia.

Bureau of Meteorology, Australian Government (2006) *Living with Drought*. http://www.bom.gov.au/climate/drought/livedrought.html

Burgess, D.B. (2002) Groundwater resource management in eastern England: a quest for environmentally sustainable development. In: Hiscock, K.M., Rivett, M.O. & Davison, R.M. (eds.) *Sustainable Groundwater Development. Geological Society, London, Special Publications*, **193**, 53–62.

Burke, E.J., Brown, S.J. & Christidis, N. (2006) Modeling the recent evolution of global drought and projections for the twenty-first century with the Hadley Centre climate model. *Journal of Hydrometeorology* **7**, 1113–1125.

Burkholder, J.M., Dickey, D.A., Kinder, C.A., Reed, R.E., Mallin, M.A., McIver, M.R., Cahoon, L.B., Melia, G., Brownie, C., Smith, J., Deamer, N., Springer, J., Glasgow, H.B. & Toms, D. (2006) Comprehensive trend analysis of nutrients and related variables in a large eutrophic estuary: A decadal study of anthropogenic and climatic influences. *Limnology and Oceanography* **51**, 463–487.

Burt, T.P., Arkell, B.P., Trudgill, S.T. & Walling, D.E. (1988) Stream nitrate levels in a small catchment in south west England over a period of 15 years (1970–1985). *Hydrological Processes* **2**, 267–284.

Buzan, D., Lee, W., Culbertson, J., Kuhn, N. & Robinson, L. (2009) Positive relationship between freshwater inflow and oyster abundance in Galveston Bay, Texas. *Estuaries and Coasts* **32**, 206–212.

Cai, W. & Cowan, T. (2008) Evidence of impacts from rising temperature on inflows to the Murray-Darling Basin. *Geophysical Research Letters* **35**, L07701, doi: 10.1029/2008GL033390, 2008.

Callow, J.N. & Smettem, K.R.J. (2009) The effect of farm dams and constructed banks on hydrologic connectivity and runoff estimation in agricultural landscapes. *Environmental Modelling & Software* **24**, 959–968.

Cane, M.A. (2005) The evolution of El Niño, past and future. *Earth and Planetary Science Letters* **230**, 227–240.

Canobbio, S., Mezzanotte, V., Sanfilippo, U. & Benvenuto, F. (2009) Effect of multiple stressors on water quality and macroinvertebrate assemblages in an effluent-dominated stream. *Water, Air and Soil Pollution* **198**, 359–371.

Canton, S.P., Cline, L.D., Short, R.A. & Ward, J.V. (1984) The macroinvertebrates and fish of a Colorado stream during a period of fluctuating discharge. *Freshwater Biology* **14**, 311–316.

Capone, T.A. & Kushlan, J.A. (1991) Fish community structure in dry-season stream pools. *Ecology* **72**, 983–992.

Caramujo, M-J., Mendes, C.R.B., Cartaxana, P., Brotas, V. & Boavida, M-J. (2008) Influence of drought on algal biofilms and meiofaunal assemblages of temperate reservoirs and rivers. *Hydrobiologia* **598**, 77–94.

Cardinale, B.J. & Palmer, M.A. (2002) Disturbance moderates biodiversity-ecosystem function relationships: Experimental evidence from caddisflies in stream mesocosms. *Ecology* **83**, 1915–1927.

Cardinale, B.J. Hillebrand, H. & Charles, D.F. (2006) Geographic patterns of diversity in streams are predicted by a multivariate model of disturbance and productivity. *Journal of Ecology* **94**, 609–618.

Cardinale, B.J., Wright, J.P., Cadotte, M.W., Carroll, I.T., Hector, A., Srivastava, D.S., Loreau, M. & Wels, J.J. (2007) Impacts of plant diversity on biomass production through time because of species complementarity. *Proceedings of the National Academy of Sciences* **104**, 10125–10128.

Carmouze, J.-P. (1983) Hydrochemical regulation of the lake. In: Carmouze, J.-P., Durand J.-R. & Lévêque C. (eds.) *Lake Chad. Ecology and Productivity of a Shallow Tropical Ecosystem.* pp. 95–123. Dr W. Junk, The Hague.

Carmouze, J.-P. & Lemoalle, J. (1983) The lacustrine environment. In: Carmouze, J.-P., Durand, J.-R. & Lévêque, C. (eds.) *Lake Chad. Ecology and Productivity of a Shallow Tropical Ecosystem.* pp. 27–63. Dr. W. Junk, The Hague.

Carmouze, J.-P., Chantraine, J.M. & Lemoalle, J. (1983) Physical and chemical characteristics of the waters. In: Carmouze, J.-P., Durand, J.-R. & Lévêque, C. (eds.) *Lake Chad. Ecology and Productivity of a Shallow Tropical Ecosystem.* pp. 65–94. Dr. W. Junk, The Hague.

Carpenter, S.R., Kitchell, J.F. & Hodgson, J.R. (1985) Cascading trophic interactions and lake productivity. *Bioscience* **35**, 634–639.

Caruso, B.S. (2002) Temporal and spatial patterns of extreme low flows and effects on stream ecosystems in Otago, New Zealand. *Journal of Hydrology* **257**, 115–133.

Casanova, M.T. (1994) Vegetative and reproductive response of charophytes to water-level fluctuations in permanent and temporary wetlands in Australia. *Australian Journal of Marine and Freshwater Research* **45**, 1409–1419.

Case, R.A. & MacDonald, G.M. (2003) Tree ring reconstructions of streamflow for three Canadian Prairie rivers. *Journal of the American Water Resources Association* **39**, 703–716.

Castella, E., Bickerton, M., Armitage, P.D. & Petts, G., (1995) The effects of water abstractions on the invertebrate communities of U.K. streams. *Hydrobiologia* **308**, 167–182.

Cayan, D.R., Redmond, K.T. & Riddle, L.G. (1999) ENSO and hydrologic extremes in the western United States. *Journal of Climate* **12**, 2881–2893.

Changnon, S.A. (1987) *Detecting drought conditions in Illinois.* Illinois State Water Survey, Champaign, Illinois, USA, Circular **169**, 1–36.

Chao, B.F., Wu, Y.H. & Li, Y.S. (2008) Impact of artificial reservoir water impoundment on global sea level. *Science* **320**, 212–214.

Chapman, L.J., Kaufman, L.S., Chapman, C.A. & McKenzie, F.E. (1995) Hypoxia tolerance in twelve species of east African cichlids: Potential for low oxygen refugia in Lake Victoria. *Conservation Biology* **9**, 1274–1287.

Chase, J.M. (2003) Community assembly: when does history matter? *Oecologia* **136**, 489–498.

Chase, J.M. (2007) Drought mediates the importance of stochastic community assembly. *Proceedings of the National Academy of Sciences* **104**, 17430–17434.

Chase, J.M. & Knight, T.M. (2003) Drought-induced mosquito outbreaks in wetlands. *Ecology Letters* **6**, 1017–1024.

Cherry, D.S., Scheller, J.L., Cooper, N.L. & Bidwell, J.R. (2005) Potential effects of Asian clam (*Corbicula fluminea*) die-offs on native freshwater mussels (Unionidae) 1. water-column ammonia levels and ammonia toxicity. *Journal of the North American Benthological Society* **24**, 369–380.

Chessman, B.C. (2009) Climatic changes and 13-year trend in stream macroinvertebrate assemblages in New South Wales, Australia. *Global Change Biology* **15**, 2791–2802.

Chessman, B.C. & Robinson, D.P. (1987) Some effects of the 1982–83 drought on water quality and macroinvertebrate fauna in the lower Latrobe River, Victoria. *Australian Journal of Marine and Freshwater Research* **38**, 289–299.

Chick, J.H., Ruetz, C.R. & Trexler, J.C. (2004) Spatial scale and abundance patterns of large fish communities in freshwater marshes of the Florida Everglades. *Wetlands* **24**, 652–664.

Chiew, F.H.S. & McMahon, T.A. (2002) Global ENSO-streamflow teleconnection, streamflow forecasting and interannual variability. *Hydrological Sciences Journal* **47**, 505–522.

Chiew, F.H.S., Piechota, T.C., Dracup, J.A. & McMahon, T.A. (1998) El Nino/Southern Oscillation and Australian rainfall, streamflow and drought: Links and potential for forecasting. *Journal of Hydrology* **204**, 138–149.

Chivas, A., De Deckker, P. & Shelley, J.M.G. (1986) Magnesium and strontium in nonmarine ostracod shells as indicators of paleosalinity and paleotemperature. *Hydrobiologia* **143**, 135–142.

Cho, H.J. & Poirrier, M.A. (2005) Response of submersed aquatic vegetation (SAV) in Lake Pontchartrain, Louisiana to the 1997–2001 El Niño Southern Oscillation shifts. *Estuaries* **28**, 215–225.

Christiansen, J.L. & Bickham, J.W. (1989) Possible historic effects of pond drying and winterkill on the behavior of *Kinosternon flavescens* and *Chrysemys picta*. *Journal of Herpetology* **23**, 91–94.

Churchell, M.A. & Batzer, D.P. (2006) Recovery of aquatic macroinvertebrate communities from drought in Georgia Piedmont headwater streams. *American Midland Naturalist* **156**, 259–272.

Clark, J.M., Chapman, P.J., Heathwaite, A.L. & Adamson, J.K. (2006) Suppression of dissolved organic carbon by sulfate induced acidification during simulated droughts. *Environmental Science and Technology* **40**, 1776–1783.

Claussen, M. (1998) On multiple solutions of the atmosphere-vegetation system in present-day climate. *Global Change Biology* **4**, 549–559.

Cloern, J.E., Alpine, A.E., Cole, B.E., Wong, R.L.J., Arthur, J.F. & Ball, M.D. (1983) River discharge controls phytoplankton dynamics in the northern San Francisco Bay estuary. *Estuarine, Coastal and Shelf Science* **16**, 415–429.

Closs, G. & Lake, P.S. (1994) Spatial and temporal variation in the structure of an intermittent-stream food web. *Ecological Monographs* **64**, 1–21.

Closs, G.P. & Lake, P.S. (1995) Drought, differential mortality and the coexistence of a native and an introduced fish species in a south east Australian intermittent stream. *Environmental Biology of Fishes* **47**, 17–26.

Coe, M.T. & Foley, J.A. (2001) Human and natural impacts on the water resources of the Lake Chad basin. *Journal of Geophysical Research* **106**, 3349–3356.

Coelho, C.A.S. & Goddard, L. (2009) El Niño-induced tropical droughts in climate change projections. *Journal of Climate* **22**, 6456–6476.

Cohen, A.S. (2003) *Paleolimnology. The History and Evolution of Lake Systems*. Oxford University Press, UK.

Colburn, E.A. (2004) *Vernal Pools. Natural History and Conservation.* The McDonald & Woodward Publishing Company, Blacksburg, VA.

Cole, G.A. (1968) Desert limnology. In: Brown, G.W. (ed.) *Desert Biology.* pp. 423–486. Academic Press, New York.

Collier, M & Webb, R. H. (2002) *Floods, Droughts and Climate Change.* The University of Arizona Press, Tucson, AZ.

Compère, P. & Iltis, A. (1983) The phytoplankton. In: Carmouze, J.-P. Durand, J.-R. & Lévêque C. (eds.) *Lake Chad. Ecology and Productivity of a Shallow Tropical Ecosystem.* pp. 145–198. Dr W. Junk, The Hague.

Connell, D. (2007) *Water Politics in the Murray-Darling Basin.* The Federation Press, Leichhardt, NSW.

Connell, J.H. (1978) Diversity in tropical rain forests and coral reefs. *Science* **199**, 1302–1310.

Cook, B.J., Miller, R.L. & Seager, R. (2008) Dust and sea surface temperature forcing of the 1930s "Dust Bowl" drought. *Geophysical Research Letters* **235**, Lo8710, doi: 10.1029/2008GL033486, 2008.

Cook, E.R. (2000) North American Drought Variability PDSI Reconstructions. International Tree-Ring Data Bank. IGBP PAGES/World Data Center-A for Paleoclimatology Data Contribution Series #2000–074. NOAA/NGDC Paleoclimatology Program, Boulder, CO.

Cook, E.R., Bird, T., Peterson, M., Barbetti, M., Buckley, B., D'Arrigo, R. & Francey, R. (1992) Climatic change over the last millenium in Tasmania reconstructed from tree rings. *The Holocene* **2**, 205–217.

Cook, E.R., Meko, D.M., Stahle, D.W. & Cleaveland, M.K. (1996) Tree-ring reconstructions of past drought across the coterminous United States: tests of a regression method and calibration/verification results. In: Dean, J.S. Meko, D.M. & Swetman, T.W. (eds.) *Tree Rings, Environment and Humanity.* pp. 155–169. Radiocarbon, Tucson, AZ.

Cook, E.R., Meko, D.M., Stahle, D.W. & Cleaveland, M.K. (1999) Drought reconstructions for the continental United States. *Journal of Climate* **12**, 1145–1162.

Cook, E.R., Woodhouse, C.A., Eakin, C.M., Meko, D.M. & Stahle, D.W. (2004) Long-term aridity changes in the western United States. *Science* **306**, 1015–1018.

Cook, E.R., Seager, R., Cane, M.A. & Stahle, D.W. (2007) North American drought: Reconstructions, causes, and consequences. *Earth-Science Reviews* **81**, 93–134.

Cook, E.R., Heim, R.R., Herweijer, C. & Woodhouse, C. (2010a) Megadroughts in North America: Placing IPCC projections of hydroclimatic change in a long-term paleoclimate context. *Journal of Quaterny Science* **25**, 48–61.

Cook, E.R., Anchukaitis, K.J., Buckley, B.M., D'Arrigo, R.D., Jacoby, G.C. & Wright, W.E. (2010b) Asian monsoon failure and megadrought during the last millennium. *Science* **328**, 486–489.

Cook, F.J., Hicks, W., Gardner, E.A., Carlin, G.D. & Froggatt, D.W. (2000) Export of acidity in drainage water from acid sulphate soils. *Marine Pollution Bulletin* **41**, 319–326.

Copeland, B.J. (1966) Effects of decreased river flow on estuarine ecology. *Journal of the Water Pollution Control Federation* **38**, 1831–1839.

Coughlan, M.J. (1985) Drought in Australia. In: *Natural Disasters in Australia*, pp. 127–149. Australian Academy of Technological Sciences and Engineering, Parkville, Victoria.

Covich, A.P., Crowl, T.A., Johnson, S.L. & Pyron, M. (1996) Distribution and abundance of tropical freshwater shrimp along a stream corridor: Response to disturbance. *Biotropica* **28**, 484–492.

Covich, A.P., Fritz, S.C., Lamb, P.J., Marzolf, R.D., Matthews, W.J., Poiani, K.A., Prepas, E. E., Richman, M.B. & Winter, T.C. (1997) Potential effects of climate change on aquatic ecosystems of the Great Plains of North America. *Hydrological Processes* **11**, 993–1021.

Covich, A.P., Palmer, M. A. & Crowl, T.A. (1999) The role of benthic invertebrate species in freshwater ecosystems: zoobenthic species influence energy flows and nutrient cycling. *BioScience* **49**, 119–127.

Covich, A.P., Crowl, T.A. & Scatena, F.N. (2000) Linking habitat stability to floods and droughts: effects on shrimps in montane streams, Puerto Rico. *Verhandlungen der Internationalen Vereinigung für Theoretische und Angewandte Limnologie* **27**, 2430–2434.

Covich, A.P., Crowl, T.A. & Scatena, F.N. (2003) Effects of extreme low flows on freshwater shrimps in a perennial tropical stream. *Freshwater Biology* **48**, 1199–1206.

Covich, A.P., Austin, M.C., Barlöcher, F., Chauvet, E., Cardinale, B.J., Biles, C.L., Inchausti, P., Dangles, O., Solan, M., Gessner, M.O., Statzner, B. & Moss, B. (2004) The role of biodiversity in the functioning of freshwater and marine benthic ecosystems. *BioScience* **54**, 767–775.

Covich, A.P., Crowl, T.A. & Heartsill-Scalley, T. (2006) Effects of drought and hurricane disturbances on headwater distributions of palaemonid river shrimp (*Macrobrachium* spp.) in the Luquillo Mountains, Puerto Rico. *Journal of the North American Benthological Society* **25**, 99–107.

Cowx, I.G., Young, W.O. & Hellawell, J.M. (1984) The influence of drought on the fish and invertebrate populations of an upland stream in Wales. *Freshwater Biology* **14**, 165–177.

Cox, P.M., Harris, P.P., Huntingford, C., Betts, R.A., Collins, M., Jones, C.D., Jupp, T.E., Marengo, J.A. & Nobre, C.A. (2008) Increasing risk of Amazonian drought due to decreasing aerosol pollution. *Nature* **453**, 212–215.

Crausbay, S.D., Russell, J.M. & Schnurrenberger, D.W. (2006) A ca. 800-year lithologic record of drought from sub-annually laminated lake sediment, East Java. *Journal of Paleolimnology* **35**, 641–659.

Crome, F.H.J. & Carpenter, S.M. (1988) Plankton community cycling and recovery after drought – dynamics in a basin on a flood plain. *Hydrobiologia* **164**, 193–211.

Cronin, T., Willard, D., Karlsen, A., Ishman, S., Verado, S., McGeehin, J., Kerhin, R., Holmes, C., Colman, S. & Zimmerman, A. (2000) Climatic variability in the eastern United States over the past millennium from Chesapeake Bay sediments. *Geology* **28**, 3–6.

Crook, K.E., Pringle, C.M. & Freeman, M.C. (2009) A method to assess longitudinal riverine connectivity in tropical streams dominated by migratory biota. *Aquatic Conservation: Marine and Freshwater Ecosystems* **19**, 714–723.

Crowley, T.J. (2000) Causes of climate change over the past 2000 years. *Science* **289**, 270–277.

Cucherousset, J., Paillisson, J-M., Carpentier, A. & Chapman, L.J. (2007) Fish emigration from temporary wetlands during drought: the role of physiological tolerance. *Archiv für Hydrobiologie* **168**, 169–178.

Curry, B.B. (1999) An environmental tolerance index for ostracodes as indicators of physical and chemical factors in aquatic habitats. *Palaeogeography, Palaeoclimatology, Palaeoecology* **148**, 51–63.

Cyrus, D. & Vivier, L. (2006) Status of the estuarine fish fauna in the St Lucia estuarine system, South Africa, after 30 months of mouth closure. *African Journal of Aquatic Science* **31**, 71–81.

Da Silva, G.D.S.F., Giusti, H., Sanchez, A.P., Carmo, J.M.D. & Glass, M.L. (2008) Aestivation in the South American lungfish, *Lepidosiren paradoxa*: Effects on cardiovascular function, blood gases, osmolality and leptin levels. *Respiration Physiology and Neurobiology* **164**, 380–385.

Dahm, C.N., Baker, M.A., Moore, D.I. & Thibault, M.A. (2003) Coupled biogeochemical and hydrological responses of streams and rivers to drought. *Freshwater Biology* **48**, 1219–1231.

Dai, A. (2010) Drought under global warming: a review. *Wiley Interdisciplinary Reviews: Climate Change* DOI: 10.1002/wcc.81

Dai, A., Lamb, P.J., Trenberth, K.E., Hulme, M., Jones, P.D. & Xie, P. (2004a) The recent Sahel drought is real. *International Journal of Climatology* **24**, 1323–1331.

Dai, A., Trenberth, K.E. & Qian, T. (2004b) A global dataset of Palmer Drought Severity Index for 1870–2002: Relationship with soil moisture and effects of surface warming. *Journal of Hydrometeorology* **5**, 1117–1130.

Danielson, J.A. & Qazi, A.R. (1972) Stream depletion by wells in Colorado. *Journal of the American Water Resources Association* **8**, 359–366.

Das, S.K. & Chakrabarty, D. (2006) Effect of prolonged summer on two ox-bow lakes. *Biological Rhythm Research* **37**, 191–212.

Daufresne, M. & Boët, P. (2007) Climate change impacts on structure and diversity of fish communities in rivers. *Global Change Biology* **13**, 2467–2478.

Davey, A.J.H. & Kelly, D.J. (2007) Fish community response to drying disturbances in an intermittent stream: a landscape perspective. *Freshwater Biology* **52**, 1719–1733.

Davey, A.J.H., Kelly, D.J. & Biggs, B.J.F. (2006) Refuge-use strategies of stream fishes in response to extreme low flows. *Journal of Fish Biology* **69**, 1047–1059.

David, P.G. (1994) Wading bird use of Lake Okeechobee relative to fluctuating water levels. *The Wilson Bulletin* **106**, 719–732.

Davies, A.W. (1978) Pollution problems arising from the 1975–76 drought. *Proceedings of the Royal Society of London A.* **363**, 97–107.

Davies, P.M. (2010) Climate change implications for river restoration in global biodiversity hotspots. *Restoration Ecology* **18**, 261–268.

Davis, J. & Finlayson, B.L. (2000) Sand slugs and stream degradation: the case of the Granite Creeks, North-east Victoria. Technical Report 7/2000. Cooperative Research Centre for Freshwater Ecology, Canberra, Australia.

Davis, M. (2001) *Late Victorian Holocausts. El Niño Famines and the Making of the Third World*. Verso, London.

Davis, S.M., Gunderson, L.H., Park, W.A., Richardson, J.R. & Mattson, J.E. (1994) Landscape dimension, composition and function in a changing Everglades ecosystem. In: Davis, S.M. & Ogden, J.C. (eds.) *Everglades. The Ecosystem and its Restoration*. pp. 419–444. St. Lucie Press, Delray Beach, FL.

Deacon, J.E. (1961) Fish populations, following a drought, in the Neosho and Marais des Cygnes Rivers of Kansas. *University of Kansas Publications, Museum of Natural History* **13**, 359–427.

Dean, W.E. (1997) Rates, timing, and cyclicity of Holocene eolian activity in north-central United States: Evidence from varved lake sediments. *Geology* **25**, 331–334.

Death, R.G. (2002) Predicting invertebrate diversity from disturbance regimes in forest streams. *Oikos* **97**, 18–30.

Death, R.G. & Winterbourn, M.J. (1995) Diversity patterns in stream benthic invertebrate communities: the influence of habitat stability. *Ecology* **76**, 1446–1460.

DeAngelis, D.L., Loftus, W.F., Trexler, J.C. & Ulanowicz, R.E. (1997) Modeling fish dynamics and effects of stress in a hydrologically pulsed ecosystem. *Journal of Aquatic Ecosystem Stress and Recovery* **6**, 1–13.

DeAngelis, D.L., Trexler, J.C., Cosner, C., Obaza, A. & Jopp, F. (2010) Fish population dynamics in a seasonally varying wetland. *Ecological Modelling* **221**, 1131–1137.

De Deckker, P. & Forester, R.M. (1988) The use of ostracodes to reconstruct continental paleoenvironmental records. In: De Deckker, P., Colin, J.P. & Peyouquet, J.P. (eds.) *Ostracods in the Earth Sciences*, pp. 176–199. Elsevier, Amsterdam, The Netherlands.

De Groot, C-J. & Fabre, A. (1993) The impact of desiccation of a freshwater marsh (Garcines Nord, Camargue, France) on sediment-water-vegetation interactions. 3. The fractional composition and the phosphate adsorption characteristics of the sediment. *Hydrobiologia* **252**, 105–116.

De Groot, C-J. & Van Wijck, C. (1993) The impact of desiccation of a freshwater marsh (Garcines Nord, Camargue, France) on sediment-water-vegetation interactions. Part 1: The sediment chemistry. *Hydrobiologia* **252**, 83–94.

Dekar, M.P. & Magoulick, D.D. (2007) Factors affecting fish assemblage structure during seasonal stream drying. *Ecology of Freshwater Fish* **16**, 335–42.

Delaney, R.G., Lahiri, S. & Fishman, A.P. (1974) Aestivation of the African lungfish *Protopterus aethiopicus*: cardiovascular and respiratory functions. *Journal of Experimental;1; Biology* **61**, 111–128.

Delettre, Y. (1989) Influence de la durée et de l'intensité de l'assèchement sur l'abondance et la phénologie des Chironomides (Diptera) d'une mare semi-permanente peu profonde. *Archiv für Hydrobiologie* **114**, 383–399.

deMenocal, P.B. (2001) Cultural responses to climate change during the late Holocene. *Science* **292**, 667–673.

Deo, R.C., Sytkus, J.I., McAlpine, C.A., Lawrence, P.J., McGowan, H.A. & Phinn, S.R. (2009) Importance of historical land cover change on daily indices of climate extremes including droughts in eastern Australia. *Geophysical Research Letters* **36**, L08705, doi: 10.1029/2009GL037666, 2009.

Detenbeck, N.E., De Vore, P.W., Niemi, G.J. & Lima, A. (1992) Recovery of temperate-stream fish communities from disturbance: A review of case studies and synthesis of theory. *Environmental Management* **16**, 33–53.

Devercelli, M. (2006) Phytoplankton of the middle Paraná River during an anomalous hydrological period: a morphological and functional approach. *Hydrobiologia* **563**, 465–478.

Devito, K, J., Hill, A.R. & Dillon, P.J. (1999) Episodic sulphate export from wetlands in acidified headwater catchments: prediction at the landscape scale. *Biogeochemistry* **44**, 187–203.

Dewson, Z.S., James, A.B.W. & Death, R.G. (2007) A review of the consequences of decreased flow for instream habitat and macroinvertebrates. *Journal of the North American Benthological Society* **26**, 401–415.

Diamond, J. (2005) *Collapse. How Societies choose to fail or survive*. Allen Lane, Penguin Group (Australia), Camberwell, Victoria.

Diamond, J. (2009) Maya, Khmer and Inca. *Science* **461**, 479–480.

Dilley, M. & Heyman, B.N. (1995) ENSO and disaster: Droughts, floods and El Niño/ Southern Oscillation warm events. *Disasters* **19**, 181–193.

Dillon, P.J., Molot, L.A. & Futter, M. (1997) The effect of El Niño-related drought on the recovery of acidified lakes. *Environmental Monitoring and Assessment* **46**, 105–111.

Dillon, P.J., Somers, K.M., Findeis, J. & Eimers, M.C. (2003) Coherent response of lakes in Ontario, Canada to reductions in sulphur deposition: the effects of climate on sulphate concentrations. *Hydrology and Earth System Sciences* **7**, 583–595.

Dodd, C.K. (1993) The cost of living in an unpredictable environment: The ecology of striped newts *Notophthalamus perstriatus* during a prolonged drought. *Copeia* **1993**, 605–614.

Dodd, C.K. (1995) The ecology of a sandhills population of the eastern narrow-mouthed toad *Gastrophyrne carolinensis*, during a drought. *Bulletin of the Florida Museum of Natural History* **38**, 11–41.

Dodds, W.K., Gudder, D.A. & Mollenhauer, D. (1995) The ecology of *Nostoc*. *Journal of Phycology* **31**, 2–18.

Dodds, W.K., Hutson, R.E., Eichem, A.C., Evans, M.A., Gudder, D.A., Fritz, K.M. & Gray, L. (1996) The relationship of floods, drying, flow and light to primary production and producer biomass in a prairie stream. *Hydrobiologia* **333**, 151–159.

Dolbeth, M., Martinho, F., Viegas, I., Cabral, H. & Pardal, M.A. (2008) Estuarine production of resident and nursery fish species: Conditioning by drought events? *Estuarine, Coastal and Shelf Science* **78**, 51–60.

Döll, P., Fiedler, K. & Zhang, J. (2009) Global-scale analysis of river flow alterations due to water withdrawls and reservoirs. *Hydrology and Earth Systems Sciences* **13**, 2413–2432.

Donders, T.H., Haberle, S.G., Hope, G., Wagner, F. & Visscher, H. (2007) Pollen evidence for the transition of the Eastern Australian climate system from the post-glacial to the present-day ENSO mode. *Quaternary Science Reviews* **26**, 1621–1637.

Doody, T. & Overton, I. (2009) Environmental management of riparian tree health in the Murray-Darling Basin, Australia. In: Brebbia, C.A. (ed.) *River Basin Management V*. pp. 197–206. WIT Transactions on Ecology and the Environment, WIT Press, Southampton, UK.

Dorn, N.J. (2008) Colonization and reproduction of large macroinvertebrates are enhanced by drought-related fish reductions. *Hydrobiologia* **605**, 209–218.

Dorn, N.J. & Trexler, J.C. (2007) Crayfish assemblage shifts in a large drought-prone wetland: the roles of hydrology and competition. *Freshwater Biology* **52**, 2399–2411.

Douglas, M.M., Townsend, S.A. & Lake, P.S. (2003) Stream Dynamics. In: Andersen, A., Cook, G.D. & Williams, R.J. (eds.) *Fire in Tropical Savannas: An Australian Study*, pp. 59–78. Springer-Verlag, New York.

Douglas, M.R., Brunner, P.C. & Douglas, M.E. (2003) Drought in an evolutionary context: molecular variability in Flannelmouth Sucker (*Catostomus latipinnis*) from the Colorado River basin of western North America. *Freshwater Biology* **48**, 1254–1273.

Dracup, J.A., Lee, K.S. & Paulson E.G. (1980a) On the statistical characteristics of drought events. *Water Resources Research* **16**, 289–296.

Dracup, J.A., Lee, K.S. & Paulson, E.G. (1980b) On the definition of droughts. *Water Resources Research* **16**, 297–302.

Drexler, J.Z. & Ewel, K.C. (2001) Effect of the 1997–1998 ENSO-related drought on hydrology and salinity in a Micronesian wetland complex. *Estuaries* **24**, 347–356.

Drosdowsky, W. & Chambers, L.E. (2001) Near-global sea surface temperatures as predictors of Australian seasonal rainfall. *Journal of Climate* **14**, 1677–1687.

Druyan, L.M. (1996a) Arid Climates In: Schneider S.H. (ed.) *Encyclopaedia of Climate and Weather. Vol.1*. pp. 48–50. Oxford University Press, New York.

Druyan, L.M. (1996b) Drought. In: Schneider S.H. (ed.) *Encyclopaedia of Climate and Weather. Vol.1*. pp. 256–259. Oxford University Press, New York.

Duever, M.J., Meeder, J.F., Meeder, L. C. & McCollom, J.M. (1994) The climate of south Florida and its role in shaping the Everglades ecosystem. In: Davis, S.M. & Ogden, J.C. (eds.) *Everglades. The Ecosystem and its Restoration*. pp. 225–248. St. Lucie Press, Delray Beach, USA.

Dumont, H.J. (1992) The regulation of plant and animal species and communities in African shallow lakes and wetlands. *Revue Hydrobiologie Tropicale* **25**, 303–346.

Durance, I. & Ormerod, S. J. (2007) Climate change effects on upland stream macro-invertebrates over a 25-year period. *Global Change Biology* **13**, 942–957.

Dusi, J.L. & Dusi, R.T. (1968) Ecological factors contributing to nesting failure in a heron colony. *The Wilson Bulletin* **80**, 458–466.

Eimers, M.C., Watmough, S.A., Buttle, J.M. & Dillon, P.J. (2007) Drought-induced sulphate release from a wetland in south-central Ontario. *Environmental Monitoring and Assessment* **127**, 399–407.

Eimers, M.C., Watmough, S.A., Buttle, J.M. & Dillon, P.J. (2008) Examination of the potential relationship between droughts, sulphate and dissolved organic carbon at a wetland-draining stream. *Global Change Biology* **14**, 938–948.

Eldon, G.A. (1979a) Habitat and interspecific relationships of the Canterbury mudfish, *Neochanna burrowsius* (Salmoniformes: Galaxiidae). *New Zealand Journal of Marine and Freshwater Research* **13**, 111–119.

Eldon, G.A. (1979b) Breeding, growth and aestivation of the Canterbury mudfish *Neochanna burrowsius* (Salmoniformes: Galaxiidae). *New Zealand Journal of Marine and Freshwater Research* **13**, 331–346.

Elliott, J.M. (1985) Population regulation for different life-stages of migratory trout *Salmo trutta* in a Lake District stream, 1966–83. *Journal of Animal Ecology* **54**, 617–638.

Elliott, J.M. (2000) Pools as refugia for brown trout during two summer droughts: trout responses to thermal and oxygen stress. *Journal of Fish Biology* **56**, 938–948.

Elliott, J.M. (2002) Shadow competition in wild juvenile sea-trout. *Journal of Fish Biology* **61**, 1268–1281.

Elliott, J.M. (2006) Periodic habitat loss alters the competitive coexistence between brown trout and bullheads in a small stream over 34 years. *Journal of Animal Ecology* **75**, 54–63.

Elliott, J.M. & Elliott, J.A. (1995) The critical thermal limits for the bullhead, *Cottus gobio*, from three populations in north-west England. *Freshwater Biology* **33**, 411–418.

Elliott, J.M. & Elliott, J.A. (2006) A 35-year study of stock-recruitment relationships in a small population of sea trout: Assumptions, implications and limitations for predicting targets. In Harris G. & Milner, N. (eds.) *Sea Trout: Biology, Conservation and Management*. pp. 257–278. Blackwell, Oxford, UK.

Elliott, J.M., Hurley, M.A. & Elliott, J.A. (1997) Variable effects of drought on the density of a sea-trout *Salmo trutta* population over 30 years. *Journal of Applied Ecology* **34**, 1229–1238.

Elsdon, T.S., De Bruin, M.B.N.A., Diepen, N.J. & Gillanders, B.M. (2009) Extensive drought negates human influence on nutrients and water quality in estuaries. *Science of the Total Environment* **407**, 3033–3043.

Eltahir, E.A.B. (1996) El Niño and the natural variability in the flow of the Nile River. *Water Resources Research* **32**, 131–137.

England, M.H., Ummenhofer, C.C. & Santoso, A. (2006) Interannual rainfall extremes over southwest Western Australia linked to Indian Ocean climate variability. *Journal of Climate* **19**, 1948–1969.

Enfield, D.B., Mestas-Nuñez, A.M. & Trimble, P.J. (2001) The Atlantic multidecadal oscillation and its relation to rainfall and river flows in the continental U.S. *Geophysical Research Letters* **28**, 2077–2080.

Eriksen, C.H. (1966) Diurnal limnology of two highly turbid puddles. *Verhandlungen des Internationalen Vereinigung für theoretische und angewandte Limnologie* **16**, 507–514.

Erman, N.A. & Erman, D.C. (1995) Spring permanence, Trichoptera species richness, and the role of drought. *Journal of the Kansas Entomological Society* **68**, 50–64.

Evans, R. (2007) *The impact of groundwater use on Australia's rivers.* Technical Report of Land & Water Australia, Canberra ACT, Australia. 137 pp.

Everard, M. (1996) The importance of periodic droughts for maintaining diversity in the freshwater environment. *Freshwater Forum* **7**, 33–50.

Extence, C.A. (1981) The effect of drought on benthic invertebrate communities in a lowland river. *Hydrobiologia* **83**, 217–224.

Faber, J.E. & White, M.M. (2000) Comparison of gene flow estimates between species of darters in different streams. *Journal of Fish Biology* **57**, 1465–1473.

Faulkenham, S.E., Hall, R.I., Dillon, P.J. & Karst-Riddoch, T. (2003) Effects of drought-induced acidification on diatom communities in acid-sensitive Ontario lakes. *Limnology and Oceanography* **48**, 1662–1673.

Fausch, K.D., Torgersen, C.E., Baxter, C.V. & Li, H.W. (2002) Landscapes to riverscapes: Bridging the gap between research and conservation of stream fishes. *BioScience* **52**, 483–498.

Feddema, J.J., Oleson, K.W., Bonan, G.B. Mearns, L.O., Buja, L.E., Meehl, G.A. & Washington, W.M. (2005) The importance of land-cover change in simulating future climates. *Science* **310**, 1674–1678.

Feminella, J.W. & Matthews, W.J. (1984) Intraspecific differences in thermal tolerance of *Etheostoma spectabile* (Agassiz) in constant versus fluctuating environments. *Journal of Fish Biology* **25**, 455–461.

Fenoglio, S., Bo, T. & Bosi, G., (2006) Deep interstitial habitat as a refuge for *Agabus paludosus* (Fabricius) (Coleoptera: Dytiscidae) during summer droughts. *The Coleopterists Bulletin* **60**, 37–41.

Fenoglio, S., Bo T., Cucco, M. & Malarcane, G. (2007) Response of benthic invertebrate assemblages to varying drought conditions in the Po River (NW Italy). *Italian Journal of Zoology* **74**, 191–201.

Ferrari, I., Viglioli, S., Viaroli, P. & Rossetti, G. (2006) The impact of the summer 2003 drought event on the zooplankton of the Po River (Italy). *Verhandlungen des Internationalen Vereinigung für theoretische und angewandte Limnologie* **29**, 2143–2149.

Findlay, D.L., Kasian, S.E.M., Stainton, M.P., Beaty, K. & Lyng, M. (2001) Climatic influences on algal populations of boreal forest lakes in the Experimental Lakes Area. *Limnology and Oceanography* **46**, 1784–1793.

Finlayson, B., Nevill, J. & Ladson, T. (2008) Cumulative impacts in water resource development. Proceedings of "Water Down Under 2008" Conference, Adelaide, 2008 9 pages.

Finn, M.A., Boulton, A.J. & Chessman, B.C. (2009) Ecological responses to artificial drought in two Australian rivers with differing water extraction. *Fundamental and Applied Limnology* **175**, 231–248.

Fischer, P. & Ohl, U. (2005) Effects of water-level fluctuations on the littoral benthic fish community in lakes: a mesocosm experiment. *Behavioral Ecology* **16**, 741–746.

Fisher, S.G. & Grimm, N. (1991) Streams and disturbance; are cross-ecosystems useful? In: Cole, J.J., Lovett, G. & Findlay, S. (eds.) *Comparative Analyses of Ecosystems. Patterns, Mechanisms and Theories*. pp. 196–221. Springer-Verlag, New York.

Fisher, S.G., Gray, L.G., Grimm, N.B. & Busch D.E. (1982) Temporal succession in a desert stream ecosystem. *Ecological Monographs* **52**, 92–110.

Fisher, S.G., Grimm, N.B., Marti, E., Holmes, R.M. & Jones, J.B. (1998) Material spiraling in stream corridors: A telescoping ecosystem model. *Ecosystems* **1**, 19–34.

Fisher, S.G., Sponseller, R.A. & Heffernan, J.B. (2004) Horizons in stream biogeochemistry: pathways to progress. *Ecology* **85**, 2369–2379.

Flanagan, C.M., McKnight, D.M., Liptzin, D., Williams, M.W. & Miller, M.P. (2009) Response of the phytoplankton community in an alpine lake to drought conditions: Colorado Rocky Mountain Front Range. *Arctic, Antarctic and Alpine Research* **41**, 191–203.

Fleig, A.K., Tallaksen, L.M., Hisdal, H. & Demuth, S. (2006) A global evaluation of streamflow drought characteristics. *Hydrology and Earth System Sciences* **10**, 535–552.

Foley, J.A., Coe, M.T., Scheffer, M. & Wang, G. (2003) Regime shifts in the Sahara and Sahel: interactions between ecological and climatic systems in northern Africa. *Ecosystems* **6**, 524–539.

Foley, J.A., DeFries, R., Asner, G.P., Barford, C., Bonan, G., Carpenter, S.R., Chapin, F.S., Coe, M.T., Daily, G.C., Gibbs, H.K., Helkowski, J.H., Holloway, T., Howard, E.A., Kucharik, C.J., Monfreda, C., Patz, J.A. Prentice, I.C. Ramankutty, N. & Snyder, P. K. (2005) Global consequences of land use. *Science* **309**, 570–574.

Folke, C., Carpenter, S.Y., Walker, B., Scheffer, M., Elmqvist, T., Gunderson, L. & Holling, C.S. (2004) Regime shifts, resilience and biodiversity in ecosystem management. *Annual Review in Ecology, Evolution and Systematics* **35**, 557–581.

Fonnesu, A., Sabetta, L. & Basset, A. (2005) Factors affecting macroinvertebrate distribution in a Mediterranean intermittent stream. *Journal of Freshwater Ecology* **20**, 641–647.

Forbes, A.T. & Cyrus, D.P. (1993) Biological effects of salinity reversals in a southeast African estuarine lake. *Netherlands Journal of Aquatic Ecology* **27**, 483–488.

Forman, S.L., Oglesby, R. & Webb, R.S. (2001) Temporal and spatial patterns of Holocene dune activity on the Great Plains of North America: megadroughts and climate links. *Global and Planetary Change* **29**, 1–29.

Foster, I.D.L. & Walling, D.E. (1978) The effects of the 1976 drought and autumn rainfall on stream solute levels. *Earth Surface Processes* **3**, 393–406.

Freeman, C., Greswell, R., Guasch, H., Hudson, J., Lock, M.A., Reynolds, B., Sabater, F. & Sabater, S. (1994) The role of drought in the impact of climatic change on the microbiota of peatland streams. *Freshwater Biology* **32**, 223–230.

Freeman, C., Ostle, N. & Kang, H. (2001a) An enzymic 'latch' on a global carbon store–A shortage of oxygen locks up carbon in peatlands by restraining a single enzyme. *Nature* **409**, 149.

Freeman, C., Evans, C.D., Monteith, D.T., Reynolds, B. & Fenner, N. (2001b) Export of organic carbon from peat soils. *Nature* **412**, 785.

Freeman, M.C., Crawford, M.K., Barrett, J.C., Facey, D.E., Flood, M.G., Hill, J., Stouder, D.J. & Grossman, G.D. (1988) Fish assemblage stability in a southern Appalachian stream. *Canadian Journal of Fisheries and Aquatic Sciences* **45**, 1949–1958.

Fritz, K.M. & Dodds, W.K. (2004) Resistance and resilience of macroinvertebrate assemblages to drying and flood in a tallgrass prairie stream system. *Hydrobiologia* **527**, 99–112.

Fritz, S.C. (2008) Deciphering climatic history from lake sediments. *Journal of Paleolimnology* **39**, 5–16.

Fritz, S.C., Ito, E., Yu, Z., Laird, K.R. & Engstrom, D.R. (2000) Hydrologic variation in the northern Great Plains during the last two millennia. *Quaternary Research* **53**, 175–184.

Fukami, T. (2001) Sequence effects of disturbance on community structure. *Oikos* **92**, 215–224.

Furse, M.T., Kirk, R.G., Morgan, P.R. & Tweddle, D. (1979) Fishes: Distribution and biology in relation to changes. In: Kalk, M., McLachlan, A.J. & Howard-Williams, C. (eds) *Lake Chilwa. Studies of change in a tropical ecosystem.* pp. 175–208. Dr. W. Junk, The Hague.

Gaboury, M.N. & Patalas, J.W. (1984) Influence of water level drawdown on the fish populations of Cross Lake, Manitoba. *Canadian Journal of Fisheries and Aquatic Sciences* **41**, 118–125.

Gaff, H., Chick, J., Trexler, J., DeAngelis, D., Gross, L. & Salinas, R. (2004) Evaluation of and insights from ALFISH: a spatially explicit, landscape-level simulation of fish populations in the Everglades. *Hydrobiologia* **520**, 73–87.

Gagneur, J. & Chaoui-Boudghane, C. (1991) Sur le rôle du milieu hyporhéique pendant l'assèchement des oueds de l'Ouest Algérien. *Stygologia* **6**, 77–89.

Gagnon, P.M., Golladay, S.W., Michener, W.K. & Freeman M.C. (2004) Drought responses of freshwater mussels (Unionidae) in Coastal Plain tributaries of the Flint River basin, Georgia. *Journal of Freshwater Ecology* **19**, 667–679.

Gaines, K.F., Bryan, A.L. & Dixon, P.M. (2000) The effects of drought on foraging habitat selection of breeding wood storks in coastal Georgia. *Waterbirds* **23**, 64–73.

Garcia, A.M., Vieira, J.P. & Winemiller, K.O. (2001) Dynamics of the shallow-water fish assemblage of the Patos Lagoon estuary (Brazil) during cold and warm ENSO episodes. *Journal of Fish Biology* **59**, 1218–1238.

Gasith, A. & Resh, V.H. (1999) Streams in Mediterranean climate regions: Abiotic influences and biotic responses to predictable seasonal events. *Annual Reviews of Ecology and Systematics* **30**, 51–81.

Gaston, K.J. (2003) *The Structure and Dynamics of Geographic Ranges.* Oxford University Press, Oxford, UK.

Geladof, Z., Peterson, D.L. & Mantua, N.J. (2004) Columbia River flow and drought since 1750. *Journal of the American Water Resources Association* **40**, 1579–1591.

Gérard, C. (2001) Consequences of a drought on freshwater gastropod and trematode communities. *Hydrobiologia* **459**, 9–18.

Gérard, C., Carpentier, A. & Paillisson, J-M. (2008) Long-term dynamics and community structure of freshwater gastropods exposed to parasitism and other environmental stressors. *Freshwater Biology* **53**, 470–484.

Gergis, J. & Fowler, A.M. (2005) Classification of synchronous oceanic and atmospheric El Niño-Southern Oscillation (ENSO) events for palaeoclimate reconstruction. *International Journal of Climatology* **25**, 1541–1565.

Gergis, J. & Fowler, A.M. (2006) How unusual was late 20th century El Niño-Southern Oscillation (ENSO)? Assessing evidence from tree-ring, coral, ice-core and documentary palaeoarchives, A.D. 1525–2002. *Advances in Geosciences* **6**, 173–179.

Gergis, J., Braganza, K., Fowler, A., Mooney, S. & Risbey, J. (2006) Reconstructing El Niño-Southern Oscillation (ENSO) from high-resolution palaeoarchives. *Journal of Quaternary Science* **21**, 707–722.

Gibbons, J.W., Greene, J.L. & Congdon, J.D. (1983) Drought-related responses of aquatic turtle populations. *Journal of Herpetology* **17**, 242–246.

Gibbons, J.W., Winne, C.T., Scott, D.E., Willson, J.D., Claudas, X., Andrews, K.M., Todd, B. D., Fedewas, L.A., Wilkinson, L., Tsalagios, R.N., Harper, S.J., Greene, J.L., Tubervillet, D., Metts, B.S., Dorcas, M.E., Nestor, J.P., Young, C.A., Akre, T., Reed, R.N., Buhlmann, K.A., Norman, J., Croshaw, D. A., Hagen, C. & Rothermel, B.B. (2006) Remarkable amphibian biomass and abundance in an isolated wetland: Implications for wetland conservation. *Conservation Biology* **20**, 1457–1465.

Gibbs, G.W. (1973) Cycles of macrophytes and phytoplankton in Pukepuke Lagoon following a severe drought. *Proceedings of the New Zealand Ecological Society* **20**, 13–20.

Gibbs, W.J. & Maher, J.V. (1967) Rainfall deciles as drought indicators. *Bureau of Meteorology Bulletin No. 48*. Bureau of Meteorology, Melbourne.

Gillanders, B.M. (2007) Linking terrestrial-freshwater and marine environments: an example from estuarine systems. In: Connell, S.D. & Gillanders, B.M. (eds.) *Marine Ecology*. pp. 252–277. Oxford University Press, South Melbourne, Australia.

Gillanders, B.M. & Kingsford, M.J. (2002) Impact of changes in flow of freshwater on estuarine and open coastal habitats and the associated organisms. *Oceanography and Marine Biology: an Annual Review* **40**, 233–309.

Giller, P.S. (1996) Floods and droughts: the effects of variations in water flow in streams and rivers. In: Giller, P.S. & Myers, A.A. (eds.) *Disturbance and recovery in ecological systems*. pp. 1–19. Royal Irish Academy, Dublin.

Gillson, J., Scandol, J. & Suthers, I. (2009) Estuarine gillnet fishery catch rates decline during drought in eastern Australia. *Fisheries Research* **99**, 26–37.

Glantz, M.H. (1994) *Drought follows the plow*. Cambridge University Press, Cambridge, UK.

Glantz, M.H. (2000) Drought follows the plough. A cautionary note. In: Wilhite, D.A. (ed.) *Drought: A Global Assessment. Volume 2*. pp. 285–291. Routledge, London.

Glasby, T.M. & Underwood, A.J. (1996) Sampling to differentiate between pulse and press perturbations. *Environmental Monitoring and Assessment* **42**, 241–252.

Glass, M.L. (2008) The enigma of aestivation in the African lungfish *Protopterus dolloi* –Commentary on the paper by Perry et al. *Respiration Physiology and Neurobiology* **160**, 18–20.

Glennon, R. (2002) *Water Follies. Groundwater pumping and the fate of America's fresh waters*. Island Press, Washington DC, USA.

Glodek, G.S. (1978) The importance of catfish burrows in maintaining fish populations of tropical freshwater sreams in western Ecuador. *Fieldiana Zoology* **73**, 1–8.

Golinski, M., Bauch, C. & Anand, M. (2008) The effects of endogenous ecological memory on population stability and resilience in a variable environment. *Ecological Modelling* **212**, 334–341.

Golladay, S.W. & Battle, J. (2002) Effects of flooding and drought on water quality in Gulf Coastal Plain streams in Georgia. *Journal of Environmental Quality* **31**, 1266–1272.

Golladay, S.W., Gagnon, P., Kearns, M., Battle, J.M. & Hicks, D.W. (2004) Response of freshwater mussel assemblages (Bivalvia: Unionidae) to a record drought in the Gulf Coastal Plain of southwestern Georgia. *Journal of the North American Benthological Society* **23**, 494–506.

Gordon, L., Dunlop, M. & Foran, B. (2003) Land cover change from water vapour flows: learning from Australia. *Philosophical transactions of the Royal Society of London* **B** **358**, 1973–1984.

Gordon, L., Peterson, G.D. & Bennett, E.M. (2007) Agricultural modifications of hydrological flows create ecological surprises. *Trends in Ecology and Evolution* **23**, 211–219.

Gordon, N., Adams, J.B. & Bate, G.C. (2008) Epiphytes of the St Lucia Estuary and their response to water level and salinity changes during a severe drought. *Aquatic Botany* **88**, 66–76.

Gordon, N.D., McMahon, T.A., Finlayson, B.L., Gippel, C.J. & Nathan R.J. (2004) *Stream Hydrology. An Introduction for Ecologists* 2nd ed. John Wiley & Sons, Chichester, UK.

Greening, H.S. & Gerritsen, J. (1987) Changes in macrophyte community structure following drought in the Okefenokee Swamp, Georgia, U.S.A. *Aquatic Botany* **28**, 113–128.

Greenwood, P.H. (1986) The natural history of African lungfishes. *Journal of Morphology* **190**, Supplement 1, 163–179.

Griffiths, T. (2005) Against the flow: A cultural history of European settlers and Australian water. Paper presented to the ANZSANA Conference, Harvard University, 29 April 2005.

Griswold, B.L., Edwards, C.J. & Woods, L.C. (1982) Recolonization of macroinvertebrates and fish in a channelized stream after a drought. *Ohio Journal of Science* **82**, 96–102.

Griswold, M.W., Berzinis, E.W., Crisman, T.L. & Golladay, S.W. (2008) Impacts of climatic stability on the structural and functional aspects of macroinvertebrate communities after severe drought. *Freshwater Biology* **53**, 2465–2483.

Groffman, P.M., Bain, D.J., Band, L.E., Belt, K.T., Brush, G.S., Grove, J.M., Pouyat, R.V., Yesilonis, I. C. & Zipperer, W.C. (2003) Down by the riverside: urban riparian ecology. *Frontiers in Ecology and the Environment* **1**, 315–321.

Grossman, G.D. & Ratajczak, R.E. (1998) Long-term patterns of microhabitat use by fish in a southern Appalachian stream from 1983 to 1992: effects of hydrologic period, season and fish length. *Ecology of Freshwater Fish* **7**, 108–131.

Grossman, G.D., Ratajczak, R.E., Crawford, M. & Freeman, M.C. (1998) Assemblage organization in stream fishes: effects of environmental variation and interspecific interactions. *Ecological Monographs* **68**, 395–420.

Gubiani, É.A., Gomes, L.C., Agostinho, A.A. & Okada, E.K. (2007) Persistence of fish populations in the upper Paraná River: effects of water regulation by dams. *Ecology of Freshwater Fish* **16**, 191–197.

Gunderson, L.H. (1994) Vegetation of the Everglades: Determinants of community composition. In: Davis, S.M. & Ogden, J.C. (eds.) *Everglades. The Ecosystem and its Restoration.* pp. 323–340. St. Lucie Press, Delray Beach, USA.

Guttman, N.B. (1999) Accepting the Standardized Precipitation Index: A calculation algorithm. *Journal of the American Water Resources Association* **35**, 311–322.

Haag, W. R. & Warren, M.L. (2008) Effects of severe drought on freshwater mussel assemblages. *Transactions of the American Fisheries Society* **137**, 1165–1178.

Haberzettl, T., Fey M., Lücke, A., Maidana, N., Mayr, C., Ohlendorf, C., Schäbitz, F., Schleser, G.H., Wille, M. & Zolitschka, B. (2005) Climatically induced lake level changes during the last two millennia as reflected in sediments of Laguna Potrok Aike, southern Patagonia, (Santa Cruz, Argentina). *Journal of Paleolimnology* **33**, 283–302.

Hakala, J. P & Hartman, L. J. (2004) Drought effect on stream morphology and brook trout (*Salvelinus fontinalis*) populations in forested headwater streams. *Hydrobiologia* **515**, 203–213.

Halfman, J.D., Johnson, T.C. & Finney, B. (1994) New AMS dates, stratigraphic correlations and decadal climate cycles for the past 4 ka at Lake Turkana, Kenya. *Palaeogeography, Palaeoclimatology, Palaeoecology* **111**, 83–98.

Hall, K. C., Baldwin, D.S., Rees, G.N. & Richardson, A.J. (2006) Distribution of inland wetlands with sulfidic sediments in the Murray-Darling Basin, Australia. *Science of the Total Environment* **370**, 235–244.

Hanley, D., Bourassa, M., O'Brian, J, Smith, S. & Spade, E. (2003) A quantitative evaluation of ENSO indices. *Journal of Climate* **16**, 1249–1258.

Hardison, E.C., O'Driscoll, M.A., DeLoatch, J.P., Howard, R.J. & Brinson, M.M. (2009) Urban land use, channel incision, and water table decline along coastal plain streams, North Carolina. *Journal of the American Water Resources Association* **45**, 1032–1046.

Harrel, R.C., Davis, B.J. & Dorris, T.C. (1967) Stream order and species diversity of fishes in an intermittent Oklahoma stream. *American Midland Naturalist* **78**, 428–436.

Harrington, R.W. (1959) Delayed hatching in stranded eggs of marsh killifish. *Ecology* **40**, 430–437.

Harrison, A.D. (1966) Recolonisation of a Rhodesian stream after drought. *Archiv für Hydrobiologie* **62**, 405–421.

Hastie, B.F. & Smith, S.D.A. (2006) Benthic macrofaunal communities in intermittent estuaries during a drought: Comparisons with permanently open estuaries. *Journal of Experimental Marine Biology and Ecology* **330**, 356–367.

Hatt, B.E., Fletcher, T.D., Walsh, C.J. & Taylor, S.L. (2004) The influence of urban density and drainage infrastructure on the concentrations and loads of pollutants in small streams. *Environmental Management* **34**, 112–124.

Hatton, T.J. & Nulsen, R.A. (1999) Towards achieving functional ecosystem mimicry with respect to water cycling in southern Australian agriculture. *Agroforestry Systems* **45**, 203–214.

Haug, G.H., Gunther, D., Peterson, LC., Sigman, D.M., Hughen, K.A. & Aeschlimann, B. (2003) Climate and the collapse of Maya civilization. *Science* **299**, 1731–1735.

Havens, K.E., Sharfstein, B., Brady, M.A., East, T.L., Harwell, M.C., Maki, R.P. & Rodusky, A.J. (2004) Recovery of submerged plants from high water stress in a large subtropical lake in Florida, USA. *Aquatic Botany* **78**, 67–82.

Havens, K.E., Fox, D.D., Gornak, S. & Hanlon, C. (2005) Aquatic vegetation and largemouth bass population responses to water level variations in Lake Okeechobee, Florida (USA). *Hydrobiologia* **539**, 225–237.

Havens, K. E., East, T.L. & Beaver, J.R. (2007) Zooplankton response to extreme drought in a large subtropical lake. *Hydrobiologia* **589**, 187–198.

Hayes, M.J. (2006) What is drought? Drought indices. www.drought.unl.edu/whatis/indices.htm

Heath, L. (2009) Climate factors that lead to fish kills in Australian coastal acid sulfate soil catchments. *Climate Change: Global Risks, Challenges and Decisions, IOP Conf. Series: Earth and Environmental Science* **6** (2009)352022 doi: 10.1088/1755-1307/6/5/35022

Heathcote, R.L. (1969) Drought in Australia. A problem of perception. *Geographical Review* **59**, 175–194.

Heathcote, R.L. (1988) Drought in Australia: Still a problem of perception? *GeoJournal* **16**, 387–397.

Heathcote, R.L. (2000) 'She'll be right, mate'. Coping with drought. Strategies old and new in Australia. In: Wilhite, D.A. (ed.) *Drought: A Global Assessment. Volume 2.* pp. 59–69. Routledge, London.

Heim R.R. (2002) A review of twentieth-century drought indices used in the United States. *Bulletin of the American Meteorological Society* **83**, 1149–1165.

Held, I.M., Delworth, T.L., Lu, J., Findell, K.L. & Knutson, T.R. (2005) Simulation of Sahel drought in the 20th and 21st centuries. *Proceedings of the National Academy of Sciences* **102**, 17891–17896.

Hendrix, N. & Loftus, W.F. (2000) Distribution and relative abundance of the crayfishes *Procambarus alleni* (Faxon) and *P. fallax* (Hagen) in southern Florida. *Wetlands* **20**, 194–199.

Hershey, A.E., Shannon, L., Niemi, G.J., Lima, A.R. & Regal, R.R. (1999) Prairie wetlands of south-central Minnesota. Effects of drought on invertebrate communities. In: Batzer, D.P., Rader, R.B. & Wissinger, S.A. (eds.) *Invertebrates in Freshwater Wetlands of North America: Ecology and Management.* pp. 515–541. John Wiley & Sons, New York.

Herweijer, C. & Seager, R. (2008) The global footprint of persistent extra-tropical drought in the instrumental era. *International Journal of Climatology* **28**, 1761–1774.

Herweijer, C., Seager, R., Cook, E. R. & Emile-Geay, J. (2007) North American droughts of the last millennium from a gridded network of tree-ring data. *Journal of Climate* **20**, 1353–1376.

Hess, B.J., Jackson, C. & Jennings, C.A. (1999) Use of cool-water springs as thermal refuge by striped bass in the Ogeechee River, Georgia. *Georgia Journal of Science* **57**, 123–130.

Hidalgo, H.G. (2004) Climate precursors of multidecadadal drought variability in the western United States. *Water Resources Research* **40**, W12504, doi: 10.1029/2004WR003350.

Hidore, J. (1996) *Global Environmental Change: Its Nature and Impact.* Prentice Hall, Upper Saddle River, N.J.

Hildrew A.G., Townsend C.R. & Hasham A. (1985) The predatory Chironomidae of an iron-rich stream: feeding ecology and food web structure. *Ecological Entomology* **10**, 403–413.

Hisdal, H., Tallaksen, L. M, Clausen, B., Peters, E. & Gustard, A. (2004) Hydrological drought characteristics. In: Tallkasen, L.M. & Van Lanen, H.A.J. (eds.) *Hydrological*

Drought. Processes and Estimation Methods for Streamflow and Groundwater. pp. 139–198. Elsevier, Amsterdam.

Hodell, D.A., Brenner, M., Curtis, J.H. & Guilderson, T. (2001) Solar forcing of drought frequency in the Maya lowlands. *Science* **292**, 1367–1370.

Hodell, D.A., Brenner, M. & Curtis, J.H. (2005) Terminal classic drought in the northern Maya lowlands inferred from multiple sediment cores in Lake Chichancanab (Mexico). *Quaternary Science Reviews* **24**, 1413–1427.

Hodell, D.A., Brenner, M. & Curtis, J.H. (2007) Climate and cultural history of the northeastern Yucatan Peninsula, Quintana Roo, Mexico. *Climatic Change* **83**, 215–240.

Hoeppner, S.S., Shaffer, G.P. & Perkins, T.E. (2008) Through droughts and hurricanes: Tree mortality, forest structure, and biomass production in a coastal swamp targeted for restoration in the Mississippi River Deltaic Plan. *Forest Ecology and Management* **256**, 937–948.

Hoerling, M. & Kumar, A. (2003) The perfect ocean for drought. *Science* **299**, 691–694.

Hoese, H.D. (1960) Biotic changes in a bay associated with the end of a drought. *Limnology and Oceanography* **5**, 326–336.

Hofmann, H., Lorke, A. & Peeters, F. (2008) Temporal scales of water-level fluctuations in lakes and their ecological implications. *Hydrobiologia* **613**, 85–96.

Holland, R.F. & Jain, S.K. (1984) Spatial and temporal variation in plant species diversity of vernal pools. In: Jain, S.K. & Moyle, P. (eds.) *Vernal Pools and Intermittent Streams.* pp. 198–209. Institute of Ecology Publication 28, University of California, Davis, CA.

Holling, C.S., Gunderson, L.H., and Walters, C.J. (1994) The structure and dynamics of the Everglades system: Guidelines for ecosystem restoration. In: Davis, S.M. & Ogden, J.C. (eds.) *Everglades. The Ecosystem and its Restoration.* pp. 741–756. St. Lucie Press, Delray Beach, USA.

Holmes, N.T.H. (1999) Recovery of headwater stream flora following the 1989–1992 groundwater drought. *Hydrological Processes* **13**, 341–354.

Horner, G.J., Baker, P.J., MacNally, R., Cunningham, S.C., Thomson, J.R. & Hamilton, F. (2009) Mortality of developing floodplain forests subjected to a drying climate and water extraction. *Global Change Biology* **15**, 2176–2186.

Hough, R.A., Allenson, T.E. & Dion, D.D. (1991) The response of macrophyte communities to drought-induced reduction of nutrient loading in a chain of lakes. *Aquatic Botany* **41**, 299–308.

Howard-Williams, C. (1979) The distribution of aquatic macrophytes in Lake Chilwa: Annual and long-term environmental fluctuations. In: Kalk, M., McLachlan, A.J. & Howard-Williams, C. (eds.) *Lake Chilwa. Studies of change in a tropical ecosystem.* pp. 105–122. Dr. W. Junk, The Hague.

Howitt, J, Baldwin, D.S., Rees, G.N. & Williams, J.L. (2007) Modelling blackwater: Predicting water quality during flooding of lowland river forests. *Ecological Modelling* **203**, 229–242.

Hrbek, T. & Larson, A. (1999) The evolution of diapause in the killifish family Rivulidae (Atherinomorpha, Cyprinodontiformes): a molecular phylogenetic and biogeographic perspective. *Evolution* **53**, 1200–1216.

Hubbell, S.P. (2001) *The Unified Theory of Biodiversity and Biogeography.* Princeton University Press, Princeton, NJ.

Hudson, P.J. & Cattadori, I.M. (1999) The Moran effect: a cause of population synchrony. *Trends in Ecology and Evolution* **14**, 1–2.

Hughen, K.A., Schrag, D. P, Jacobsen, S.B. & Hantoro, W. (1999) El Niño during the last interglacial period recorded by a fossil coral from Indonesia. *Geophysical Research Letters* **26**, 3129–3132.

Hughes, M.K. & Diaz, H.F. (1994) *The Medieval Warm Period*. Kluwer Academic Publishers, Dordrecht, The Netherlands.

Hughes, S., Reynolds, B., Hudson, J.A. & Freeman, C. (1997) Effect of summer drought on peat soil solution chemistry in an acid gully mire. *Hydrology and Earth System Sciences* **1**, 661–669.

Humphries, P. & Baldwin, D.S. (2003) Drought and aquatic ecosystems: an introduction. *Freshwater Biology* **48**, 1141–1146.

Hunt, B.G. & Elliott, T.I. (2005) A simulation of the climatic conditions associated with the collapse of the Maya civilization. *Climatic Change* **69**, 393–407.

Huntsman, A.G. (1942) Death of salmon and trout with high temperature. *Journal of the Fisheries Research Board of Canada* **5**, 485–501.

Hurrell, J.W., Kushnir, Y., Ottersen, G. & Visbeck, M. (2003) An overview of the North Atlantic Oscillation. In: Hurrell, J.W. (ed.) *The North Atlantic Oscillation-Climatic significance and environmental impact*. pp. 1–35. American Geophysical Union, Washington DC, USA.

Hynes, H.B.N. (1958) The effect of drought on the fauna of a small mountain stream in Wales. *Verhandlungen der Internationalen Vereinigung für Theoretische und Angewandte Limnologie* **13**, 826–833.

Hynes, H.B.N. (1961) The invertebrate fauna of a Welsh mountain stream. *Archiv für Hydrobiologie* **57**, 344–388.

Iltis, A. & Lemoalle, J. (1983) The aquatic vegetation of Lake Chad. In: Carmouze, J.-P., Durand, J.-R. & Lévêque, C. (eds.) *Lake Chad. Ecology and Productivity of a Shallow Tropical Ecosystem*. pp. 125–144. Dr W. Junk, The Hague.

Institute of Hydrology (1980) *Low Flow Studies* (1–4), Wallingford, UK.

Ip, Y.K., Chew, S.F. & Randall, D.J. (2004) Five tropical air-breathing fishes, six different strategies to defend against ammonia toxicity on land. *Physiological and Biochemical Zoology* **77**, 768–782.

IPCC (2007) Climate Change 2007: Synthesis Report. Contribution of Working Groups I, II and III to the Fourth Assessment Report of the Intergovernmental Panel on Climate Change [Core Writing Team, Pachauri, R.K and Reisinger, A. (eds.)]. IPCC, Geneva, Switzerland, 104 pp.

Ito, E. (2001) Application of stable isotope techniques to inorganic and biogenic carbonates. In: Last, W.M. & Smol, J.P. (eds.) *Tracking Environmental Change using Lake Sediments. Volume 2: Physical and Geochemical Methods*. pp. 351–371. Kluwer Academic Publishers, Dordrecht, The Netherlands.

Iversen, T.M., Wiberg-Larsen, Hansen S.B. & Hansen, F.S. (1978) The effect of partial and total drought on the macroinvertebrate communities of three small Danish streams. *Hydrobiologia* **60**, 235–242.

Jacobs, S.M., Bechtold, J.S., Biggs, H.C., Grimm, N.B., Lorentz, S., McClain, M.E., Naiman, R.J., Perakis, S.S., Pinay, G. & Scholes, M.C. (2007) Nutrient vectors and riparian processing: A review with special reference to African semiarid savanna ecosystems. *Ecosystems* **10**, 1231–1249.

James, M.C. (1934) Effect of 1934 drought on fish life. *Transactions of the American Fisheries Society* **64**, 57–62.

James, R.T. (1991) Microbiology and chemistry of acid lakes in Florida: Effects of drought and post-drought conditions. *Hydrobiologia* **213**, 205–225.

Janovy, J., Snyder, S.D. & Clopton, R.E. (1997) Evolutionary constraints on population structure: the parasites of *Fundulus zebrinus* (Pisces: Cyprinodontidae) in the South Platte River of Nebraska. *Journal of Parasitology* **83**, 584–592.

Jassby, A.D., Kimmerer, W.J., Monismith, S.G., Armor, C., Cloern, J.E., Powell, T.M., Schubel, J.R. & Vendlinski, T.J. (1995) Isohaline positions as a habitat indicator for estuarine populations. *Ecological Applications* **5**, 272–289.

Jeffries, M. (1994) Invertebrate communities and turnover in wetland ponds affected by drought. *Freshwater Biology* **32**, 603–612.

Jellyman, D. (1989) The Canterbury drought and effects on fish. *Freshwater Catch* **39**, 22–24.

Jenkins, K.M. & Boulton, A.J. (1998) Community dynamics of invertebrates emerging from reflooded lake sediments: Flood pulse and aeolian influences. *International Journal of Ecology and Environmental Sciences* **24**, 179–192.

Jenkins, K.M. & Boulton, A.J. (2003) Connectivity in a dryland river: Short-term aquatic microinvertebrate recruitment flowing floodplain inundation. *Ecology* **84**, 2708–2727.

Jenkins, K.M. & Boulton, A.J. (2007) Detecting impacts and setting restoration targets in arid-zone rivers: aquatic micro-invertebrate responses to reduced floodplain inundation. *Journal of Applied Ecology* **44**, 833–842.

John, K.R. (1964) Survival of fish in intermittent streams of the Chiricahua Mountains, Arizona. *Ecology* **45**, 112–119.

Johnels, A.G. & Svensson, G.S.O. (1954) On the biology of *Protopterus annectens* (Owen). *Arkiv för Zoologi* **7**, 131–164.

Jöhnk, K.D., Huisman, J., Sharples, J., Sommeijer, B., Visser, P.M. & Strooms, J.M. (2008) Summer heatwaves promote blooms of harmful cyanobacteria. *Global Change Biology* **14**, 495–512.

Jones, A.R. (1990) Zoobenthic variability associated with a flood and drought in the Hawkesbury Estuary, New South Wales: Some consequences for environmental monitoring. *Environmental Monitoring and Assessment* **14**, 185–195.

Jordan, F., Babbitt, K.J. & McIvor, C.C. (1998) Seasonal variation in habitat use by marsh fishes. *Ecology of Freshwater Fish* **7**, 159–166.

Jordan, F., Babbitt, K.J., McIvor, C.C. & Miller, S.J. (2000) Contrasting patterns of habitat use by prawns and crayfish in a headwater marsh of the St. Johns River, Florida. *Journal of Crustacean Biology* **20**, 769–776.

Junk, W.J., Bayley, P.B. & Sparks, R.E. (1989) The flood- pulse concept in river-floodplain systems. In: Dodge, D.P. (ed.) Proceedings of the International Large River Symposium. pp. 110–127. *Canadian Special Publication Fisheries and Aquatic Sciences* **106**.

Junk, W.J., Nuna da Cunha, C., Wantzen, K.M., Petermann, P., Strüssman, C., Marques, M.I. & Adis, J. (2006) Biodiversity and its conservation in the Pantanal of Mato Grosso, Brazil. *Aquatic Sciences* **68**, 278–309.

Kalk, M. (1979a) Introduction: Perspectives of research at Lake Chilwa. In: Kalk, M., McLachlan, A.J. & Howard-Williams, C. (eds.) *Lake Chilwa. Studies of change in a tropical ecosystem*. pp. 7–16, Dr. W. Junk. The Hague.

Kalk, M. (1979b) Zooplankton in Lake Chilwa: adaptations to changes. In: Kalk, M., McLachlan, A.J. & Howard-Williams, C. (eds.) *Lake Chilwa. Studies of change in a tropical ecosystem.* pp. 125–141. Dr. W. Junk, The Hague.

Kalk, M., McLachlan, A.J. & Howard-Williams, C. (eds.) (1979) *Lake Chilwa. Studies of change in a tropical ecosystem.* Dr. W. Junk. The Hague.

Kamler, E. & Riedel, W. (1960) The effect of drought on the fauna *Ephemeroptera*, *Plecoptera* and *Trichoptera* of a mountain stream. *Polskie Archiwum Hydrobiologii* **8**, 87–94.

Kangas, R.S. & Brown, T.J. (2007) Characteristics of US drought and pluvials from a high-resolution spatial dataset. *International Journal of Climatology* **27**, 1303–1325.

Karoly, D.J., Risbey, J., Reynolds, A. & Braganza, K. (2003) Global warming contributes to Australia's worst drought. *Australasian Science, April 2003 issue*, 14–17.

Katz, G.L., Stromberg, J.C. & Denslow, M.W. (2009) Streamside herbaceous vegetation response to hydrologic restoration on the San Pedro River, Arizona. *Ecohydrology* **2**, 213–225.

Keaton, M., Haney, D. & Andersen, C.B. (2005) Impact of drought upon fish assemblage structure in two South Carolina Piedmont streams. *Hydrobiologia* **545**, 209–223.

Keating, J. (1992) *The Drought walked through. A History of Water Shortage in Victoria.* Department of Water Resources, Melbourne.

Keller, W. & Yan, N.D. (1991) Recovery of crustacean zooplankton species in Sudbury area lakes following water quality improvements. *Canadian Journal of Fisheries and Aquatic Sciences* **48**, 1635–1644.

Keller, W. & Yan, N.D. (1998) Biological recovery from lake acidification: zooplankton communities as a model of patterns and processes. *Restoration Ecology* **6**, 364–375.

Kelsch, S.W. (1994) Lotic fish-community structure following transition from severe drought to high discharge. *Journal of Freshwater Ecology* **9**, 331–341.

Kemp, N.E. (1987) The biology of the Australian lungfish, *Neoceratodus forsteri* (Krefft, 1870). *Journal of Morphology* **190**, Supplement 1, 181–198.

Ker, P. (2010) Murray River alert as toxic algae spreads. *The Age*, **24** February, 2010.

Kerr, R.A. (1998) Sea-floor dust shows drought felled Akkadian Empire. *Science* **279**, 325–326.

Kerr, R.A. (2001) A north Atlantic climate pacemaker for the centuries. *Science* **288**, 1984–1985.

Keyantash, J. & Dracup, J.A. (2002) The quantification of drought: An evaluation of drought indices. *Bulletin of the American Meteorological Society* **83**, 1167–1180.

Khan, S., Hafeez, M., Abbas, A. & Ahmad, A. (2009) Spatial assessment of water use in an environmentally sensitive wetland. *Ambio* **38**, 157–165.

Kiem, A.S. & Franks, S.W. (2004) Multi-decadal variability of drought risk, eastern Australia. *Hydrological Processes* **18**, 2039–2050.

Kim, D-W. & Byun, H-R. (2009) Future pattern of Asian drought under global warming scenario. *Theoretical and Applied Climatology* **98**, 137–150.

Kingsford, R.T. (1989) The effects of drought on ducking survival of Maned Ducks. *Australian Wildlife Research* **16**, 405–412.

Kingsford, R.T. & Norman, F.I. (2002) Australian waterbirds – product of the continent's ecology. *Emu* **102**, 47–69.

Kingsford, R.T. & Thomas, R.F. (2004) Destruction of wetlands and waterbird populations by dams and irrigation on the Murrumbidgee River in arid Australia. *Environmental Management* **34**, 383–396.

Kingsford, R.T., Curtin, A.L. & Porter, J.L. (1999) Water flows on Cooper Creek in arid Australia determine 'boom' and 'bust' periods for waterbirds. *Biological Conservation* **88**, 231–248.

Kingsford, R.T., Jenkins, K.M. and Porter, J.L. (2004). Imposed hydrological stability on lakes in arid Australia and effect on waterbirds. *Ecology* **85**, 2478–2492.

Kingsford, R.T., Roshier, D.A. & Porter, J.L. (2010) Australian waterbirds – time and space travelers in dynamic desert landscapes. *Marine and Freshwater Research* **61**, 875–884.

Kinninmonth, W.R., Voice, M.E., Beard, G.S., de Hoelt, G.C. & Mullen, C.E. (2000). Australian climate services for drought management. In: Wilhite, D.A. (ed.) *Drought; A Global Assessment.* Volume 1 pp. 210–222. Routledge, London.

Kneitel, J.M. & Chase, J.M. (2004) Disturbance, predator, and resource interactions alter container community composition. *Ecology* **85**, 2088–2093.

Knowles, N. (2002) Natural and management influences on freshwater inflows and salinity in the San Francisco Estuary at monthly to interannual scales. *Water Resources Research* **38** (12), 1289, doi: 10.1029/2001WR000360, 2002.

Kobayashi, D., Suzuki, K. & Nomura, M. (1990) Diurnal fluctuation in stream flow and in specific electric conductance during drought periods. *Journal of Hydrology* **115**, 105–114.

Kobza, R.M., Trexler, J.C., Loftus, W.F. & Perry, S.A. (2004) Community structure of fishes inhabiting aquatic refuges in a threatened karst wetland and its implications for ecosystem management. *Biological Conservation* **116**, 153–165.

Koch, M. (2004) Faunal survey. I. The distribution of aquatic gastropods of the New England Tablelands. *Memoirs of the Queensland Museum* **49**, 653–658.

Konikow, L.F. & Kendy, E. (2005) Groundwater depletion: a global problem. *Hydro-geological Journal* **13**, 317–320.

Koster, R.D., Dirmeyer, P.A., Guo, Z, Bonan, G., Chan, E., Cox, P., Gordon, C.T., Kanae, S., Kowalczyk, E., Lawrence, D., Liu, P., Lu, C-H., Malyshev, S., McAvaney, B., Mitchell, K., Mocko, D., Oki, T., Oleson, K., Pitman, A., Sud, Y.C., Taylor, C.M., Verseghy, D., Vaic, T., Xue, Y. & Yamada, T. (2004) Regions of strong coupling between soil moisture and precipitation. *Science* **305**, 1138–1140.

Kramer, D.L. (1983) The evolutionary ecology of respiratory mode in fishes: an analysis based on the costs of breathing. *Environmental Biology of Fishes* **9**, 145–158.

Kramer, D.L. (1987) Dissolved oxygen and fish behavior. *Environmental Biology of Fishes* **18**, 81–92.

Kratz, T.K., Webster, K.E., Bowser, C.J., Magnuson, J.J. & Benson, B.J. (1997) The influence of landscape position on lakes in northern Wisconsin. *Freshwater Biology* **37**, 209–217.

Kundzewicz, Z.W., Mata, L.J., Arnell, N.W., Döll, P., Kabat, P., Jiménez, B., Miller, K.A., Oki, T., Şen, Z. & Shiklomanov, I.A. (2007) Freshwater resources and their management. Climate Change 2007: Impacts, Adaptation and Vulnerability. In: Parry, M.L., Canziani, O.F., Palutikof, J.P., van der Linden, P.J. & Hanson C.E. (eds.) *Contribution of Working Group II to the Fourth Assessment Report of the Intergovernmental Panel on Climate Change.* pp. 173–210. Cambridge University Press, Cambridge, UK.

Kundzewicz, Z. W., Mata, L.J., Arnell, N.W., Döll, P., Jiminez, B., Miller, K., Oki, T., Şen, Z. & Shilkomanov, I. (2008). The implications of projected climate change for freshwater resources and their management. *Hydrological Sciences Journal* **53**, 3–10.

Kushlan, J.A. (1973) Differential responses to drought in two species of *Fundulus*. *Copeia* **1973**, 808–809.

Kushlan, J.A. (1974a) Effects of a natural fish kill on the water quality, plankton, and fish populations of a pond in the Big Cypress Swamp, *Florida. Transactions of the American Fisheries Society* **103**, 235–243.

Kushlan, J.A. (1974b) Observations on the role of the American alligator (*Alligator mississippiensis*) in the southern Florida wetlands. *Copeia* **1974**, 993–996.

Kushlan, J.A. (1976a) Environmental stability and fish community diversity. *Ecology* **57**, 821–825.

Kushlan, J.A. (1976b) Wading bird predation in a seasonally-fluctuating pond. *Auk*, **93**, 464–476.

Kushlan J.A. (1980) Population fluctuations of Everglades fishes. *Copeia* **1980**, 870–874.

Kushlan, J.A. & Kushlan, M.S. (1980) Population fluctuations in the prawn, *Palaemonetes paludosus*, in the Everglades. *American Midland Naturalist* **103**, 401–403.

Labbe, T.R. & Fausch, K.D. (2000) Dynamics of intermittent stream habitat regulate persistence of a threatened fish at multiple scales. *Ecological Applications* **10**, 1774–1791.

Ladle, M. & Bass, J.A.B. (1981) The ecology of a small chalk stream and its responses to drying during drought conditions. *Archiv für Hydrobiologie* **90**, 448–466.

Laë, R. (1994) Effect of drought, dams and fishing pressure on the fisheries of the Central Delta on the Niger River. *International Journal of Ecology and Environmental Sciences* **20**, 119–128.

Laë, R. (1995) Climatic and anthropogenic effects on fish diversity and fish yields in the Central Delta of the Niger River. *Aquatic Living Resources* **8**, 43–58.

Laird, K.R. & Cumming, B.F. (1998) A diatom based reconstruction of drought intensity, duration and frequency from Moon Lake, North Dakota: A sub-decadal records for the last 2300 years. *Journal of Palaeoliminology* **19**, 161–179.

Laird, K.R., Fritz, S.C., Grimm, E.C. & Mueller, P.G. (1996a) Century-scale paleoclimatic reconstruction from Moon Lake, a closed-basin lake in the northern Great Plains. *Limnology & Oceanography* **41**, 890–902.

Laird, K.R., Fritz, S.C., Maasch, K.A. & Cumming, B.F. (1996b) Greater drought intensity and frequency before AD 1200 in the Northern Great Plains, USA. *Nature* **384**, 552–554.

Laird, K.R., Cumming, B.F., Wunsam, S., Rusak, J.A., Oglesby, R.J., Fritz, S.C. & Leavitt, P.R. (2003) Lake sediments record large-scale shifts in moisture regimes across the northern prairies of North America during the past two millennia. *Proceedings of the National Academy of Sciences* **100**, 2483–2488.

Lake, P.S. (1977) Pholeteros, the faunal assemblage found in crayfish burrows. *Newsletter of the Australian Society for Limnology* **15**, 57–60.

Lake, P.S. (2000) Disturbance, patchiness, and diversity in streams. *Journal of the North American Benthological Society* **19**, 573–592.

Lake, P.S. (2003) Ecological effects of perturbation by drought in flowing waters. *Freshwater Biology* **48**, 1161–1172.

Lake, P.S. (2007) Flow-generated disturbances and ecological responses: Floods and droughts. In: Wood, P.J., Hannag, D.M. & Sadler, J.P. (eds.) *Hydroecology and Ecohydrology. Past, Present and Future.* pp. 75–92. John Wiley & Sons. Ltd, Chichester, UK.

Lake, P.S. (2008) *Drought, the "creeping disaster". Effects on aquatic ecosystems.* Land & Water Australia, Canberra, Australia.

Lake, P.S., Bayly, I.A.E. & Morton, D.W. (1989) The phenology of a temporary pond in western Victoria, Australia, with special reference to invertebrate succession. *Archiv für Hydrobiologie* **115**, 171–202.

Lake, P.S., Bond, N. & Reich, P. (2006) Floods down rivers: From damaging to replenishing forces. *Advances in Ecological Research* **39**, 41–62.

Lake, P.S., Bond, N.R. & Reich, P. (2007) Linking ecological theory with stream restoration. *Freshwater Biology* **52**, 597–615.

Lake, P.S., Bond, N.R. & Reich, P. (2008) An appraisal of studies on the impacts of drought on aquatic ecosystems: knowledge gaps and future directions. *Verhandlungen der Internationale Vereinigung für Theoretische und Angewandte Limnologie* **30**, 506–508.

Lamb, H.H. (1977). *Climate: Present, Past and Future – Volume 2: Climatic History and the Future.* Methuen, London, UK.

Lamb, H.H. (1995) *Climate History and the Modern World.* 2nd edition. Routledge, London.

Lamontagne, S., Hicks, W., Fitzpatrick, R.W. & Rogers, S. (2006) Sulfidic materials in dryland river wetlands. *Marine and Freshwater Research* **57**, 775–788.

Lancaster, N. (1979) The physical environment of Lake Chilwa. In: Kalk, M., McLachlan, A.J. & Howard-Williams, C. (eds.) *Lake Chilwa. Studies of change in a tropical ecosystem.* pp. 17–40. Dr. W. Junk, The Hague.

Lancaster, J. & Belyea, L.R. (1997) Nested hierarchies and scale-dependence of mechanisms of flow refugium use. *Journal of the North American Benthological Society* **16**, 221–238.

Lancaster, J. & Hildrew, A.G. (1993) Characterizing in-stream flow refugia. *Canadian Journal of Fisheries and Aquatic Sciences* **50**, 1663–1675.

Larimore, R.W., Childers, W.F. & Heckrotte, C. (1959) Destruction and re-establishment of stream fish and invertebrates affected by drought. *Transactions of the American Fisheries Society* **88**, 261–285.

Larson, E.R., Magoulick, D. D., Turner, C. & Laycock, K.H. (2009) Disturbance and species displacement: different tolerances to stream drying and desiccation in a native and an invasive crayfish. *Freshwater Biology* **54**, 1899–1908.

Laudon, H. (2008) Recovery from episodic acidification delayed by drought and high sea salt deposition. *Hydrology and Earth System Sciences* **12**, 363–370.

Laudon, H.M. & Bishop, K. (2002) Episodic stream water pH decline during autumn storms following a summer drought in northern Sweden. *Hydrological Processes* **16**, 1725–1733.

Laudon, H., Dillon, P.J., Eimers, M.C., Semkin, R.G. & Jeffries, D.S. (2004) Climate-induced episodic acidification of streams in central Ontario. *Environmental Science & Technology* **38**, 6009–6015.

Lawrence, P.J. & Chase, T.N. (2010) Investigating the climate impacts of global land cover change in the community climate system model. *International Journal of Climatology* DOI: 10.1002/joc.2061

LaZerte, B.D. (1993) The impact of drought and acidification on the chemical exports from a minerotrophic conifer swamp. *Biogeochemistry* **18**, 153–175.

Leblanc, M.J., Favreau, G., Massuel, S., Tweed, S.O., Loireau, M. & Capelaere, B. (2008) Land clearance and hydrological change in the Sahel: SW Niger. *Global and Planetary Change* **61**, 135–150.

Leblanc, M. J., Tregoning, P., Ramillien, G., Tweed, S.O. & Fakes, A. (2009) Basin-scale, integrated observations of the early 21st century multiyear drought in southeast Australia. *Water Resources Research* **45**, W04408, doi: 10.1029/2008WR007333.

Leck, M.A. (1989) Wetland seed banks. In: Leck, M.A., Parker, V.T. & Simpson, R.L. (eds.) *Ecology of Soil Seed Banks* pp. 283–308. Academic Press, San Diego, CA.

Leck, M.A. & Brock, M.A. (2000) Ecological and evolutionary trends in wetlands: evidence from seeds and seed banks in New South Wales, Australia and New Jersey, USA. *Plant Species Biology* **15**, 97–112.

Ledger, M.E. & Hildrew, A.G. (2001) Recolonization by the benthos of an acid stream following a drought. *Archiv für Hydrobiologie* **152**, 1–17.

Ledger, M.E., Harris, R.M.L., Armitage, P.D. & Milner, A.M. (2008) Disturbance frequency influences patch dynamics in stream benthic algal communities. *Oecologia* **155**, 809–819.

Legier, P. & Talin, J. (1975) Recolonisation d'un ruisseau temporaire et évolution du degré de stabilité de la zoocénose. *Ecologia Mediterranea*. **1**, 149–164.

Lehman, P.W. & Smith, R.W. (1991) Environmental factors associated with phytoplankton succession for the Sacramento-San Jaoquin Delta and Suisun Bay Estuary, California. *Estuarine, Coastal and Shelf Science* **32**, 105–128.

Lehner, B., Döll, P., Alcamo, J., Henrichs, T. & Kaspar, F. (2006) Estimating the impact of global change on flood and drought risks in Europe: a continental, integrated analysis. *Climatic Change* **75**, 273–299.

Leng, M.J., Lamb, A.L., Heaton, T.H.E., Marshall, J.D., Wolfe, B.B., Jones, M.D. Holmes, J.A. & Arrowsmith, C. (2006). Isotopes in lake sediments. In: Leng, M.J. (ed.) *Isotopes in Palaeoenvironmental Research*, pp. 147–184. Springer, Dordrecht, The Netherlands.

Lenton, T.M., Held, H., Kriegler, E., Hall, J.W., Lucht, W., Rahmstorf, S. & Schellnuber, H.J. (2008) Tipping elements in the Earth's climate system. *Proceedings of the National Academy of Sciences* **105**, 1786–1793.

Lévêque, C. (1997) *Biodiversity Dynamics and Conservation. The Freshwater Fish of Tropical Africa.* Cambridge University Press, Cambridge, UK.

Lévêque, C., Dejoux, C. & Lausanne, L. (1983) The benthic fauna: ecology, biomass and communities. In: Carmouze, J.-P. . Durand, J.-R. & Lévêque, C. (eds.) *Lake Chad. Ecology and Productivity of a Shallow Tropical Ecosystem.* pp. 233–272. Dr. W. Junk, The Hague.

Light, S.S. & Dineen, J.W. (1994) Water control in the Everglades: a historical perspective. In: Davis, S.M. & Ogden, J.C. (eds.) *Everglades. The Ecosystem and its Restoration.* pp. 47–84. St. Lucie Press, Delray Beach, FL.

Lind, P.R., Robson, B.J. & Mitchell, B.D. (2006) The influence of reduced flow during drought on patterns of variation in macroinvertebrate assemblages across a spatial hierarchy in two lowland rivers. *Freshwater Biology* **51**, 2282–2295.

Lindesay, J.A. (2003) Climate and drought in Australia. In: Botterill, L. C & Fisher, M. (eds.) *Beyond Drought. People, Policy and Perspectives.* pp. 21–47. CSIRO Publishing, Collingwood, Australia.

Lindström, S. (1990) Submerged tree stumps as indicators of mid-Holocene activity in the Lake Tahoe region. *Journal of California and Great Basin Anthropology* **12**, 146–157.

Lindström, S. (1997) Lake Tahoe case study: Lake levels. Appendix 7.1. Sierra Nevada Ecosystem Project Final report to Congress. Status of the Sierra Nevada. pp. 265–268. Wildland Resources Center Report No. 40, Centers for Water and Wildland resources, University of California, Davis, CA.

Lite, S.J. & Stromberg, J.C. (2005) Surface water and ground-water thresholds for maintaining *Populus-Salix* forests, San Pedro River, Arizona. *Biological Conservation* **125**, 153–167.

Little, C. (2000) *The biology of soft shores and estuaries*. Oxford University Press, Oxford, UK.

Livingston, R.J. (1997) Trophic response of estuarine fishes to long-term changes of river runoff. *Bulletin of Marine Science* **60**, 984–1004.

Livingston, R.J. (2002) *Trophic Organization in Coastal Systems*. CRC Press, Boca Raton, Florida, USA.

Livingston, R.J., Niu, X., Lewis, F.G. & Woodsum, G.C. (1997) Freshwater input to a Gulf estuary: Long-term control of trophic organization. *Ecological Applications* **7**, 277–299.

Lloyd-Hughes, B. & Saunders, M.A. (2002) A drought climatology for Europe. *International Journal of Climatology* **22**, 1571–1592.

Lobón-Cerviá, J. (2009a) Why, when and how do fish populations decline, collapse and recover? The example of brown trout (*Salmo trutta*) in Rio Chaballos (northwestern Spain). *Freshwater Biology* **54**, 1149–1162.

Lobón-Cerviá, J. (2009b) Recruitment as a driver of production dynamics in stream-resident brown trout (*Salmo trutta*). *Freshwater Biology* **54**, 1692–1704.

Lobón-Cerviá, J., Utrilla, C.G., Querol, E. & Puig, M.A. (1993) Population ecology of pike-cichlid, *Crenicichla lepidota*, in two streams in the Brazilian Pampa subject to a severe drought. *Journal of Fish Biology* **43**, 537–557.

Loftus, W.F. & Eklund, A-M. (1994) Long-term dynamics of an Everglades small-fish assemblage. In: Davis, S.M. & Ogden, J.C. (eds.) *Everglades. The Ecosystem and its Restoration*. pp. 461–483. St. Lucie Press, Delray Beach, USA.

Lomholt, J.P. (1993) Breathing in the aestivating African lungfish, *Protopterus amphibious*. *Advances in Fish Research* **1**, 17–34.

Lopez, B. (1990). Drought. In *Desert notes: reflections in the eye of a raven; River notes: the dance of herons*. pp. 133–138. Avon Books, New York.

Lopez, O.R. & Kursar, T.A. (2007) Interannual variation in rainfall, drought stress and seedling mortality may mediate monodominance in tropical flooded forests. *Oecologia* **154**, 35–43.

Los, S.O., Weedon, G.P., North, P.R.J., Kaduk, J.D., Taylor, C.M. & Cox, P.M. (2006) An observation-based estimate of the strength of rainfall-vegetation interactions in the Sahel. *Geophysical Research Letters* **33**, L16402, doi: 10.1029/2006GL27065, 2006.

Love, J.W., Taylor, C.M. & Warren, M.P. (2008) Effects of summer drought on fish and macroinvertebrate assemblage properties in upland Ouachita Mountain streams, USA. *American Midland Naturalist* **160**, 265–277.

Lowe-McConnell, R.H. (1975) *Fish Communities in Tropical Freshwaters*. Longman, London, UK.

Lucassen, E.C.H.T., Smolders, A.J.P. & Roelofs, J.G.M. (2002) Potential sensitivity of mires to drought, acidification and mobilization of heavy metals: the sediment S/(Ca + Mg) ratio as diagnostic tool. *Environmental Pollution* **120**, 635–646.

Luo, W., Song, F. & Xie, Y. (2008) Trade-off between tolerance to drought and tolerance to flooding in three wetland plants. *Wetlands* **28**, 866–873.

Lytle, D.A. (2008) Life-history and behavioural adaptations to flow regime in aquatic insects. In: Lancaster, J. & Briers, R.A. (eds.) *Aquatic Insects. Challenges to Populations*. pp. 122–138. CABI, Wallingford, UK.

Lytle, D.A. & Merritt, D.M. (2004) Hydrologic regimes and riparian forests: a structured population model for cottonwood. *Ecology* **85**, 2493–2503.

Lytle, D.A. & Poff, N.L. (2004) Adaptation to natural flow regimes. *Trends in Ecology and Evolution* **19**, 94–100.

Lytle, D.A., Olden, J.D. & McMullen, L.E. (2008) Drought-escape behaviors of aquatic insects may be adaptations to highly variable flow regimes characteristic of desert rivers. *The Southwestern Naturalist* **53**, 399–402.

Maamri, A., Chergui, C. & Pattee, E. (1997a) Dynamique de apports de litière végétale et la faune invertébrée dans une rivière mediterranéenne temporaire: l'Oued Cheraa, au Maroc. *Ecologie* **28**, 251–264.

Maamri, A., Chergui, H. & Pattee, E. (1997b) Leaf litter processing in a temporary northeastern Moroccan river. *Archiv für Hydrobiologie* **140**, 513–531.

Maceda-Viega, A., Salvadó, H., Vinyoles, D. & de Sostoa, A. (2009) Outbreaks of *Ichthyophthirius multifilis* in redtail barbs *Barbus haasei* in a Mediterranean stream during drought. *Journal of Aquatic Animal Health* **21**, 189–194.

Maceina, M.J. (1993) Summer fluctuations in planktonic chlorophyll-a concentrations in Lake Okeechobee, Florida: The influence of lake levels. *Lake and Reservoir Management* **8**, 1–11.

Mackay, C.F. & Cyrus, D.P. (2001) Is freshwater quality adequately defined by physico-chemical components? Results from two drought-affected estuaries on the east coast of South Africa. *Marine and Freshwater Research* **52**, 267–281.

Mac Nally, R., Molyneux, G., Thomson, J.R., Lake, P.S. & Read, J. R. (2008) Variation in widths of riparian-zone vegetation of higher-elevation streams and implications for conservation management. *Plant Ecology* **198**, 89–100.

Magalhães, M.F., Beja, P., Canas, C. & Collares-Pereira, M.J. (2002) Functional heterogeneity of dry-season fish refugia across a Mediterranean catchment: the role of habitat and predation. *Freshwater Biology* **47**, 1919–1934.

Magalhães, M.F., Schlosser, I.J. & Collares-Pereira, M.J. (2003) The role of life history in the relationship between population dynamics and environmental variability in two Mediterranean stream fishes. *Journal of Fish Biology* **63**, 300–317.

Magalhães, M.F., Beja, P., Schlosser, I.J. & Collares-Pereira, M.J. (2007) Effects of multi-year droughts on fish assemblages of seasonally drying Mediterranean streams. *Freshwater Biology* **52**, 1494–1510.

Magalhães, V.F.D., Soares, R.M. & Azevedo, S.M.F.O. (2001) Microcystin contamination in fish from the Jacarepaguá Lagoon (Rio de Janeiro, Brazil): ecological implication and human health risk. *Toxicon* **39**, 1077–1085.

Magnuson, J.J., Benson, B.J. & Kratz, T.K. (2004) Patterns of coherent dynamics within and between lake districts at local to intercontinental scales. *Boreal Environment Research* **9**, 359–369.

Magoulick, D.D. (2000) Spatial and temporal variation in fish assemblages of drying stream pools: The role of abiotic and biotic factors. *Aquatic Ecology* **34**, 29–41.

Magoulick, D.D. & Kobza, R.M. (2003) The role of refugia for fishes during drought: a review and synthesis. *Freshwater Biology* **48**, 1186–1198.

Maher, M. (1984) Benthic studies of waterfowl breeding habitat in south-western New South Wales. 1. The fauna. *Australian Journal of Marine and Freshwater Research* **35**, 85–96.

Maher, M. & Carpenter, S.M. (1984) Benthic studies of waterfowl breeding habitat in south-western New South Wales. II. Chironomid populations. *Australian Journal of Marine and Freshwater Research* **35**, 97–110.

Makarieva, A.M. & Gorshkov, V.G. (2007) Biotic pump of atmospheric moisture as driver of the hydrological cycle on land. *Hydrology and Earth System Sciences* **11**, 1013–1033.

Makarieva, A.M., Gorshkov, V.G. & Li, B-L. (2009) Precipitation on land versus distance from the ocean: Evidence for a forest pump of atmospheric moisture. *Ecological Complexity* **6**, 302–307.

Maki, R.P., Sharfstein, B., East, T.L. & Rodusky, A.J. (2004) Phytoplankton photosynthesis-irradiance relationships in a large, managed, eutrophic, subtropical lake: The influence of lake stage on ecological homogeneity. *Archiv für Hydrobiologie* **161**, 159–180.

Malbrouck, C. & Kestemont, P. (2006) Effects of microcystins on fish. *Environmental Toxicology and Chemistry* **25**, 72–86.

Mantua, N.J., Hare, S.R., Zhang, Y., Wallace, J.M. & Francis, R.C. (1997) A Pacific decadal climate oscillation with impacts on salmon. *Bulletin of the American Meteorological Society.* **78**, 1069–1079.

Marques, S.C., Azeiteiro, U.M., Martinho, F. & Pardal, M.A. (2007) Climate variability and planktonic communities: The effect of an extreme event (severe drought) in a southern European estuary. *Estuarine, Coastal and Shelf Science* **73**, 725–734.

Marshall, B.E. (1988) Seasonal and annual variations in the abundance of pelagic sardines in Lake Kariba, with special reference to the effects of drought. *Archiv für Hydrobiologie* **112**, 399–409.

Martinho, F., Leitão, R., Viegas, I., Dolbeth, M., Neto, J.M., Cabral, H.N. & Pardal, M.A. (2007) The influence of an extreme drought event in the fish community of a southern Europe temperate estuary. *Estuarine, Coastal and Shelf Science* **75**, 537–546.

Mason, I.M., Guzkowska, M. A.J., Rapley, C.G. & Street-Perrott, F.A. (1994). The response of lake levels and areas to climatic change. *Climatic Change* **27**, 161–197.

Masson-Delmotte, V., Raffalli-Delerce, G., Danis, P.A., Yiou, P., Stievenard, M., Guibal, F., Mestre, O., Bernard, V., Goosse, H., Hoffmann, G. & Jouzel, J. (2005) Changes in European precipitation seasonality and in drought frequencies revealed by a four-century-long tree-ring isotopic record from Brittany, western France. *Climate Dynamics* **24**, 57–69.

Matthews, T.G. (2006) Spatial and temporal changes in abundance of the infaunal bivalve *Soletellina alba* (Lamarck 1818) during a time of drought in the seasonally-closed Hopkins River estuary, Victoria, Australia. *Estuarine, Coastal and Shelf Sciences* **66**, 13–20.

Matthews, T.G. & Constable, A. J. (2004) Effect of flooding on estuarine bivalve populations near the mouth of the Hopkins River, Victoria, Australia. *Journal of the Marine Biological Association of the United Kingdom* **84**, 633–639.

Matthews, W.J. (1998) *Patterns in Freshwater Fish Ecology.* Chapman and Hall, New York.

Matthews, W.J. & Marsh-Matthews, E. (2003) Effects of drought on fish across axes of space, time and ecological complexity. *Freshwater Biology* **48**, 1232–1253.

Matthews, W.J. & Marsh-Matthews, E. (2007) Extirpation of red shiner in direct tributaries of Lake Texoma (Oklahoma-Texas): a cautionary case history from a fragmented river-reservoir system. *Transactions of the American Fisheries Society* **136**, 1041–1062.

Matthews, W.J. & Styron, J.T. (1981) Tolerance of headwater vs. mainstream fishes for abrupt physicochemical changes. *American Midland Naturalist* **105**, 149–158.

Matthews, W.J., Surat, E. & Hill, L.G. (1982) Heat death of the orangethroat darter *Etheostoma spectabile* (Percidae) in a natural environment. *The Southwestern Naturalist* **27**, 216–217.

Mawdsley, J.A., Petts, G.E. & Walker, S. (1994) Assessment of drought Severity. *British Hydrological Society Occasional Paper No.3*.

Mayewski, P.A., Rohling, E.E., Stager, J.C., Karlén, W., Maasch, K.A., Meeker, L.D., Meyerson, E.A., Gasse, F., van Kreveld, S., Holmgren, K., Lee-Thorp, J., Rosqvist, G., Rack, F., Staubwasser, M., Schneider, R.R. & Steig, E.J. (2004) Holocene climate variability. *Quaternary Research* **62**, 243–255.

McAlpine, C.A., Syktus, J., Deo, R.C., Lawrence, P.J., McGowan, H. A, Watterson, I.G. & Phinn, S.R. (2007) Modeling the impact of historic land cover change on Australia's regional climate. *Geophysical Research Letters* **34**, L22711, doi: 10.1029/2007GL031524

McAlpine C.A., Syktus J., Ryan J.G., Deo R.C., McKeon G.M., McGowan H.A. & Phinn S.R. (2009) A continent under stress: interactions, feedbacks and risks associated with impact of modified land cover on Australia's climate. *Global Change Biology* **15**, 2206–2223.

McCabe, G.J., Palecki, M.A. & Betancourt, J.L. (2004) Pacific and Atlantic Ocean influences on multidecadal drought frequency. *Proceedings of the National Academy of Sciences* **101**, 4136–4141.

McCully, P. (2001) *Silenced Rivers: The Ecology and Politics of Large Dams* (revised edition). Zed Books, London, UK.

McDowall, R.M. (2006) Crying wolf, crying foul, or crying shame: alien salmonids and a biodiversity crisis in the south cool-temperate galaxioid fishes? *Reviews in Fish Biology and Fisheries* **16**, 233–422.

McElravy, E.P., Lamberti, G.A. & Resh, V.H. (1989) Year-to-year variation in the aquatic macroinvertebrate fauna of a northern Californian stream. *Journal of the North American Benthological Society* **8**, 51–63.

McGowan, S., Leavitt, P.R. & Hall, R.I. (2005) A whole-lake experiment to determine the effects of winter droughts on shallow lakes. *Ecosystems* **8**, 694–708.

McKee, K.L., Mendelssohn, I.A. & Materne, M.D. (2004) Acute salt marsh dieback in the Mississippi River deltaic plain: a drought-induced phenomenon? *Global Ecology and Biogeography* **13**, 65–73.

McKee, T.B., Doeken, N.J. & Kleist, J. (1993) The relationship of drought frequency and duration to time scales. *Preprints of Eighth Conference on Applied Climatology, Anaheim, California*. American Meteorological Society, 179–184.

McKee, T.B., Doesken, N.J., Kleist, J. & Shrier, C.J. (2000) *A history of drought in Colorado. Lessons learned and what lies ahead*. Bulletin # 9 pp. 1–20. Colorado Water Resources Research Institute, Colorado State University, Fort Collins, Colorado, USA.

McKernan, M. (2005) *Drought. The Red Marauder*. Allen and Unwin, Crows Nest, NSW.

McKenzie-Smith, F.J., Bunn, S.E. & House, A.P.N. (2006) Habitat dynamics in the bed sediment of an intermittent upland stream. *Aquatic Sciences* **68**, 86–99.

McLachlan, A.J. (1979a) The aquatic environment: 1. Chemical and physical characteristics of Lake Chilwa. In: Kalk, M., McLachlan, A.J. & Howard-Williams, C. (eds.) *Lake Chilwa. Studies of change in a tropical ecosystem.* pp. 59–78. Dr. W. Junk, The Hague.

McLachlan, A.J. (1979b) Decline and recovery of benthic invertebrate communities. In: Kalk M., McLachlan A.J. & Howard-Williams C. (eds.) *Lake Chilwa. Studies of change in a tropical ecosystem.* pp. 143–160. Dr. W. Junk, The Hague.

McMahon, T.A. & Finlayson, B.H. (2003) Droughts and anti-droughts: the low flow hydrology of Australian rivers. *Freshwater Biology* **48**, 1147–1160.

McMaster, D. & Bond, N.R. (2008) A field and experimental study on the tolerances of fish to *Eucalyptus camaldulensis* leachate and low dissolved oxygen concentrations. *Marine and Freshwater Research* **59**, 177–185.

McMenamin, S.K., Hadly, E.A. & Wright, C.K. (2008) Climate change and wetland desiccation cause amphibian decline in Yellowstone National Park. *Proceedings of the National Academy of Sciences* **105**, 16988–16993.

McNeil, D. G. & Closs, G.P. (2007). Behavioural responses of south-east Australian floodplain fish community to gradual hypoxia. *Freshwater Biology* **52**, 412–420.

McPhail, J.D. (1999) A fish out of water: observations on the ability of black mudfish, *Neochanna diversus,* to withstand hypoxic water and drought. *New Zealand Journal of Marine and Freshwater Research* **33**, 417–424.

Me-Bar, Y. & Valdez, F. (2003) Drought as random events in the Maya lowlands. *Journal of Archeological Science* **30**, 1599–1606.

Medeiros, E.S.F. & Maltchik, L. (1999) The effects of hydrological disturbance on the intensity of infestation of *Lernaea cyprinacea* in an intermittent stream fish community. *Journal of Arid Environments* **43**, 351–356.

Medeiros, E.S.F. & Maltchik, L. (2001) Fish assemblage stability in an intermittently flowing stream from the Brazilian semiarid region. *Austral Ecology* **26**, 156–164.

Meesters, A.G.C.A., Dolman, A.J. & Bruijnzeel, (2009) Comment on "Biotic pump of atmospheric moisture as driver of the hydrological cycle on land" by A.M. Makarieva and V.G. Gorshov, Hydrol. Earth Sys. Sci., 11, 1013–1033, 2007". *Hydrology and Earth System Sciences* **13**, 1299–1305.

Meko, D.M. & Woodhouse, C.A. (2005) Tree-ring footprint of joint hydrologic drought in Sacramento and Upper Colorado river basins, western USA. *Journal of Hydrology* **308**, 196–213.

Meko, D.M., Therrell, M.D., Baisan, C.H. & Hughes, M.K. (2001) Sacramento River flow reconstructed to A.D. 869 from tree rings. *Journal of the American Water Resources Association* **37**, 1029–1039.

Meko, D.M., Woodhouse, C.A., Baisan, C.A., Knight, T., Lukas, J.J., Hughes, M.K. & Salzer, M.W. (2007) Medieval drought in the upper Colorado River Basin. *Geophysical Research Letters* **34**, L10705, doi: 10.1029/2007GL029988, 2007.

Mensing, S.A., Benson, L.V., Kashgarian, M. & Lund, S. (2004) A Holocene pollen record of persistent droughts from Pyramid Lake, Nevada, USA. *Quaternary Research* **62**, 29–38.

Merron, G., la Hausse de Lalouvière, P. & Bruton, M. (1985) The recovery of the fishes of the Pongolo floodplain after a severe drought. *J.L.B. Smith Institute of Ichthyology Investigation Report No. 13*, 1–48.

Merron, G., Bruton, M. & la Hausse de Lalouvière, P. (1993) Changes in fish communities of the Phongolo floodplain, Zululand (South Africa) before, during and after a severe drought. *Regulated Rivers Research & Management* **8**, 335–344.

Mesquita, N., Coelho, M.M. & Filomena, M.M. (2006) Spatial variation in fish assemblages across small Mediterranean drainages: effects of habitat and landscape context. *Environmental Biology of Fish* **77**, 105–120.

Mesquita-Saad, L.S.B., Leitão, M.A.B., Paula-Silva, M.N., Chippari-Gomez, A.R. & Almeida-Val, V.M.F. (2002) Specialized metabolism and biochemical suppression during aestivation of the extant South American Lungfish-*Lepidosiren paradoxa*. *Brazilian Journal of Biology* **62**, 495–501.

Meyer, A., Meyer, E.I. & Meyer, C. (2003) Lotic communities of two small temporary karstic stream systems (East Westphalia, Germany) along a longitudinal gradient of hydrological intermittency. *Limnologica* **33**, 271–279.

Michels, A., Laird, K. R., Wilson, S.E., Thomson, D., Leavitt, P.R., Oglesby, R.J. & Cumming, B.F. (2007) Multidecadal to millennial-scale shifts in drought conditions on the Canadian prairies over the past six millennia: implications for future drought assessment. *Global Change Biology* **13**, 1295–1307.

Mikhailov, V.N. & Isupova, M.V. (2008) Hypersalinization of river estuaries in West Africa. *Water Resources* **35**, 367–385.

Miller, A.M. & Golladay, S.W. (1996) Effects of spates and drying on macroinvertebrate assemblages of an intermittent and a perennial prairie stream. *Journal of the North American Benthological Society* **15**, 670–689.

Miller, R.L. & Tegen, I. (1998) Climate response to soil dust aerosols. *Journal of Climate* **11**, 3247–3267.

Miller, S.W., Wooster, D. & Li, J. (2007) Resistance and resilience of macroinvertebrates to irrigation water withdrawls. *Freshwater Biology* **52**, 2494–2510.

Milly, P.C.D., Dunne, K.A. & Vecchia, A.V. (2005) Global pattern of trends in streamflow and water availability in a changing climate. *Nature* **438**, 347–350.

Milly, P.C.D., Betancourt, J., Falkenmark, M, Hirsch, R.M., Kundzewicz, Z.W., Lettenmaier, D.P. & Stouffer, R.J. (2008) Stationarity is dead: Whither water management? *Science* **319**, 573–574.

Mitchell, A.M. & Baldwin, D.S. (1999) The effects of sediment desiccation on the potential for nitrification, denitrification, and methanogenesis in an Australian reservoir. *Hydrobiologia* **392**, 1–11.

Mol, J.H., Resida, D., Ramlal, J.S. & Becker, C.R. (2000) Effects of El Niño-related drought on freshwater and brackish-water fishes in Suriname, South America. *Environmental Biology of Fishes* **59**, 429–440.

Montalto, L. & Marchese, M. (2005) Cyst formation in Tubificidae (Naidinae) and Opistocystidae (Annelida, Oligochaeta) as an adaptive strategy for drought tolerance in fluvial wetlands of the Paraná River, Argentina. *Wetlands* **25**, 488–494.

Montalto, L. & Paggi, A.C. (2006) Diversity of chironomid larvae in a marginal fluvial wetland of the middle Paraná River floodplain, Argentina. *Annales de Limnologie* **42**, 289–300.

Morrison, B.R.S. (1990) Recolonisation of four small streams in central Scotland following drought conditions in 1984. *Hydrobiologia* **208**, 261–267.

Mosisch, T.D. (2001) Effects of desiccation on stream epilithic algae. *New Zealand Journal of Marine and Freshwater Research* **35**, 173–179.

Moss, B. (1979) The Lake Chilwa ecosystem – a limnological overview. In: Kalk M., McLachlan A.J. & Howard-Williams C. (eds.) *Lake Chilwa. Studies of change in a tropical ecosystem.* pp. 399–415. Dr. W. Junk, The Hague.

Mourguiart, Ph., Corrège, T., Wirrmann, D., Argollo, J., Montenegro, M.E., Pourchet, M. & Carbonel, P. (1998) Holocene palaeohydrology of Lake Titicaca estimated from an ostracod-based transfer function. *Palaeogeography, Palaeoclimatology, Palaeoecology* **143**, 51–72.

Moy, C.M., Seitzer, G.D., Rodbell, D.T. & Anderson, D.M. (2002) Variability of El Niño/ Southern Oscillation activity at millennial timescales during the Holocene epoch. *Nature* **420**, 162–165.

Mpelasoka, F., Hennessy, K., Jones, R. & Bates, B. (2008) Comparison of suitable drought indices for climate change impacts assessment over Australia towards resource management. *International Journal of Climatology* **28**, 1283–1292.

Mundahl, N.D. (1990) Heat death of fish in shrinking stream pools. *American Midland Naturalist* **123**, 40–46.

Munger, T.T. (1916) Graphic method of representing and comparing drought intensities. *Monthly Weather Review* **44**, 642–643.

Murdock, J., Gido, K.B., Dodds, W.K., Bertrand, K.N. & Whiles, M.R. (2010) Consumer return chronology alters recovery trajectory of stream ecosystem structure and function following drought. *Ecology* **91**, 1048–1062.

Murrell, M.C., Hagy, III J.D., Lores, E.M. & Greene, R.M. (2007) Phytoplankton production and nutrient distributions in a subtropical estuary: Importance of freshwater flow. *Estuaries and Coasts* **30**, 390–402.

Muskó, I.B., Balogh, C., Tóth, A.P. Varga, E. & Lakatos, G. (2007) Differential response of invasive malacostracan species to lake level fluctuations. *Hydrobiologia* **590**, 65–74.

Muteveri, T. & Marshall, B.E. (2007) The impact of fish and drought on frog breeding in temporary waters of Zimbabwe. *African Zoology* **42**, 124–130.

Nadai, R. & Henry, R. (2009) Temporary fragmentation of a marginal lake and its effects on zooplankton community structure and organization. *Brazilian Journal of Biology* **69**, 819–835.

Naiman, R.J. & Décamps, H. (1997) The ecology of interfaces: riparian zones. *Annual Review of Ecology and Systematics* **28**, 621–658.

Naiman, R.J., Décamps, H. & McClain, M.E. (2005) *Riparia: ecology, conservation and management of streamside communities*. Elsevier/Academic Press, San Diego, USA.

Naselli-Flores, L. (2003) Man-made lakes in Mediterranean semi-climate: the strange case of Dr Deep Lake and Mr Shallow Lake. *Hydrobiologia* **506–509**, 13–21.

Nathan, R.J. & McMahon, T.A. (1990) Practical aspects of low-flow frequency analysis. *Water Resources Research*, **26**, 2135–2141.

Nathan, R.J. & McMahon, (1992) Estimating flow characteristics in ungauged catchments. *Water Resources Management* **6**, 85–100.

Nathan, R.J. & Weinmann, P.E. (1993) *Low Flow Atlas for Victorian Streams*. Department of Conservation and Natural Resources, Melbourne.

National Drought Mitigation, Center., (2005) What is drought? Understanding and defining drought. www.drought.unl.edu/

National Drought Policy Commission (2000) *Preparing for Drought in the 21st Century*. U.S. Department of Agriculture, Washington, USA. pp. 1–48.

National Land and Water Resources Audit (2001) Australian dryland salinity assessment 2000. Extents, impacts, processes, monitoring and management options. Land and Water Australia, Canberra, ACT, Australia.

Neiland, A.E., Goddard, J.P. & McGregor Reid, G. (1990) The impact of damming, drought and over-exploitation on the conservation of marketable fish stocks of the River Benue, Nigeria. *Journal of Fish Biology* **37** (Supplement A) 203–205.

Nichols, F.H. (1985) Increased benthic grazing: An alternative explanation for low phytoplankton biomass in northern San Francisco Bay during the 1976–1977 drought. *Estuarine, Coastal and Shelf Science* **21**, 379–388.

Nicholls, N. (1985) Towards the prediction of major Australian drought. *Australian Meteorological Magazine* **33**, 161–166.

Nicholls, N. (1988) More on early ENSOs: Evidence from Australian documentary sources. *Bulletin of the American Meteorological Society* **69**, 4–6.

Nicholls, N. (1989a) Sea surface temperatures and Australian winter rainfall. *Journal of Climate* **2**, 965–973.

Nicholls, N. (1989b) How old is ENSO? *Climatic Change* **14**, 111–115.

Nicholls, N. (1997) The centennial drought. In: Webb, E.K. (ed.) *Windows on Meteorology: Australian Perspective*, pp. 118–126. CSIRO Publishing, Melbourne.

Nicholls, N. (2004) The changing nature of Australian droughts. *Climatic Change* **63**, 323–336.

Nicholls, N. (2010) Local and remote causes of the southern Australian autumn-winter rainfall decline. *Climate Dynamics* **34**, 835–845.

Nicholson, A.J. (1954) An outline of the dynamics of animal populations. *Australian Journal of Zoology* **2**, 9–65.

Nicola, G.G., Almodóvar, A. & Elvira, B. (2009) Influence of hydrologic attributes on brown trout recruitment in low-latitude range margins. *Oecologia* **160**, 515–524.

Nielsen, D.L. & Brock, M.A. (2009) Modified water regime and salinity as a consequence of climate change: prospects for wetlands of southern Australia. *Climatic Change* **95**, 523–533.

Nilsson, C., Reidy, C.A., Dynesius, M. & Revenga, C. (2005) Fragmentation and flow regulation of the World's large river systems. *Science* **308**, 405–408.

Nippert, J.B., Butler, J.J., Kluitenberg, G.J., Whittemore, D.O., Arnold, D., Spal, S.E. & Ward, J.K. (2010) Patterns of *Tamarix* water use during a record drought. *Oecologia* **162**, 283–292.

Nõges, T. & Nõges, P. (1999) The effect of extreme water level decrease on hydrochemistry and phytoplankton in a shallow eutrophic lake. *Hydrobiologia* **409**, 277–283.

Nõges, T., Nõges, P. & Laugaste, R. (2003) Water level as the mediator between climate change and phytoplankton in a shallow temperate lake. *Hydrobiologia* **506–509**, 257–263.

Nybakken, J.W. & Bertness, M.D. (2005) *Marine biology: an ecological approach*. 6th edition. Pearson Education Inc., San Francisco, CA.

O'Connell, M., Baldwin, D.S., Robertson, A.I. & Rees, G. (2000) Release and availability of dissolved organic matter from floodplain litter: influence of origin and oxygen levels. *Freshwater Biology* **45**, 333–342.

O'Connor, T.G. (2001) Effect of small catchment dams on downstream vegetation of a seasonal river in semi-arid African savanna. *Journal of Applied Ecology* **38**, 1314–1325.

Odebrecht, C., Abreu, P.C., Möller, O.O., Niencheski, L.F., Proença, L.A. & Torgan, L.C. (2005) Drought effects on pelagic properties in the shallow and turbid Patos Lagoon, Brazil. *Estuaries* **28**, 675–685.

Odum, E.P. (1969) The strategy of ecosystem development. *Science* **164**, 262–270.

Oglesby, R.J. (1991) Springtime soil moisture, natural climatic variability, and the North American drought as simulated by the NCAR community climate model. *Journal of Climate* **4**, 890–897.

Ogurtsov, M.G. (2007) Cosmogenic isotopes and their role in present-day solar paleoastrophysics. *Geomagetism and Aeronomy* **47**, 85–93.

Ostman, O, Kneitel, J.M. & Chase, J.M. (2006) Disturbance alters habitat isolation's effect on biodiversity in aquatic microcosms. *Oikos* **114**, 360–366.

Ostrand, K.G. & Marks, D.E. (2000) Mortality of prairie stream fishes confined in an isolated pool. *Texas Journal of Science* **52**, 255–258.

Ostrand, K.G. & Wilde, G.R. (2001) Temperature, dissolved oxygen, and salinity tolerances of five prairie stream fishes and their role in explaining fish assemblage structure. *Transactions of the American Fisheries Society* **130**, 742–749.

Overpeck, J.T. & Cole, J.E. (2006) Abrupt change in Earth's climate systems. *Annual Review of Environment and Resources.* **31**, 1–31.

Padisák, J. (1992) Seasonal succession of phytoplankton in a large shallow lake (Balaton, Hungary)-a dynamic approach to ecological memory, its possible role and mechanisms. *Journal of Ecology* **80**, 217–230.

Paine, R.T., Tegner, M.J. & Johnson, E.A. (1998) Compounded perturbations yield ecological surprises. *Ecosystems* **1**, 535–545.

Palis, J.G., Aresco, M.J. & Kilpatrick, S. (2006) Breeding biology of a Florida population of *Ambystoma cingulatum* (Flatwoods Salamander) during a drought. *Southeastern Naturalist* **5**, 1–8.

Palmer, M.A., Reidy Liermann, C.A., Nilsson, C., Florke, M., Alcamo, J., Lake, P.S. & Bond, N. (2008) Climate change and the world's river basins: anticipating management options. *Frontiers in Ecology and the Environment* **6**, 81–89.

Palmer, M.A., Lettenmaier, D.P., Poff, N.L., Postel, S.L., Richter, B. & Warner, R. (2009) Climate change and river ecosystems: protection and adaptation options. *Environmental Management* **44**, 1053–1068.

Palmer, W.C. (1965) Meteorological Drought. Research Paper No. 45, U. S. Department of Commerce Weather Bureau, Washington, DC.

Paloumpis, A.A. (1956) Stream havens save fish. *Iowa Conservationist* **15** (8),60.

Paloumpis, A.A. (1957) The effects of drought conditions on the fish and bottom organisms of two small oxbow ponds. *Transactions of the Illinois State Academy of Science* **50**, 60–64.

Paloumpis, A.A. (1958) Responses of some minnows to flood and drought conditions in an intermittent stream. *Iowa State College Journal of Science* **32**, 547–561.

Palumbi, S.R., McLeod K.L. & Grünbaum D. (2008) Ecosystems in action: Lessons from marine ecology about recovery, resistance, and reversibility. *BioScience* **58**, 33–42.

Panter, J. & May, A. (1997) Rapid changes in the vegetation of a shallow pond in Epping Forest, related to recent droughts. *Freshwater Forum* **8**, 55–64.

Parsons, W.T. & Cuthbertson, E.G. (2001) *Noxious Weeds of Australia*. CSIRO Publishing, Collingwood, Australia.

Pearce, F. (2007) *When the rivers run dry*. Eden Project Books, Transworld Publishers, London, UK.

Pelicice, F.M. & Agostinho, A.A. (2008) Fish-passage facilities as ecological traps in large Neotropical rivers. *Conservation Biology* **22**, 180–188.

Perry, G.L.W. & Bond, N.R. (2009) Spatially explicit modeling of habitat dynamics and fish population persistence in an intermittent lowland stream. *Ecological Applications* **19**, 731–746.

Perry, S.F., Euverman, R., Wang, T., Loong, A.M., Chew, S.F., Ip, Y.K. & Gilmour, K.M. (2008) Control of breathing in African lungfish (*Protopterus dolloi*): A comparison of aquatic and cocooned (terrestrialized) animals. *Respiratory Physiology and Neurobiology* **160**, 8–17.

Peters, E., van Lanen, H.A.J., Torfs, P.J.J.F. & Bier, G. (2005) Drought in groundwater–drought distribution and performance indicators. *Journal of Hydrology* **306**, 302–317.

Peterson, G.D. (2002) Contagious disturbance, ecological memory, and the emergence of landscape pattern. *Ecosystems* **5**, 329–338.

Petranka, J.W., Murray, S.S. & Kennedy, C.A. (2003) Responses of amphibians to restoration of a southern Appalachian wetland: Perturbations confound post-restoration assessment. *Wetlands* **23**, 278–290.

Philander, S.G.H. (1985) El Niño and La Niña. *Journal of Atmospheric Science* **42**, 2652–2662.

Philbrick, C.T. & Les, D.H. (1996) Evolution of aquatic angiosperm reproductive systems. *BioScience* **46**, 813–826.

Phlips, E.J., Cichra, M., Havens, K., Hanlon, C., Badylak, S., Rueter, B., Randall, M. & Hansen, P. (1997) Relationship between phytoplankton dynamics and the availability of light and nutrients in a shallow sub-tropical lake. *Journal of Plankton Research* **19**, 319–342.

Piet, G.J. (1998) Impact of environmental perturbation on a tropical fish community. *Canadian Journal of Fisheries and Aquatic Sciences* **55**, 1842–1853.

Piha, H., Luoto, M., Piha, M. & Merilä, J. (2007) Anuran abundance and persistence in agricultural landscapes during a climatic extreme. *Global Change Biology* **13**, 300–311.

Pillay, D. & Perissinotto, R. (2008) The benthic macrofauna of the St Lucia Estuary during the 2005 drought year. *Estuarine, Coastal and Shelf Science* **77**, 35–46.

Pillay, D. & Perissinotto, R. (2009) Community structure of epibenthic meiofauna in the St Lucia Lake (South Africa) during a drought phase. *Estuarine, Coastal and Shelf Science* **81**, 94–104.

Pinay, G., Clément, J.C. & Naiman, R.J. (2002) Basic principles and ecological consequences of changing water regimes on nitrogen cycling in fluvial systems. *Environmental Management* **30**, 481–491.

Pinna, M. & Basset, A. (2004) Summer drought disturbance on plant detritus decomposition processes in three River Tirso (Sardinia, Italy) sub-basins. *Hydrobiologia* **522**, 311–319.

Pinna, M., Fonessu, A., Sangiorgio, F. & Basset, A. (2004) Influence of summer drought on spatial patterns of resource availability and detritus processing in Mediterranean stream sub-basins (Sardinia, Italy). *Internationale Revue der Gesamten Hydrobiologie* **89**, 484–499.

Pires, A.M., Cowx, I.G. & Coelho, M.M. (2000) Benthic macroinvertebrate communities of intermittent streams in the middle reaches of the Guadiana Basin (Portugal). *Hydrobiologia* **435**, 167–175.

Pires, D.F., Pires, A.M., Collares-Pereira, M.J. & Magalhães, M.F. (2010) Variation in fish assemblages across dry-season pools in a Mediterranean stream: effects of pool morphology, physicochemical factors and spatial context. *Ecology of Freshwater Fish* **19**, 74–86.

Pitman, A.J., Narisma, G.T., Pielke, R.A. & Holbrook, N.J. (2004) Impact of land cover change on the climate of southwest Western Australia. *Journal of Geophysical Research* **109**, D18109, doi: 10.1029/2003JD004347

Planchon, O., Dubreuil, V., Bernard, V. & Blain, S. (2008) Contribution of tree-ring analysis to the study of droughts in northwestern France (XIX–XXth century). *Climate of the Past Discussions* **4**, 249–270.

Pociask-Karteczka, J. (2006) River hydrology and the North Atlantic Oscillation: A general review. *Ambio* **35**, 312–314.

Poff, N.L. & Zimmerman, J.K.H. (2010) Ecological responses to altered flow regimes: a literature review to inform the science and management of environmental flows. *Freshwater Biology* **55**, 194–205.

Poff, N.L., Allan, J.D., Bain, M.B., Karr, J.R., Prestegaard, K. L, Richter, B.D. Sparks, R.E. & Stromberg, J.C. (1997) The natural flow regime: a paradigm for river conservation and restoration. *BioScience* **47**, 769–784.

Poff, N.L., Angermeier, P.L., Cooper, S.D., Lake, P.S., Fausch, K.D., Winemiller, K. O., Mertes, L.A.K., Oswood, M. W., Reynolds, J. & Rahel, F.J. (2001) Fish diversity in streams and rivers. In: Chapin, F.S., Sala, O.E. & Huber-Sannwald, E. (eds.) *Global Diversity in a Changing Environment*. pp. 316–349. Springer-Verlag, New York.

Poff, N.L., Olden, J.D., Vieira, K.M., Finn, D.S., Simmons, M.P. & Kondratieff, B.C. (2006) Functional trait niches of North American lotic insects: trait-based ecological applications in light of phylogenetic relationships. *Journal of the North American Benthological Society* **25**, 730–755.

Pompeu, P.D.S. & Godinho, H.P. (2006) Effects of extended absence of flooding on the fish assemblages of three floodplain lagoons in the middle São Francisco River, Brazil. *Neotropical Ichthyology* **4**, 427–433.

Postel, S & Richter, B. (2003) *Rivers for life. Managing water for people and nature.* Island Press, Washington, DC.

Power, M. E (1990) Effects of fish in river food webs. *Science* **250**, 811–814.

Power, M.E. (1992) Hydrologic and trophic controls of seasonal algal blooms in northern California rivers. *Archiv für Hydrobiologie* **125**, 385–410.

Power, M.E., Matthew, W.J. & Stewart, A.J. (1985) Grazing minnows, piscivorous bass and stream algae: dynamics of a strong interaction. *Ecology* **66**, 1448–1456.

Power, M.E., Parker, M.S. & Dietrich, W.E. (2008) Seasonal reassembly of a river food web: floods, droughts and impacts of fish. *Ecological Monographs* **78**, 263–282.

Power, S., Casey, T., Folland, C., Colman, A. & Mehta, V. (1999) Inter-decadal modulation of the impact of ENSO in Australia. *Climate Dynamics* **15**, 319–324.

Pringle, C.M., Freeman, M.C. & Freeman, B.J. (2000) Regional effects of hydrologic alterations on riverine macrobiota in the New World: Tropical-temperate comparisons. *BioScience* **50**, 807–822.

Pritchard, D.W. (1967) What is an estuary: physical viewpoint. In: Lauf, G.H. (ed.) *Estuaries*, pp. 3–5. American Association for Advancement of Science Publ. No. 83, Washington, DC.

Puckridge, J.T., Sheldon, F., Walker, K.F. & Boulton, A.J. (1998) Flow variability and the ecology of large rivers. *Marine and Freshwater Research* **49**, 55–72.

Pusey, B.J. (1990) Seasonality, aestivation and the life history of the salamanderfish (*Lepidogalaxias salamandroides*) Pisces: Lepidogalaxiidae. *Environmental Biology of Fishes* **29**, 15–26.

Pusey, B.J. & Edward, D.H.D. (1990) Structure of fish assemblages in waters of the southern acid peat flats, south-western Australia. *Australian Journal of Marine and Freshwater Research* **41**, 721–734.

Qiu, S. & McComb, A.J. (1995) Planktonic and microbial contributions to phosphorus release from fresh and air-dried sediments. *Marine and Freshwater Research* **46**, 1039–1045.

Qiu, S. & McComb, A.J. (1996) Drying-induced stimulation of ammonium release and nitrification in reflooded lake sediment. *Marine and Freshwater Research* **47**, 531–536.

Quade, J., Forester, R.M., Pratt, W.L. & Carter, C. (1998) Black mats, spring-fed streams, and late Glacial-Age recharge in the southern Great Basin. *Quaternary Research* **49**, 129–148.

Quinn, J.M., Steele, G.L., Hickey, C.W. & Vickers, M.L. (1994) Upper thermal tolerances of twelve New Zealand stream invertebrate species. *New Zealand Journal of Marine and Freshwater Research* **28**, 391–397.

Rasmusson, E.M. & Carpenter, T.H. (1982) Variations in tropical sea surface temperatures and surface wind field associated with the Southern Oscillation-El Niño. *Monthly Weather Review* **110**, 354–384.

Reid, D.J., Quinn, G.P., Lake, P.S. & Reich, P. (2008) Terrestrial detritus supports the food webs in lowland intermittent streams of south-eastern Australia: a stable isotope study. *Freshwater Biology* **53**, 2036–2050.

Rejmánková, E., Rejmánek, M., Djohan, T. & Goldman, C.R. (1999) Resistance and resilience of subalpine wetlands with respect to prolonged drought. *Folia Geobotanica* **34**, 175–188.

Resh, V.H. (1992) Year-to-year changes in the age structure of a caddisfly population following loss and recovery of a springbrook habitat. *Ecography* **15**, 314–317.

Resh, V.H., Jackson, J.K. & McElravy, E.P. (1990) Disturbance, annual variability, and lotic benthos: Examples from a California stream influenced by a Mediterranean climate. *Memorie dell'Istituto Italiano du Idrobiologia* **47**, 309–329.

Resilience Alliance (2008) Resilience Alliance. The basis for sustainability. http://www.resalliance.org/576.php

Reynolds, B. & Edwards, A. (1995) Factors influencing dissolved nitrogen concentrations and loading in upland streams of the UK. *Agricultural Water Management* **27**, 181–202.

Reynolds, C.S., Padisák, J. & Sommer, U. (1993) Intermediate disturbance in the ecology of phytoplankton and the maintenance of diversity: a synthesis. In: Padisák, J., Reynolds, C.S. & Sommer, U. (eds.) *Intermediate Disturbance Hypothesis in Phytoplankton Ecology*. pp. 157–171. Developments in Hydrobiology 81, Kluwer Academic Publishers, Dordrecht, Netherlands.

Riebsame, W. E, Changnon, S.A. & Karl, T.R. (1991) *Drought and Natural Resources Management in the United States. Impacts and Implications of the 1987–89 Drought.* Westview Press, Boulder, Co.

Rincon, J. & Cressa, C. (2000) Temporal variability of macroinvertebrate assemblages in a neotropical intermittent stream in northwestern Venezuela. *Archiv für Hydrobiologie* **148**, 421–432.

Robson, B.J. (2000) Role of residual biofilm in the recolonization of rocky intermittent streams by benthic algae. *Marine and Freshwater Research* **51**, 725–732.

Robson, B.J. & Matthews, T.G. (2004) Drought refuges affect algal colonization in intermittent streams. *River Research and Applications* **20**, 753–763.

Robson, B.J., Matthews, T.G., Lind, P.R. & Thomas, N.A. (2008a) Pathways for algal recolonization in seasonally-flowing streams. *Freshwater Biology* **53**, 2385–2401.

Robson, B.J., Chester, E.T., Mitchell, B.D. & Matthews, T.G. (2008b) *Identification and management of refuges for aquatic organisms*. National Water Commission, Canberra, Australia.

Rodda, J.C. (2000) Drought and water Resources. In: Wilhite, D.A. (ed.) *Drought: A Global Assessment. Volume 1*. pp. 241–263. Routledge, London.

Roe, J.H. & Georges, A. (2007) Heterogeneous wetland complexes, buffer zones, and travel corridors: Landscape management for freshwater reptiles. *Biological Conservation* **135**, 67–76.

Romanello, G.A., Chuchra-Zbytniuk, K.L., Vandermer, J.L. & Touchette, B.W. (2008) Morphological adjustments promote drought avoidance in the wetland plant Acorus americanus. *Aquatic Botany* **89**, 390–396.

Romani, A.M. & Sabater, S. (1997) Metabolism recovery of a stromatolitic biofilm after drought in a Mediterranean stream. *Archiv für Hydrobiologie* **140**, 261–271.

Romani, A.M., Vázquez, E. & Butturini, A. (2006) Microbial availability and size fractionation of dissolved organic carbon after drought in an intermittent stream: Biogeochemical link across the stream-riparian interface. *Microbial Ecology* **52**, 501–512.

Romme, W.H., Everham, E.H., Frelich, L.E., Moritz, M.A. & Sparks, R.E. (1998) Are large, infrequent disturbances qualitatively different from small, frequent disturbances? *Ecosystems* **1**, 524–534.

Rood, S.B., Braatne, J.H. & Hughes, F.M.R. (2003) Ecophysiology of riparian cottonwoods: stream flow dependency, water relations and restoration. *Tree Physiology* **23**, 1113–1124.

Ropelewski, C.F. & Halpert, M.S. (1987) Global and regional scale precipitation patterns associated with the El Niño/Southern Oscillation (ENSO). *Monthly Weather Review* **115**, 1606–1626.

Ropelewski, C.F. & Halpert, M.S. (1989) Precipitation patterns associated with the high index phase of the Southern Oscillation. *Journal of Climate* **2**, 268–284.

Roshier, D.A., Robertson, A.I., Kingsford, R.T. & Green, D.G. (2001) Continental-scale interactions with temporary resources may explain the paradox of large populations of desert waterbirds in Australia. *Landscape Ecology* **16**, 547–556.

Rose, P., Metzeling, L & Catzikiris, S. (2008) Can macroinvertebrate rapid bioassessment methods be used to assess river health during drought in south eastern Australian streams? *Freshwater Biology* **53**, 2626–2638.

Ross, S.T., Matthews, W.J. & Echelle, A.A. (1985) Persistence of stream fish assemblages: effects of environmental change. *American Naturalist* **126**, 24–40.

Ruegg, J. & Robinson, C.T. (2004) Comparison of macroinvertebrate assemblages of permanent and temporary streams in an alpine flood plain, *Switzerland. Archiv für Hydrobiologie* **161**, 489–510.

Ruetz, C.R., Trexler, J.C., Jordan, F., Loftus, W.F. & Perry, S.A. (2005) Population dynamics of wetland fishes: spatio-temporal patterns synchronized by hydrological disturbance? *Journal of Animal Ecology* **74**, 322–332.

Rusak, J.A., Leavitt, P.R., McGowan, S., Chen, G., Olson, O. & Wunsam, S. (2004) Millenial-scale relationships of diatom species richness and production in two prairie lakes. *Limnology and Oceanography* **49**, 1290–1299.

Russell, J.M. & Johnson, T.C. (2005). A high-resolution geochemical record from Lake Edward, Uganda–Congo and the timing and causes of tropical African drought during the late Holocene. *Quaternary Science Reviews* **24**, 1375–1389.

Russell, J.M. & Johnson, T.C. (2007) Little Ice Age drought in equatorial Africa: Intertropical Convergence Zone migrations and El Niño-Southern Oscillation variability. *Geology* **35**, 21–24.

Russell, J.M., Johnson, T.C., Kelts K.R., Laerdal T. & Talbot M.R. (2003) An 11,000-year lithostratigraphic and paleohydrologic record from equatorial Africa: Lake Edward, Uganda-Congo. *Palaeogeography, Palaeoclimatology, Palaeoecology* **193**, 25–49.

Rutherford, M.C., Marsh, N.A., Davies, P.M. & Bunn, S.E. (2004) Effects of patchy shade on stream water temperatures: how quickly do small streams heat and cool? *Marine and Freshwater Research* **55**, 737–748.

Rutledge, C.J., Zimmerman, E.G. & Beitinger, T.L. (1990) Population genetic responses of two minnow species (Cyprinidae) to seasonal stream intermittency. *Genetica* **80**, 209–219.

Ryder, D.S. (2004) Response of epixylic biofilm metabolism to water level variability in a regulated floodplain river. *Journal of the North American Benthological Society* **23**, 214–223.

Saint-Jean, L. (1983) The zooplankton. In: Carmouze, J.-P., Durand, J.-R. & Lévêque, C. (eds.) *Lake Chad. Ecology and Productivity of a Shallow Tropical Ecosystem.* pp. 199–232. Dr. W. Junk, The Hague.

Saji, N.H. & Yamagata, T. (2003) Structure of SST and surface wind variability during the Indian Ocean Dipole mode events: COADS observations. *Journal of Climate* **16**, 2735–2751.

Saji, N.H., Goswami, B.N., Vinayachandran, P.N. & Yamagata, T. (1999) A dipole mode in the tropical Indian Ocean. *Nature* **401**, 360–363.

Sala, O.E., Chapin, F.S., Armesto, J.J., Berlow, E., Bloomfield, J., Dirzo, R., Huber-Sanwald, E., Huenneke, L.F., Jackson, R.B., Kinzig, A., Leemans, R., Lodge, D.M., Money, H.A., Oesterheld, M., Poff, N.L., Sykes, M.T., Walker, B.H., Walker, M. & Wall, D.H. (2000) Global biodiversity scenarios for the year 2100. *Science* **287**, 1770–1774.

Sandgren, P. & Snowball, I. (2001) Application of mineral magnetic techniques to paleolimnology. In: Last, W.M. & Smol, J.P. (eds.) *Tracking Environmental Change using Lake Sediments. Volume 2: Physical and Geochemical Methods.* pp. 217–237. Kluwer Academic Publishers, Dordrecht, The Netherlands.

Sangiorgio, F., Fonnesu, A., Pinna, M., Sabetta, L. & Basset, A. (2006) Influence of drought and abiotic factors on *Phragmites australis* leaf decomposition in the River Pula, Sardinia, Italy. *Journal of Freshwater Ecology* **21**, 411–420.

Santos, A.M.R. & Thomaz, A.S.M. (2007) Aquatic macrophytes diversity in lagoons of a tropical floodplain: The role of connectivity and water level. *Austral Ecology* **32**, 177–190.

Sarr, D.A. (2002) Riparian livestock exclosure research in the western United States: a critique and some recommendations. *Environmental Management* **30**, 516–526.

Savadamuthu, K. (2002) *Impact of farm dams on streamflow in the Upper Marne catchment.* Report DWR 02/01/0003 Department of Water Resources, Adelaide, South Australia.

Scanlon, B.R., Keese, K.E., Flint, A.L., Flint, L.E., Gaye, C.B., Edmunds, W.M. & Simmers, I. (2006) Global synthesis of groundwater recharge in semiarid and arid regions. *Hydrological Processes* **20**, 3335–3370.

Schaefer, J. (2001) Riffles as barriers to interpool movements by three cyprinids (*Notropis boops*, *Campostoma anomalum* and *Cyprinella venusta*). *Freshwater Biology* **46**, 379–388.

Scheffer, M. (1998) *Ecology of Shallow Lakes*. Chapman & Hall, London.

Schindler, D.W. (1997) Widespread effects of climatic warming on freshwater ecosystems in North America. *Hydrological Processes* **11**, 225–251.

Schindler, D.W. (1998) A dim future for boreal waters and landscapes. *BioScience* **48**, 157–164.

Schindler, D.W. & Curtis, P.J. (1997) The role of DOC in protecting freshwaters subjected to climate warming and acidification from UV exposure. *Biogeochemistry* **36**, 1–8.

Schindler, D.W., Beaty, K.G., Fee, E.J., Cruikshank, D.R., DeBruyn, E.D., Findlay, D.L., Linsey, G.A., Shearer, J.A., Stainton, M.P. & Turner, M.A. (1990) Effects of climate warming on lakes of the central boreal forest. *Science* **250**, 967–970.

Schindler, D.W., Curtis, P.J., Bayley, S.E., Parker, B.R., Beaty, K.G. & Stainton, M.P. (1997) Climate-induced changes in the dissolved organic carbon budgets of boreal lakes. *Biogeochemistry* **36**, 9–28.

Schlosser, I.J. (1995) Critical landscape attributes that influence fish population dynamics in headwater streams. *Hydrobiologia* **303**, 71–81.

Schlosser, I.J. & Angermeier, P.L. (1995) Spatial variation in demographic processes of lotic fishes: Conceptual models, empirical evidence, and implications for conservation. *American Fisheries Society Symposium* **17**, 392–401.

Schneider, D.W. (1997) Predation and food web structure along a habitat duration gradient. *Oecologia* **110**, 567–575.

Schneider, D.W. (1999) Snowmelt ponds in Wisconsin. Influence of hydroperiod on invertebrate community structure. In: Batzer, D.P., Rader, R.B. & Wissinger, S.A. (eds.) *Invertebrates in Freshwater Wetlands of North America: Ecology and Management.* pp. 299–318. John Wiley & Sons, New York.

Schneider, D.W. & Frost, T.M. (1996) Habitat duration and community structure in temporary ponds. *Journal of the North American Benthological Society* **15**, 64–86.

Schreider, S.Y., Jakeman, A.J., Letcher, R.A., Nathan, R.J., Neal, B.P. & Beavis, S.G. (2002) Detecting changes in streamflow response to changes in non-climatic catchment conditions: farm dam development in the Murray-Darling basin, Australia. *Journal of Hydrology* **262**, 84–98.

Schubert, S.D., Suarez, T.C., Pegion, P.J., Koster, R.D. & Bacmeister, J.T. (2004) Causes of long-term drought in the United States Great Plains. *Journal of Climate* **17**, 485–503.

Schubert, S.D., Suarez, M.J., Pegion, P.J., Koster, R.D. & Bacmeister, J. T. (2008) Potential predictability of long-term drought and pluvial condition in the U.S. Great Plains. *Journal of Climate* **21**, 802–816.

Scoppettone, G.G., Rissler, P.H. & Buettner, M.E. (2000) Reproductive longevity and fecundity associated with nonannual spawning in Cui-ui. *Transactions of the American Fisheries Society* **129**, 658–669.

Scott, M.L., Shafroth, P.B. & Auble, G.T. (1999) Responses of riparian cottonwoods to alluvial water table declines. *Environmental Management* **23**, 347–358.

Scully, N.M. & Lean, D.R.S. (1994) The attenuation of ultraviolet radiation in temperate lakes. *Archiv für Hydrobiologie* **43**, 135–144.

Seager, R. (2007) The turn of the century North American drought: global context, dynamics and past analogs. *Journal of Climate* **20**, 5527–5552.

Seager, R., Kushnir, Y., Herweijer, C., Naik, N. & Miller, J. (2005) Modeling of tropical forcing of persistent droughts and pluvials over western North America: 1856–2000. *Journal of Climate* **18**, 4068–4091.

Seager, R., Ting, M., Held, I., Kushnir, Y., Lu, J., Vecchi, G., Huang, H-P., Harnik, N., Leetmaa, A., Lau, N-C., Li, C., Velez, J. & Naik, N. (2007) Model projections of an imminent transition to a more arid climate in southwestern North America. *Science* **316**, 1181–1184.

Seager, R., Kushnir, Y., Ting, M., Cane, M., Naik, N. & Miller, J. (2008) Would advance knowledge of 1930s SSTs have allowed prediction of the Dust Bowl drought. *Journal of Climate* **21**, 3261–3281.

Sebestyén, O., Entz, B. & Felföldy, L. (1951). Alacsony Vízállással kapcsolatos biológiai jelenségek a Balatonon 1949 őszén. *Annales Instituti Biologici (Tihany) Hungaricae Academiae Scientiarum* **20**, 127–160.

Sedell, J.R., Reeves, G.H., Hauer, F.R., Stanford, J.A. & Hawkins, C.P. (1990) Role of refugia in recovery from disturbance: modern fragmented and disconnected river systems. *Environmental Management* **14**, 711–724.

Semlitsch, R.D. (1987) Relationship of pond drying to the reproductive success of the salamander *Ambystoma talpoideum*. *Copeia* **1987**, 61–69.

Shafer, B.A. & Dezman, L.E. (1982) Development of a Surface Water Supply Index (SWSI) to assess the severity of drought conditions in snowpack runoff areas. *Proceedings of the 50th Western Snow Conference, Reno, NV*. 164–175.

Shanahan, T.M., Overpeck, J.T., Anchukaitis, K.J., Beck, J.W., Cole, J.E., Dettman, D.L., Peck, J.A., Scholz, C.A. & King, J.W. (2009) Atlantic forcing of persistent drought in West Africa. *Science* **324**, 377–380.

Shapley, M.D., Johnson, W.C., Engstrom, D.R. & Osterkamp, W.R. (2005) Late-Holocene flooding and drought in the Northern Great Plains, USA, reconstructed from tree rings, lake sediments and ancient shorelines. *The Holocene* **15**, 29–41.

Shepley, M.G., Streetly, M., Voyce, K. & Bamford, F. (2009) Management of stream compensation for a large conjunctive use scheme, Shropshire, UK. *Water and Environment Journal* **23**, 263–271.

Shi, G., Ribbe, J., Cai, W. & Cowan, T. (2008) An interpretation of Australian rainfall projections. *Geophysical Research Letters* 35, doi: 10.1029/2007GL032436, 2008.

Shiel, D. & Murdiyarso, D. (2009) How forests attract rain: an examination of a new hypothesis. *BioScience* **59**, 341–347.

Shili, A., Maïz, N. B., Boudouresque, C.F. & Trabelsi, E.B. (2007) Abrupt changes in *Potamogeton* and *Ruppia* beds in a Mediterranean lagoon. *Aquatic Botany* **87**, 181–188.

Shorthouse, C.A. & Arnell, N.W. (1997) Spatial and temporal variability in European river flows and the North Atlantic Oscllation. In: Gustard, A., Blazkova, S., Brilly, M., Delmuth, S., Dixon, J., Van Lanen, H., Llasat, C., Mkhandi, S. & Servat, E. (eds.) *FRIEND'97: regional hydrology: concepts and models of sustainable water resource management*. pp. 77–85. IAHS Press, Wallingford, Oxfordshire, UK.

Shrestha, A. & Kostaschuk, R. (2005) El Niño/Southern Oscillation (ENSO)-related variability in mean-monthly streamflow in Nepal. *Journal of Hydrology* **308**, 33–49.

Siegel, R.A., Gibbons, J.W. & Lynch, T.K. (1995) Temporal changes in reptile populations: effects of a severe drought on aquatic snakes. *Herpetologica* **51**, 424–434.

Sigleo, A.C. & Frick, W.E. (2007) Seasonal variations in river discharge and nutrient export to a northeastern Pacific estuary. *Estuarine, Coastal and Shelf Science* **73**, 368–378.

Silliman, B.R. & Newell, S.Y. (2003) Fungal farming in a snail. *Proceedings of the National Academy of Sciences* **100**, 16543–15648.

Silliman, B.R., van de Koppel, J., Bertness, M.D., Stanton, L.E. & Mendelssohn, I.A. (2005) Drought, snails, and large-scale die-off of southern U.S. salt marshes. *Science* **310**, 1803–1806.

Simpson, H.J., Cane, M.A., Herczeg, A.L. Zebiak, S.E. & Simpson, J.H. (1993) Annual river discharge in southeastern Australia related to El Niño-Southern Oscillation forecasts of sea surface temperatures. *Water Resources Research* **29**, 3671–3680.

Simpson, S.L., Fitzpatrick, R.W., Shand, P., Angel, B.M., Spadaro, D.A. & Mosley, L. (2010) Climate-driven mobilization of acid and metals from acid sulfate soils. *Marine and Freshwater Research* **61**, 129–138.

Sinclair, P. (2001) *The Murray. A River and its People.* Melbourne University Press, Melbourne, Australia.

Sklar, F.H., Chimney, M.J., Newman, S., McCormick, P., Gawlik, D., Miao, S., McVoy, C., Said W., Newman, J., Coronado, C., Crozier, G., Korvela, M., & Rutchey, K. (2005) The ecological-societal underpinnings of Everglades restoration. *Frontiers in Ecology and the Environment* **3**, 161–169.

Slack, K.V. (1955) A study of factors affecting stream productivity by the comparative method. *Investigations of Indiana Lakes and Streams* **4** (1), 3–47.

Slack, J.G. (1977) River water quality in Essex during and after the 1976 drought. *Effluent and Water Treatment Journal* **17**, 575–578.

Sloman, K.A., Taylor, A.C., Metcalfe, N.B. & Gilmour, K.M. (2001) Effects of an environmental perturbation on the social behaviour and physiological function of brown trout. *Animal Behaviour* **61**, 325–333.

Smakhtin, V.U. (2001) Low flow hydrology: A review. *Journal of Hydrology* **240**, 147–186.

Smale, M.A. & Rabeni, C.F. (1995) Hypoxia and hyperthermia tolerances of headwater stream fishes. *Transactions of the American Fisheries Society* **124**, 698–711.

Smith, A.J. (1993) Lacustrine ostracodes as hydrochemical indicators in lakes of the north-central Unites States. *Journal of Paleolimnology* **8**, 121–134.

Smock, L.A., Smith, L.C., Jones, J.B. & Hooper, S.M. (1994) Effects of drought and a hurricane on a coastal headwater stream. *Archiv für Hydrobiologie* **131**, 25–38.

Smolders, A.J.P., Van der Velde, G. & Roelofs, J.G.M. (2000) *El Niño* caused collapse of the Sábalo fishery (*Prochilodus lineatus*, Pisces: Prochilodontidae) in a South American river. *Naturwissenschaften* **87**, 30–32.

Snodgrass J.W., Bryan A.L., Lide R.F. & Smith G.M. (1996) Factors affecting the occurrence and structure of fish assemblages in isolated wetlands of the upper Coastal Plain, U.S.A. *Canadian Journal of Fisheries and Aquatic Science* **53**, 443–454.

Sobek, S, Tranvik L.J., Prairie Y.T., Kortelainen P. & Cole J.J. (2007) Patterns and regulation of dissolved organic carbon: An analysis of 7,500 widely distributed lakes. *Limnology and Oceanography* **52**, 1208–1219.

Sommer, B. & Horwitz, P. (2001) Water quality and macroinvertebrate response to acidification following intensified summer droughts in a West Australian wetland. *Marine and Freshwater Research* **52**, 1015–1021.

Søndergaard, M., Kristensen, P. & Jeppesen, E. (1992) Phosphorus release from resuspended sediment in the shallow and wind-exposed Lake Arresø, Denmark. *Hydrobiologia* **228**, 91–99.

Sophocleous, M. (2002) Interactions between groundwater and surface water: the state of the Science. *Hydrogeology Journal* **10**, 52–67.

Soranno, P.A., Webster, K.E., Riera, J.L., Kratz, T.K., Baron, J.S., Bukaveckas, P.A., Kling, G.W., White, D. S., Caine, N., Lathrop, R.C. & Leavitt, P.R. (1999) Spatial variation among lakes within landscapes: ecological organization along lake chains. *Ecosystems* **2**, 395–410.

Sprague, L.A. (2005) Drought effects on water quality in the South Platte River Basin, Colorado. *Journal of the American Water Resources Association* **41**, 11–24.

Spranza, J.J. & Stanley, E.H. (2000) Condition, growth, and reproductive styles of fishes exposed to different environmental regimes in a prairie drainage. *Environmental Biology of Fishes* **59**, 99–109.

Srivastava, D.S. (2005). Do local processes scale to global patterns? The role of drought and the species pool in determining treehole insect diversity. *Oecologia* **145**, 205–215.

Stager, J.C., Cumming, B.F. & Meeker, L.D. (2003) A 10,000-year high-resolution diatom record from Pilkington Bay, Lake Victoria, East Africa. *Quaternary Research* **59**, 172–181.

Stahle, D.W., Cleaveland, M.K., Blanton, D.B., Therrell, M.D. & Gay, D.A. (1998a) The Lost Colony and Jamestown droughts. *Science* **280**, 564–567.

Stahle, D.W., D'Arrigo, R.D., Krusic, P.J., Cleaveland, M.K., Cook, E.R., Allan, R.J., Cole, J.E., Dunbar, R.B., Therrell, M.D., Gay, D.A., Moore, M.D., Stokes, M.A., Burns, B.T., Villanueva-Diaz, J. & Thompson, L.G. (1998b) Experimental dendroclimatic reconstruction of the Southern Oscillation. *Bulletin of the American Meteorological Society* **79**, 2137–152.

Stahle, D.W., Cook, E. R., Cleaveland, M.K., Therrell, M.D., Meko, D.M., Grissino-Mayer, H.D., Watson, E. & Luckman, B.H. (2000) Tree-ring data document 16th century megadrought over North America. *Eos, Transactions, American Geophysical Union* **81**, 121–123.

Stahle, D.W., Fye, F.K., Cook, E.R. & Griffin, R.D. (2007) Tree-ring reconstructed megadroughts over North America since A.D. 1300. *Climatic Change* **83**, 133–149.

Stanley, E.H., Buschman, D.L., Boulton, A.J., Grimm, N.B. & Fisher, S.G. (1994) Invertebrate resistance and resilience to intermittency in a desert stream. *American Midland Naturalist* **131**, 288–300.

Stanley, E.H. & Boulton, A.J. (1995) Hyporheic processes during flooding and drying in a Sonoran Desert stream. I. Hydrological and chemical dynamics. *Archiv für Hydrobiologie* **134**, 27–52.

Stanley, E.H., Fisher, S.G. & Grimm, N.B. (1997) Ecosystem expansion and contraction in streams. *BioScience* **47**, 427–435.

Stanley, E.H., Fisher, S.G. & Jones, J.B. (2004) Effects of water loss on primary production: a landscale-scale model. *Aquatic Sciences* **66**, 130–138.

Stark, J. (1993) A macroinvertebrate community index of water quality for stony streams. *New Zealand Journal of Marine and Freshwater Research* **27**, 463–478.

Staubwasser, M. & Weiss, H. (2006) Holocene climate and cultural evolution in late prehistoric-early historic West Asia. *Quaternary Research* **66**, 372–387.

Stenseth, N. C., Ottersen, G., Hurrell, J. W., Mysterud, A., Lima, M., Chan, K-S., Yoccoz, N. G. & Adlandsvik, B. (2003) Studying climate effects on ecology through the use of climate indices: the North Atlantic Oscillation, El Niño Southern Oscillation and beyond. *Proceedings of the Royal Society London* **B 270**, 2087–2096.

Steward, J.R. & Lister, A. (2001) Cryptic northern refugia and the origin of the modern biota. *Trends in Ecology and Evolution* **16**, 608–613.

Stine, S. (1994) Extreme and persistent drought in California and Patagonia during mediaeval time. *Nature* **369**, 546–549.

Stockton, C.W. & Meko, D.M. (1975) A long-term history of drought occurrence in western United States as inferred from tree rings. *Weatherwise* **28**, 244–249.

Stone, J.R. & Fritz, S.C. (2006) Multidecadal drought and Holocene climate instability in the Rocky Mountains. *Geology* **34**, 409–412.

Straille, D., Livingstone, D.M., Weyhenmeyer, G.A. & George, D.G. (2003) The response of freshwater ecosystems to climate variability associated with the North Atlantic Oscillation. In: Hurrell, J.W. (ed.) *The North Atlantic Oscillation-Climatic significance and environmental impact.* pp. 263–279. American Geophysical Union, Washington DC, USA.

Strauch, A.M., Kapust, A.R. & Jost, C.C. (2009) Impact of livestock management on water quality and streambank structure in a semi-arid, African ecosystem. *Journal of Arid Environments* **73**, 795–803.

Stromberg, J.C. (1998) Dynamics of Fremont cottonwood (*Populus fremontii*) and saltcedar (*Tamarix chinensis*) populations along the San Pedro River, Arizona. *Journal of Arid Environments* **40**, 133–155.

Stromberg, J.C. Tiller, R. & Richter, B. (1996) Effects of ground water decline on riparian vegetation of semi-arid regions: The San Pedro River, AZ. *Ecological Applications* **6**, 113–131.

Stromberg, J.C. Bagstad, K.J., Leenhouts, J.M., Lite, S.J. & Makings, E. (2005) Effects of stream flow intermittency on riparian vegetation of a semiarid region river (San Pedro River, Arizona). *River Research and Applications* **21**, 925–938.

Stubbington, R., Wood, P.J. & Boulton, A.J. (2009a) Low flow controls on benthic and hyporheic macroinvertebrate assemblages during supra-seasonal drought. *Hydrological Processes* **23**, 2252–2263.

Stubbington, R., Greenwood, A.M., Wood, P.J., Armitage, P.D., Gunn, J. & Robertson, A.L. (2009b) The response of perennial and temporary headwater stream invertebrate communities to hydrological extremes. *Hydrobiologia* **630**, 299–312.

Svoboda, M. (2000) An introduction to the Drought Monitor. *Drought Network News* **12**, 15–20.

Svoboda, M., Lecomte, D., Hayes, D., Hem, R., Gleason, K., Angel, J.M., Rippey, B., Tinker, R., Palecki, M., Stocksbury, D., Miskle, D. & Stephens, S. (2002) The Drought Monitor. *Bulletin of the American Meteorological Society* **83**, 1181–1190.

Swales, S., Storey, A.W., Roderick, I.D. & Figa, B.S. (1999) Fishes of floodplain habitats of the Fly River system, Papua New Guinea, and changes associated with El Niño droughts and algal blooms. *Environmental Biology of Fishes* **54**, 389–404.

Sykora, K.V. (1979) The effects of the severe drought of 1976 on the vegetation of some moorland pools in the Netherlands. *Biological Conservation* **16**, 145–162.

Tallaksen, L.M. & Van Lanen, H.A.J. (2004) Introduction. In: Tallaksen, L.M. & Van Lanen, H.A.J. (eds.) *Hydrological Drought. Processes and Estimation Methods for Stream-flow and Groundwater*. pp. 3–17. Elsevier, Amsterdam.

Tallaksen, L.M., Madsen, H. & Clausen, B. (1997) On the definition and modeling of streamflow drought duration and deficit volume. *Hydrological Sciences Journal* **42**, 15–33.

Tannehill, I.R. (1947) *Drought: Its causes and effects*. Princeton University Press, Princeton, NJ.

Tarboton, D.G. (1995) Hydrologic scenarios for severe sustained drought in the south-western United States. *Water Resources Bulletin* **31**, 803–813.

Tate, E.L. & Gustard, A. (2000) Drought definition: A hydrological perspective. In: Vogt, J.V. & Somma, F. (eds.) *Drought and Drought Mitigation in Europe*. pp. 23–48. Kluwer, Dordrecht.

Taylor, R., Adams, J.B. & Haldorsen, S. (2006) Primary habitats of the St Lucia Estuarine System, South Africa, and their responses to mouth management. *African Journal of Aquatic Sciences* **31**, 31–41.

Taylor R.C. (1983) Drought-induced changes in crayfish populations along a stream continuum. *American Midland Naturalist* **110**, 286–298.

Taylor, R.C. (1988) Population dynamics of the crayfish *Procambarus spiculifer* observed in different-sized streams in response to two droughts. *Journal of Crustacean Biology* **8**, 401–409.

Teoh, K.S. (2002) *Estimating the impact of current farm development on the surface water resources of the Onkaparinga River catchment*. Report DWR 2002/22 Department of Water, Land and Biodiversity Conservation, Adelaide, South Australia.

Theissen, K.M., Dunbar, R.B., Rowe, H.D. & Mucciarone, D.A. (2008) Multidecadal- to century-scale arid episodes on the northern Altiplano during the middle Holocene. *Palaeogeography, Palaeolimnology, Palaeoecology* **257**, 361–376.

Thibault, K.M. & Brown, J.H. (2008) Impact of an extreme climatic event on community assembly. *Proceedings of the National Academy of Sciences* **105**, 3410–3415.

Thomas, S., Cecchi, P., Corbin, D. & Lemoalle, J. (2000) The different primary producers in a small African tropical reservoir during a drought: temporal changes and inter-actions. *Freshwater Biology* **45**, 43–56.

Thomé-Souza, M.J.F. & Chao, N.L. (2004) Spatial and temporal variation of benthic fish assemblages during the extreme drought of 1997–98 (El Niño) in the middle Rio Negro, Amazonia, Brazil. *Neotropical Ichthyology* **2**, 127–136.

Thompson, D.W.J. & Wallace, J.M. (2000) Annual modes in the extratropical circulation. Part 1: Month-to-month variability. *Journal of Climate* **13**, 1000–1016.

Tibby, J., Gell, P.A., Fluin, J. & Sluiter, I.R.K. (2007) Diatom-salinity relationships in wetlands: assessing the influence of salinity variability on the development of inference models. *Hydrobiologia* **591**, 207–218.

Timbal, B. & Jones, D.A. (2008) Future projections of winter rainfall in southeast Australia using a statistical downscaling technique. *Climatic Change* **86**, 165–187.

Timilsena, J., Piechota, T.C., Hidalgo, H. & Tootle, G. (2007) Five hundred years of hydrological drought in the Upper Colorado River Basin. *Journal of the American Water Resources Association* **43**, 798–812.

Tipping, E., Smith, E.J., Lawlor, A.J. Hughes, S. & Stevens, P.A. (2003) Predicting the release of metals from ombrotrophic peat due to drought-induced acidification. *Environmental Pollution* **123**, 239–253.

Titus, R.G. & Mosegaard, H. (1992) Fluctuating recruitment and variable life history of migratory brown trout, *Salmo trutta* L., in a small, unstable stream. *Journal of Fish Biology* **41**, 239–255.

Tockner, K., Malard, F. & Ward, J.V. (2000) An extension of the flood pulse concept. *Hydrological Processes* **14**, 2861–2883.

Tockner, K., Bunn, S.E., Gordon, C., Naiman, R.J., Quinn, G.P. & Stanford, J.A. (2008) Floodplains: Critically threatened ecosystems. In: Polunin, N.V.C. (ed.) *Aquatic Ecosystems. Trends and Global Prospects.* pp. 45–64. Cambridge University Press, Cambridge, UK.

Touchette, B.W., Iannacone, L.R., Turner, G.E. & Frank, A.R. (2007a) Drought tolerance versus drought avoidance: A comparison of plant-water relations in herbaceous wetland plants subjected to water withdrawl and repletion. *Wetlands* **27**, 656–667.

Touchette, B.W., Burkholder, J.M., Allen, E.H., Alexander, J.L., Kinder, C.A., Brownie, C., James, J. & Britton, C.H. (2007b) Eutrophication and cyanobacteria blooms in run-of-river impoundments in North Carolina, USA. *Lake and Reservoir Management* **23**, 179–192.

Towns, D.R. (1985) Limnological characteristics of a South Australian intermittent stream, Brown Hill Creek. *Australian Journal of Marine and Freshwater Research* **36**, 821–837.

Towns, D.R. (1991) Ecology of leptocerid caddisfly larvae in an intermittent South Australian stream receiving *Eucalyptus* litter. *Freshwater Biology* **25**, 117–129.

Townsend, C.R., Scarsbrook, M.R. & Dolédec, A. (1997) The intermediate disturbance hypothesis, refugia, and biodiversity in streams. *Limnology and Oceanography* **42**, 938–949.

Tracey, C.R., Reynolds, S.J., McArthur, L., Tracy, C.R. & Christian, K.A. (2007). Ecology of aestivation in a cocoon-forming frog, *Cyclorana australis* (Hylidae). *Copeia* **2007**, 901–912.

Tramer, E.J. (1977) Catastrophic mortality of stream fishes trapped in shrinking pools. *American Midland Naturalist* **97**, 469–478.

Trauth, J.B., Raith, S.E. & Johnson, R.L. (2006) Best management practices and drought combine to silence the Illinois chorus frog in Arkansas. *Wildlife Society Bulletin* **34**, 514–518.

Trenberth, K. (1997) The definition of El Niño. *Bulletin of the American Meteorological Society* **78**, 2771–2777.

Trexler, J.C., Loftus, W.F., Jordan, F., Chick, J.H., Kandl, K.L., McElroy, T.C. & Bass, O.L. (2002) Ecological scale and its implications for freshwater fishes in the Florida Everglades. In: Porter, J.W. & Porter, K.G. (eds.) *The Everglades, Florida Bay, and coral reefs of the Florida Keys.* pp. 153–181. CRC Press, Boca Raton, USA.

Trexler, J.C., Loftus, W.F. & Perry, S. (2005) Disturbance frequency and community structure in a twenty-five year intervention study. *Oecologia* **145**, 140–152.

Turner, M.G. & Dale, V.H. (1998) Comparing large, infrequent disturbances: what have we learned? *Ecosystems* **1**, 493–496.

Turney, C.S.M., Kershaw, A.P., Clemens, S.C., Branch, N., Moss, P.T. & Fifield, L.K. (2004) Millenial and orbital variations of El Niño/Southern Oscillation and high-latitude climate in the last glacial period. *Nature* **428**, 306–310.

Tweed, S., Leblanc, M. & Cartwright, I. (2009) Groundwater-surface water interaction and the impact of a multi-year drought on lakes conditions in South-East Australia. *Journal of Hydrology* **379**, 41–53.

UN News Centre (2009) Shrinking Lake Chad could trigger humanitarian disaster, UN agency warns. http://www.un.org/apps/news/story.asp?NewsID=32555. (15 October, 2009).

Usseglio-Polatera, P., Bournaud, M., Richoux, P. & Tachet, H. (2000) Biological and ecological traits of benthic macroinvertebrates: relationships and definition of groups with similar traits. *Freshwater Biology* **43**, 175–205.

Valett, H.M., Baker, M.A., Morrice, J.A., Crawford, C.S., Molles, M.C., Dahm, C.N., Moyer, D.L., Thobault, J.R. & Ellis, L.M. (2005) Biogeochemical and metabolic responses to the flood pulse in a semiarid floodplain. *Ecology* **86**, 220–234.

Van Dam, H. (1988) Acidification of three moorland pools in The Netherlands by acid precipitation and extreme drought periods over seven decades. *Freshwater Biology* **20**, 157–176.

Van Der Valk, A.G. (1994) Effects of prolonged flooding on the distribution and biomass of emergent species along a freshwater wetland coenocline. *Vegetatio* **110**, 185–196.

Van Der Valk, A.G. & Davis, C.B. (1978) The role of seed banks in the vegetation dynamics of prairie glacial marshes. *Ecology* **59**, 322–335.

Van Geest, G.J., Wolters, H., Roozen, F.C. J. M., Roijackers, R.M.M., Buijse, A.D. & Scheffer, M. (2005) Water-level fluctuations affect macrophyte richness in floodplain lakes. *Hydrobiologia* **539**, 239–248.

Van Lanen, H.A.J. & Peters, E. (2000) Definition, effects and assessment of groundwater droughts. In: Vogt, J.V. & Somma, F. (eds.) *Drought and Drought Mitigation in Europe.* pp. 49–61. Kluwer, Dordrecht.

Van Lanen, H.A.J., Fendeková, M., Kupczyk, E., Kaspryzk, A. & Pokojski, W. (2004) Flow generating processes. In: Tallaksen, L.M. & Van Lanen, H.A.J. (eds) *Hydrological Drought. Processes and Estimation Methods for Streamflow and Groundwater.* pp. 53–96. Elsevier, Amsterdam.

Van Vliet, M.T.H. & Zwolsman, J.J.G. (2008) Impact of summer drought on the water quality of the Meuse River. *Journal of Hydrology* **353**, 1–17.

Vandersande, M.W., Glenn E.P. & Walworth J.L. (2001) Tolerance of five riparian plants from the lower Colorado River to salinity drought and inundation. *Journal of Arid Environments* **49**, 147–159.

Velz, C.J. & Gannon, J.J. (1953) Low flow characteristics of streams. *Ohio StateUniversity Engineering Series* **22**, 138–157.

Verdon-Kidd, D.C. & Kiem, A.S. (2009) Nature and causes of protracted droughts in southeast Australia: Comparison between the Federation, WWII, and Big Dry droughts. *Geophysical Research Letters* **36**, L22707, doi: 10.1029/2009GL041067, 2009.

Verschuren, D., Laird, K.R. & Cumming, B.F. (2000) Rainfall and drought in equatorial east Africa during the past 1,100 years. *Nature* **403**, 410–414.

Villar-Argaiz, M., Medina-Sánchez, J.M., Cruz-Pizarro, L. & Carillo, P. (2001) Inter- and intra-annual variability in the phytoplankton community of a high mountain lake: the influence of external (atmospheric) and internal (recycled) sources of P. *Freshwater Biology* **46**, 1017–1034.

Villar-Argaiz, M., Medina-Sánchez, J.M. & Carillo, P. (2002) Microbial plankton response to contrasting climatic conditions: insights from community structure, productivity and fraction stoichiometry. *Aquatic Microbial Ecology* **29**, 253–266.

Visser, J.M., Sasser, C.E., Chabreck, R.H. & Linscombe, R.G. (2002) The impact of a severe drought on the vegetation of a subtropical estuary. *Estuaries* **25**, 1184–1195.

Wade, A.J., Whitehead, P.G., Hornberger, G.M. & Snook, D.L. (2002) On modelling the flow controls on macrophyte and epiphyte dynamics in a lowland permeable catchment: the River Kennet, southern England. *The Science of the Total Environment* **282–283**, 375–393.

Wahl, E.R. & Morrill, C. (2010) Toward understanding and predicting monsoon patterns. *Science* **328**, 437–438.

Waldhoff, D., Junk, W.J. & Furch, B. (1998) Responses of three central Amazonian tree species to drought and flooding under controlled conditions. *International Journal of Ecology and Environmental Sciences* **24**, 237–252.

Walker, B. & Salt, D. (2006) *Resilience Thinking. Sustaining ecosystems and people in a Changing World.* Island Press, Washington, DC.

Walker, B.H. (1995) Conserving biological diversity through ecosystem resilience. *Conservation Biology* **9**, 747–752.

Walling, D.E. & Foster, I.D.L. (1978) The 1976 drought and nitrate levels in the River Exe basin. *Journal of the Institution of Water Engineers and Scientists* **32**, 341–352.

Walsh, C.J., Roy, A.H., Feminella, J.W., Cottinham, P.D., Groffman, P.M. & Morgan, R.P. (2005) The urban stream syndrome: current knowledge and the search for a cure. *Journal of the North American Benthological Society* **24**, 706–723.

Wang, G. (2004) A conceptual modeling study of biosphere-atmosphere interactions and its implications for physically based climate modeling. *Journal of Climate* **17**, 572–2583.

Wang, G. & Eltahir, E.A.B. (2000) Role of vegetation dynamics in enhancing the low-frequency variability of the Sahel rainfall. *Water Resources Research* **36**, 1013–1021.

Wang, G., Eltahir, E.A.B., Foley, J.A., Pollard, D. & Levis, S. (2004) Decadal variability of rainfall in the Sahel: results from the coupled GENESIS-IBIS atmosphere-biosphere model. *Climate Dynamics* **22**, 625–637.

Wantzen, K.M., Junk, W.J. & Rothhaupt K-O. (2008) An extension of the floodpulse concept (FPC) for lakes. *Hydrobiologia* **613**, 151–170.

Ward, J.V. (1989a) The four-dimensional nature of lotic ecosystems. *Journal of the North American Benthological Society* **8**, 2–8.

Ward, J.V. (1989b) Riverine-wetland interactions. In: Sharitz, R.R. & Gibbons, J.W. (eds.) *Freshwater Wetlands and Wildlife, 1989. Conf 8603101, DOE Symposium Series No. 61.* pp. 385–400. USDOE Office of Scientific and Technical Information, Oak Ridge, TN.

Warren, F.J., Waddington, J.M., Bourbonniere, R.A. & Day, S.M. (2001) Effect of drought on hydrology and sulphate dynamics in a temperate swamp. *Hydrological Processes* **15**, 3133–3150.

Watkins, A.B. (2005) The Australian drought of 2005. *Bulletin of the World Meteorological Organization.* **54**, 156–162.

Watmough, S.A., Eimers, M.C., Aherne, J. & Dillon, P.J. (2004) Climate effects on stream nitrate concentrations at 16 forested catchments in south central Ontario. *Environmental Science & Technology* **38**, 2383–2388.

Webb, B.W. (1996) Trends in stream and river temperature. *Hydrological Processes* **10**, 205–226.

Webb, B.W. & Nobilis, F. (2007) Long-term changes in river temperature and the influences of climatic and hydrological factors. *Hydrological Sciences Journal* **52**, 74–85.

Webster, J.R., Gurtz, M.E., Hains, J.J., Meyer, J.L., Swank, W.T., Waide, J.B. & Wallace, J.B. (1983) Stability of stream ecosystems. In: Barnes, J.R. & Minshall, G.W. (eds.) *Stream Ecology: Applications and Testing of General Ecological Theory*. pp. 355–395. Plenum Press, New York.

Webster, K.E., Kratz, T.K., Bowser, C.J. & Magnuson, J.J. (1996) The influence of landscape position on lake chemical responses to drought in northern Wisconsin. *Limnology and Oceanography* **41**, 977–984.

Weiss, H. & Bradley, R.S. (2001) What drives societal collapse? *Science* **291**, 609–610.

Welcomme, R.L. (1979) *Fisheries Ecology of Floodplain Rivers*. Longman, London, UK.

Welcomme, R.L. (1985) *River Fisheries*. FAO Fisheries Technical Paper 262, FAO Rome.

Welcomme, R.L. (1986) The effects of the Sahelian drought on the fishery of the central delta of the Niger River. *Aquaculture and Fisheries Management* **17**, 147–154.

Welcomme, R.L. (2001) *Inland Fisheries Ecology and Management*. Blackwell Science Ltd, Oxford, UK.

Wellborn, G.A., Skelly, D.K. & Werner, E.E. (1996) Mechanisms creating community structure across a freshwater habitat gradient. *Annual Review in Ecology and Systematics* **27**, 337–363.

Wells, F., Metzeling, L. & Newall, P. (2002) Macroinvertebrate regionalization for use in the management of aquatic ecosystems in Victoria, Australia. *Environmental Monitoring and Assessment* **74**, 271–294.

Werner, E.E., Relyea, R.A., Yurewicz, K.L., Skelly, D.K. & Davis, C.J. (2009) Comparative landscape dynamics of two anuran species: climate-driven interaction of local and regional processes. *Ecological Monographs* **79**, 503–521.

Werner, S. & Rothhaupt, K-O. (2008) Mass mortality of the invasive bivalve *Corbicula fluminea* induced by a severe low-water event and associated low water temperatures. *Hydrobiologia* **613**, 143–150.

Westwood, C.G., Teeuw, R.M., Wade, P.M. & Holmes, N.T.H. (2006) Prediction of macrophyte communities in drought-affected groundwater-fed headwater streams. *Hydrological Processes* **20**, 127–145.

Wetzel, R.G. (2001) *Limnology. Lake and River Ecosystems*. 3rd Edition. Academic Press, San Diego, CA.

Whetton, P. (1997) Floods, droughts and the Southern Oscillation. In: Webb, E.K. (ed.). *Windows on Meteorology: Australian Perspective*. pp. 180–199. CSIRO Publishing, Melbourne, Australia.

Whetton, P. & Rutherfurd, I. (1994) Historical ENSO teleconnections in the eastern hemisphere. *Climatic Change* **28**, 221–253.

Whillans, T.H. (1996) Historic and comparative perspectives on rehabilitation of marshes as habitat for fish in the lower Great Lakes basin. *Canadian Journal of Fisheries and Aquatic Sciences* **53** (Supplement 1) 58–66.

Whitaker, D.W., Wasim, S.A. & Islam, S. (2001) The El Niño-Southern Oscillation and long-range forecasting of flows in the Ganges. *International Journal of Climatology* **21**, 77–87.

White, I., Melville, M., MacDonald, B., Quirk, R., Hawken, R., Tunks, M., Buckley, D., Beattie, R., Williams, J. & Heath, L. (2007) From conflicts to wise practice agreement and national strategy: cooperative learning and coastal stewardship in estuarine floodplain management, Tweed River, eastern Australia. *Journal of Cleaner Production* **15**, 1545–1558.

White, P.K., Merron, G.S., Quick, A.J.R. & la Hausse de Lalouvière, P. (1984) The impact of sustained drought conditions on the fishes of the Phongolo floodplain based on a survey in September 1983. *J.L.B. Smith Institute of Ichthyology Investigation Report 7*, 1–39.

White, P.S. & Pickett, S.T.A. (1985) Natural disturbance and patch dynamics: an introduction. In: Pickett, S.T.A. & White, P.S. (eds.) *The Ecology of Natural Disturbance and Patch Dynamics*. pp. 3–13. Academic Press, New York.

White, S.M. & Rahel, F.J. (2008) Complementation of habitats for Bonneville cutthroat trout in watersheds influenced by beavers, livestock and drought. *Transactions of the American Fisheries Society* **137**, 881–894.

White, S.N. & Alber, M. (2009) Drought-associated shifts in *Spartina alterniflora* and *S. cynosuroides* in the Altamaha River estuary. *Wetlands* **29**, 215–224.

Whitfield, A.K., Taylor, R.H., Fox, C. & Cyrus, D.P. (2006) Fishes and salinity in the St Lucia estuarine system-a review. *Review of Fish Biology and Fisheries* **16**, 1–20.

Whittington, J. (1999) Blue green algal blooms: a preventable emergency? *Australian Journal of Emergency Management* **14**, 20–22.

Wiggins, G.B., Mackay, R.J. & Smith, I.M. (1980) Evolutionary and ecological strategies of animals in annual temporary ponds. *Archiv für Hydrobiologie* **58** (Suppl.) 97–206.

Wilber, D.H. (1992) Associations between freshwater inflows and oyster productivity in Apalachicola Bay, Florida. *Estuarine, Coastal and Shelf Science* **35**, 179–190.

Wilber, D.H., Tighe, R.E. & O'Neil, (1996) Associations between changes in agriculture and hydrology in the Cache River basin, Arkansas, USA. *Wetlands* **16**, 366–378.

Wilbers, G.J., Zwolsman, G., Klaver, G. & Hendriks, A.J. (2009) Effects of a drought period on physico-chemical surface water quality in a regional catchment area. *Journal of Environmental Monitoring* **11**, 1298–1302.

Wilhite, D.A. (1992) Drought: its physical and social dimensions. In: Majumdar, S.K., Forbes, G.S., Miller, E.W. & Schmalz, R.F. (eds.). *Natural and Technological Disasters: Causes, Effects and Preventive Measures*. pp. 239–253. Pennsylvania Academy of Science, Easton, Pennsylvania.

Wilhite, D.A. (2000) Drought as a natural hazard: Concepts and definitions. In: Wilhite D.A. (ed.) *Drought: A Global Assessment. Volume 1*. pp. 3–18. Routledge, London.

Wilhite, D.A. (2003) Drought policy and preparedness: The Australian experience in an international context. In: Botterrill, L.C. & Fisher, M. (eds.) *Beyond Drought. People, Policy and Perspectives*. pp. 175–195. CSIRO Publishing, Melbourne, Australia.

Wilhite, D.A. & Buchanan-Smith, M. (2005) Drought as hazard: Understanding the natural and social context. In: Wilhite, D.A. (ed.) *Drought and Water Crises. Science, Technology and Management Issues*. pp. 4–29. Taylor & Francis, CRC Press, Boca Raton, FL.

Wilhite, D.A. & Glantz, M.H. (1985) Understanding the drought phenomenon: The role of definitions. *Water International* **10**, 111–120.

Wilhite, D.A., Svoboda, M.D. & Hayes, M.J. (2007) Understanding the complex impacts of drought: A key to enhancing drought mitigation and preparedness. *Water Resources Management* **21**, 763–774.

Williams, A.J. & Trexler, J.C. (2006) A preliminary analysis of the correlation of food-web characteristics and nutrient gradients in the southern Everglades. *Hydrobiologia* **569**, 493–504.

Williams, D.D. (2006) *The Biology of Temporary Waters*. Oxford University Press, Oxford, UK.

Williams, D.D. & Hynes, H.B.N. (1976) The ecology of temporary streams. I. The faunas of two Canadian streams. *Internationale Revue gesamten Hydrobiologie* **61**, 761–787.

Williams, D.D., Nalewajko, C. & Magnusson, A.K. (2005) Temporal variation in algal communities in an intermittent pond. *Journal of Freshwater Ecology* **20**, 165–170.

Williams, M.R. & Melack, J.M. (1997) Effects of prescribed burning and drought on the solute chemistry of mixed-conifer forest streams of the Sierra Nevada, California. *Biogeochemistry* **39**, 225–253.

Williams, W.D. (2001) Anthropogenic salinisation of inland waters. *Hydrobiologia* **466**, 329–337.

Williamson, C.E., Saros, J.E., Vincent, W.F. & Smol, J.P. (2009) Lakes and reservoirs as sentinels, integrators, and regulators of climate change. *Limnology and Oceanography* **54**, 2273–2282.

Willson, J.D., Winne, C.T., Dorcas, M.E. & Gibbons, J.W. (2006). Post-drought responses of semi-aquatic snakes inhabiting an isolated wetland: Insights on different strategies for persistence in a dynamic habitat. *Wetlands* **26**, 1071–1078.

Wilson, A.E., Sarnelle, O. & Tillmanns, A.R. (2006) Effects of cyanobacterial toxicity and morphology on the population growth of freshwater zooplankton: Meta-analyses of laboratory experiments. *Limnology and Oceanography* **51**, 1915–1924.

Winemiller, K.O. & Jepsen, D.B. (1998) Effects of seasonality and fish movement on tropical river food webs. *Journal of Fish Biology*, **53** (Supplement) A 267–296.

Winne, C.T., Willson, J.D. & Gibbons, J.W. (2006) Income breeding allows an aquatic snake *Seminatrix pygaea* to reproduce normally following prolonged drought-induced aestivation. *Journal of Animal Ecology* **75**, 1352–1360.

Wissinger, S.A. & Gallagher, L.J. (1999) Beaver pond wetlands in northwestern Pennsylvania. Mode of colonization and succession after drought. In: Batzer, D.P., Rader, R.B. & Wissinger, S.A. (eds.) *Invertebrates in Freshwater Wetlands of North America: Ecology and Management.* pp. 333–362. John Wiley & Sons, New York.

Wolter, C. & Menzel, R. (2005) Using commercial catch statistics to detect habitat bottlenecks in large lowland rivers. *River Research and Applications* **21**, 245–255.

Woo, M.K. & Tarhule A. (1994) Streamflow droughts of northern Nigerian rivers. *Hydrological Sciences Journal* **39**, 19–34.

Wood, P.J. (1998) The ecological impact of the 1995–1996 drought on a small groundwater-fed stream. In: Wheaten, H. & Kirby, C. (eds) *Hydrology in a Changing Environment. Volume 1.* pp. 303–311. John Wiley & Sons, Chichester, UK.

Wood, P.J. & Armitage, P.D. (2004) The response of the macroinvertebrate community to low-flow variability and supra-seasonal drought within a groundwater dominated stream. *Archiv für Hydrobiologie* **161**, 1–20.

Wood, P.J. & Petts, G.E. (1994) Low flows and recovery of macroinvertebrates in a small regulated chalk stream. *Regulated Rivers: Research & Management* **9**, 303–316.

Wood, P.J. & Petts, G.E. (1999) The influence of drought on chalk stream macroinvertebrates. *Hydrological Processes* **13**, 387–399.

Wood, P.J., Agnew, M.D. & Petts, G.E. (2000) Flow variations and macroinvertebrate community responses in a small groundwater-dominated in south-east England. *Hydrological Processes* **14**, 3133–3147.

Wood, P.J., Boulton, A.J., Little, S. & Stubbington, R. (2010) Is the hyporheic zone a refugium for aquatic macroinvertebrates during low flow conditions? *Archiv für Hydrobiologie* **176**, 377–390.

Woodhouse, C.A. (2004) A paleo perspective on hydroclimatic variability in the western United States. *Aquatic Sciences* **66**, 346–356.

Woodhouse, C.A. & Overpeck, J.T. (1998) 2000 years of drought variability in the central United States. *Bulletin of the American Meteorological Society* **79**, 2693–2714.

Woodhouse, C.A., Gray, S.T. & Meko, D.M. (2006) Updated streamflow reconstructions for the Upper Colorado River Basin. *Water Resources Research* **42**, W05415, doi: 10,1029/2005WR004455

World Meteorological Organization (1992) *International Meteorological Vocabulary*. 2nd Edition. World Meteorological Organization No.182, World Meteorological Organization, Geneva, Switzerland.

Worrall, F. & Burt, T.P. (2007) Flux of dissolved organic carbon from UK rivers. *Global Biogeochemical Cycles* **21**, GB1013. doi: 10.1029/2006GB002709, 2007.

Worrall, F., Burt, T. P & Adamson, J. K. (2006) Trends in drought frequency – the fate of DOC export from British peatlands. *Climatic Change* **76**, 339–359.

Worrall, F., Burt, T.P. & Adamson, J. (2008) Long-term records of dissolved organic carbon flux from peat-covered catchments: evidence for a drought effect? *Hydrological Processes* **22**, 3181–3193.

Worster, D. (1979) *Dust Bowl. The Southern Plains in the 1930s*. Oxford University Press, New York.

Wourma, J.P. (1972) The developmental biology of annual fishes. 3. Pre-embryonic and embryonic diapause of variable duration in the eggs of annual fishes. *Journal of Experimental Zoology* **182**, 389–414.

Wright, J.F. & Berrie, A.D. (1987) Ecological effects of groundwater pumping and a natural drought on the upper reaches of a chalk stream. *Regulated Rivers: Research & Management* **1**, 145–160.

Wright, J.F. & Symes, K.L. (1999) A nine-year study of the macroinvertebrate fauna of a chalk stream. *Hydrological Processes* **13**, 371–385.

Wright, J.F., Gunn, R.J.M., Winder, J.M., Wiggers, R., Vowles, K., Clarke, R.T. & Harris, I. (2002a) A comparison of the macrophyte cover and macroinvertebrate fauna at three sites on the River Kennet in the mid 1970s and late 1990s. *The Science of the Total Environment* **282–283**, 121–142.

Wright, J.F., Gunn, R. J.M., Winder, J.M. Wiggers, R., Kneebone, N.T. & Clarke, R.T. (2002b) The impact of drought events in 1976 and 1997 on the macroinvertebrate fauna of a chalk stream. *Verhandlungen des Internationalen Vereinigung für theoretische und angewandte Limnologie* **28**, 948–952.

Wright, J.F., Clarke, R.T., Gunn, R.J.M., Kneebone, N.T. & Davy-Bowker, J. (2004) Impact of major changes in flow regime on the macroinvertebrate assemblages of four chalk stream sites, 1997–2001. *River Research & Applications* **20**, 775–794.

Wyrtki, K. (1976) Predicting and observing El Niño. *Science* **191**, 343–346.

Wyrtki, K. (1977) Sea level during the 1973 El Niño. *Journal of Physical Oceanography* **7**, 779–787.

Xenopoulos, M. A., Lodge, D.M., Alcamo, J., Märker, M., Schulze, K. & van Vuuren, D.P. (2005) Scenarios of freshwater fish extinctions from climate change and water withdrawl. *Global Change Biology* **11**, 1557–1564.

Yan, N.D., Keller, W., Scully, N.M., Lean, D.R.S. & Dillon, P.J. (1996) Increased UV-B penetration in a lake owing to drought-induced acidification. *Nature* **381**, 141–143.

Yevjevich, V. M (1967) An objective approach to definitions and investigations of continental hydrologic droughts. Colorado State University, Hydrology Paper 23, Fort Collins CO, 18pp.

Yevjevich, V.M. (1972) *Stochastic Processes in Hydrology*. Water Resource Publications, Fort Collins, CO.

Yiou, P. & Nogaj, M. (2004) Extreme climatic events and weather regimes over the North Atlantic: When and where? *Geophysical Research Letters* **31**, L07202, doi: 10.1029/2003GL019119, 2004.

Yu, Z. & Ito, E. (1999) Possible solar forcing of century-scale drought frequency in the northern Great Plains. *Geology* **27**, 263–266.

Yu, Z., Ito, E., Engstrom, D.R. & Fritz, S.C. (2002) A 2100-year trace-element and stable-isotope record at decadal resolution from Rice Lake in the Northern Great Plains, USA. *The Holocene* **12**, 605–617.

Zaret, T. M. & Rand, A.S. (1971) Competition in tropical stream fishes: support for the competitive exclusion principle. *Ecology* **52**, 336–342.

Zeldis, J. R., Howard-Williams, C., Carter, C.M. & Schiel, D.R. (2008) ENSO and riverine control of nutrient loading, phytoplankton biomass and mussel aquaculture yield in Pelorus Sound, New Zealand. *Marine Ecology Progress Series* **371**, 131–142.

Zeng, N., Neelin, J.D., Lau, K.M. & Tucker, C.J. (1999) Enhancement of interdecadal climate variability in the Sahel by vegetation interaction. *Science* **286**, 1537–1540.

Zielinski, P., Gorniak, A. & Piekarski, M.K. (2009) The effect of hydrological drought on chemical quality of water and dissolved organic carbon concentrations in lowland rivers. *Polish Journal of Ecology* **57**, 217–227.

Zohary, T., Pais-Madeira, A.M., Robarts, R. & Hambright, K.D. (1996). Interannual phytoplankton dynamics of a hypertrophic African lake. *Archiv für Hydrobiologie* **136**, 105–126.

Zwolsman, J.J.G. & van Bokhoven, A.J. (2007) Impact of summer droughts on water quality of the Rhine River-a preview of climate change? *Water Science & Technology* **56**, 45–55.

Index

Note: page numbers in *italics* refer to figures and tables

Drought and Aquatic Ecosystems: Effects and Responses, First Edition. P. Sam Lake.
© 2011 P. Sam Lake. Published 2011 by Blackwell Publishing Ltd.